ARSENIC

EXPOSURE AND HEALTH

ARSENIC

EXPOSURE AND HEALTH

Edited by

Willard R. Chappell

Charles O. Abernathy

C. Richard Cothern

SCIENCE AND TECHNOLOGY LETTERS

British Library Cataloguing in Publication Data

Arsenic exposure and health

A catalogue record of this book is available from the British Library

ISBN 0-905927-49-4

Special issue of *Environmental Geochemistry and Health*, Volume 16

ISSN 0269-4042

Production editor: Sara Nash
Copy editor: Pam Tamblyn
Typeset by: DL Technology, PO Box 9, Buckhurst Hill, Essex, IG9 5BE
Printed by: Laws & Stimson Associates, 31 Malden Way, New Malden, Surrey, KT3 6EB

CONTENTS

Preface

The Society of Environmental Geochemistry and Health (SEGH) International Conference on Arsenic Exposure and Health Effects was held July 28-30, 1993 in New Orleans, Louisiana. It was attended by approximately 160 people representing academia, government, industry and environmental action groups. There were many attendees from outside the USS indicating that arsenic in the environment is truly an international issue.

The Conference had its origins at a meeting of the Executive Board of the SEGH in December 1991 when the Board was considering possible efforts similar to its successful Lead in Soil Task Force. It was noted that EPA was faced with a law suit requiring it to establish a Maximum Contaminant Level (MCL) for arsenic in drinking water. This effort was going to necessarily result in a risk assessment for arsenic. As a result, not only drinking water, but other media and programs could also be potentially impacted because arsenic is an important contaminant in many Superfund Sites and represents approximately half of the estimated carcinogenic potential in coal fly ash. Any adjustment of the carcinogenic potency, therefore, could have wide-ranging impacts on drinking water, superfund remediation efforts, and fly ash control technology. Because of the broad impact, the Executive Board agreed that it might be appropriate to launch an Arsenic Task Force. Willard Chappell and Charles Abernathy were asked to co-chair the effort and to suggest names of the members of the group.

After discussions with numerous experts in the field, it was decided that an international conference could be the appropriate vehicle to begin the efforts of the Task Force. Luckily there were was widespread interest in the conference and funding was obtained from a diverse group of government agencies, companies and industry groups (these are listed below).

The organizers were also fortunate in that there was interest in the leading researchers in participating in the conference. As a result, the speakers represent the top researchers in the world on the subject. Because of the considerable possible impact of regulations on local and state government, as well as industry, we also invited participation from representatives of these sectors.

In addition to excellent presentation by the invited speakers, there were also many fine poster presentations about a wide variety of topics. In all, the meeting was apparently successful in meeting its objectives. The contents of this volume represent most of the talks presented at the conference. We wish to thank the authors for their fine work and the sponsors of the conference who made it all possible.

Co-editors:

Willard R . Chappell *University of Colorado at Denver*
Charles O. Abernathy *US Environmental Protection Agency*
C. Richard Cothern *US Environmental Protection Agency*

Sponsors:

Agency for Toxic Substances and Disease Registry (ATSDR)
US Environmental Protection Agency (EPA)
American Water Works Association (AWWA)
American Mining Congress (AMC)
International Council on Metal in the Environment (ICME)
Electric Power Research Institute (EPRI)
US Borax

We are also grateful for assistance from Health Canada.

1 The value of research on health effects of ingested inorganic arsenic

D. Warner North, Frank Selker and Tom Guardino

Decision Focus Incorporated, 650 Castro Street, Suite 300, Mountain View, CA 94041, USA

Abstract

A method for estimating the potential "value of information" from alternative research opportunities is presented. The method is introduced with a simple example using a coin-toss game. Four potential research topics are then evaluated in the context of arsenic drinking water standards: (1) uncertainty in effective delivered dose, (2) uncertainty in the dose-response model at low doses; (3) the risk of internal organ cancers, and (4) the applicability of existing epidemiology results to the US population. Data and assumptions are based on the authors' review of available information. Where information was not available, the authors sought informal views of others in the field and selected values they considered realistic. Results suggest that: (1) value-of-information calculations can be useful for understanding uncertainties and estimating the value of research opportunities, (2) further biologically-based arsenic research appears likely to have value in the hundreds of millions of dollars, which is considerably higher than its expected cost, (3) research regarding the risk of internal organ cancer may have the greatest value. We also observe that food appears to account for about 3/4 of ingested inorganic arsenic exposure, and may thus present both a significant public health risk and an opportunity to reduce arsenic exposures. The calculated values of research depend upon the range of uncertainty, exposure assumptions, expected costs of reducing arsenic in water, the monetary equivalent value of avoiding health impacts, and other factors. However, the value of research appears to remain high for a variety of scenarios considered.

This paper was condensed from "Estimating the Value of Research: An Illustrative Calculation for Ingested Inorganic Arsenic", prepared by the same authors with funding from the National Science Foundation, the Electric Power Research, the Edison Electric Institute, Monsanto, ARCO, and FMC. Copies of the full report are available from the authors at Decision Focus Inc., 650 Castro Streeet, Suite 300, Mountain View, California 94041, USA.

Introduction

Ingested inorganic arsenic has been associated with increased risk of skin cancer, and is classified by the US Environmental Protection Agency (US EPA) as a known human carcinogen (category A). A primary challenge of arsenic risk management is to select cost-effective control levels even though arsenic's potency and mechanism of action remain highly uncertain. A second challenge is determining what, if any, research is cost-effective in helping us improve regulations in the future. This question regarding arsenic has been taken up by several groups, including the US EPA Science Advisory Board (US EPA SAB, 1989; US EPA SAB, 1992; Fowle, 1992). The 1992 SAB report recommends research expected to be possible in 3-5 years and research that may take longer.

With the help of a decision tree we model decision making regarding the arsenic drinking water standard, including uncertainties regarding arsenic's potency and effects. We use this tree to compare *expected effects* of alternative drinking water standards in terms of costs and human health. We then carry out value-of-information calculations to estimate the potential value of research addressing each of the uncertainties considered. Finally, we consider the impacts of varying our assumptions in several sensitivity analyses.

Methods

Decision analysis offers a method of calculating the value of obtaining new information. The method is based on the observation that information is valuable to the extent that it helps us make better decisions. Therefore we can value information by comparing a decision before and after hypothetical information is available. For

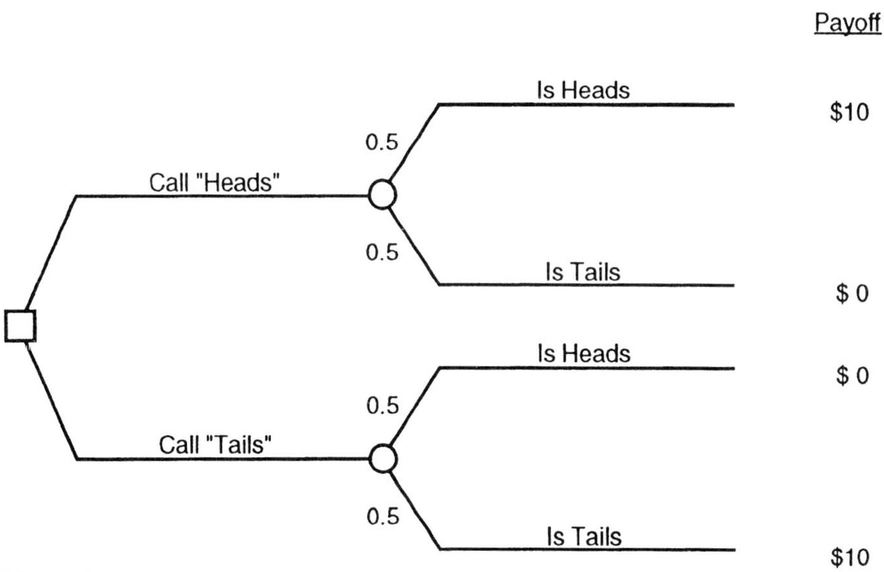

Figure 1 *Decision tree representing coin-toss game.*

background and further discussion of value-of-information methods, see Raiffa (1968), Howard and Matheson (1983), and Clemen (1991).

We explore how methods for calculating the value of information can be used to help evaluate and prioritize research opportunities. The results are in dollars, allowing us to consider the expected cost-effectiveness in evaluating and prioritizing research opportunities. The methods could be similarly applied to research regarding other carcinogens.

Calculating the value of information: coin-toss game example

Imagine that someone offers to pay you $10 if you correctly call a coin toss. Assuming the coin toss is fair, there is a 50% probability that your call will be correct. This situation is shown in Figure 1. Note that the first, square node of the decision tree is your decision of whether to call heads or tails. After your call, the uncertainty of the toss is resolved, as depicted by the second, circular node. Whether you call head or tail, you have a 50% chance of getting $10 and a 50% chance of getting nothing. Thus the "expected value" of this opportunity is (0.5 x $10) + (0.5 x $0) = $5.

Now imagine that you could obtain information revealing whether the toss was a head or tail before you had to make your call. In this case you could always call the toss correctly, and you would be guaranteed of winning the $10. This situation is shown in Figure 2. Notice that we have now reversed the order of the nodes in the decision tree. This is because your information source will now resolve the uncertainty before you need to make the decision.

The coin-toss game is worth $5 without the information and $10 with the information. The value of information is the increase in value, or $5. To summarize,

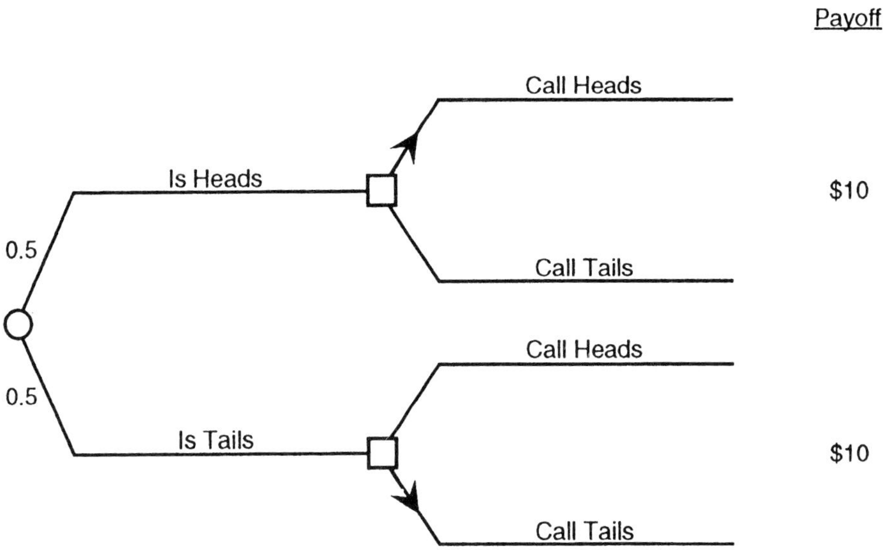

Figure 2 *Coin-toss decision tree with information regarding the toss outcome (arrows indicate preferred decisions).*

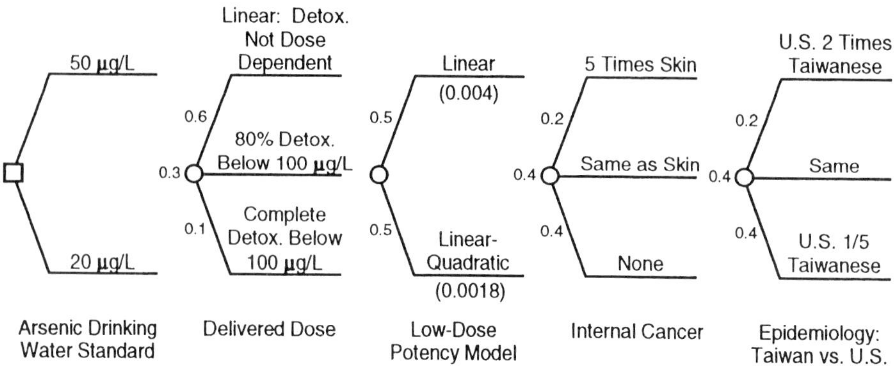

Figure 3 *Decision tree depicting choice between arsenic drinking water standard of 50 µg L^{-1} and 20 µg L^{-1} with uncertainties regarding arsenic's potency and action. All values are illustrative.*

we evaluated a decision with current information, and then evaluated the same decision with the opportunity to acquire better information before the decision is made. The difference in expected values is the value of information. This same approach can be used with much more complex, realistic decisions, and is the basis for the calculated values of information that we estimate below for research pertaining to arsenic.

The actual value of research may be either less than or greater than the calculated value of perfect information. On one hand, the $5 value of information in our coin-toss game should be considered an upper bound, since we have presumed that the information is correct. The value would be less than $5 if you were provided with information that might be in error. Value-of-information calculations can readily accommodate imperfect information (information that is not definitive and completely reliable), although the calculations are slightly more involved (Raiffa, 1968; Howard and Matheson, 1983; and Clemen, 1991). In the case of scientific research, however, such calculated values of perfect information are *not* upper bounds. This is because research often yields unexpected results that are helpful with unrelated uncertainties and goals, thereby offering value beyond the initial decision context. More detailed analysis could go further in the evaluation of both imperfect information and results that may have value in a broader context.

Ingested inorganic arsenic

We begin by constructing a "decision tree" reflecting the arsenic regulatory decision and important uncertainties (Figure 3). The decision tree depicts a regulatory decision and four uncertainties related to the decision.

The first node, designated with a square, represents the choice to regulate arsenic at the level of 50 µg L^{-1} or 20 µg L^{-1}. 50 µg L^{-1} is the current US EPA drinking water standard, and 20 μg L^{-1} was selected as one of several possible new standards. Other potential standards could be used equally well as the basis for value-of-information calculations, and would yield different results. We explore this further with a sensitivity analysis of a 5 μg L^{-1} standard later in this paper.

The second through fifth nodes, designated with circles, represent four uncertainties regarding risks from arsenic ingestion. Each uncertainty corresponds to

Table 1 *Low-dose risks for arsenic ingestion.*

	Linear low-dose (10^{-3} risk at 1 μg kg^{-1} day^{-1})	Quadratic low-dose (10^{-3} risk at 1 μg kg^{-1} day^{-1})
US EPA, 1988, Males	5.0	2.3
US EPA, 1988, Females	3.4	1.0
Brown *et al.*, 1989, Males	3.0	1.3
Brown *et al.*, 1989, Females	2.1	0.6
Value used (Average for Males):	4.0	1.8

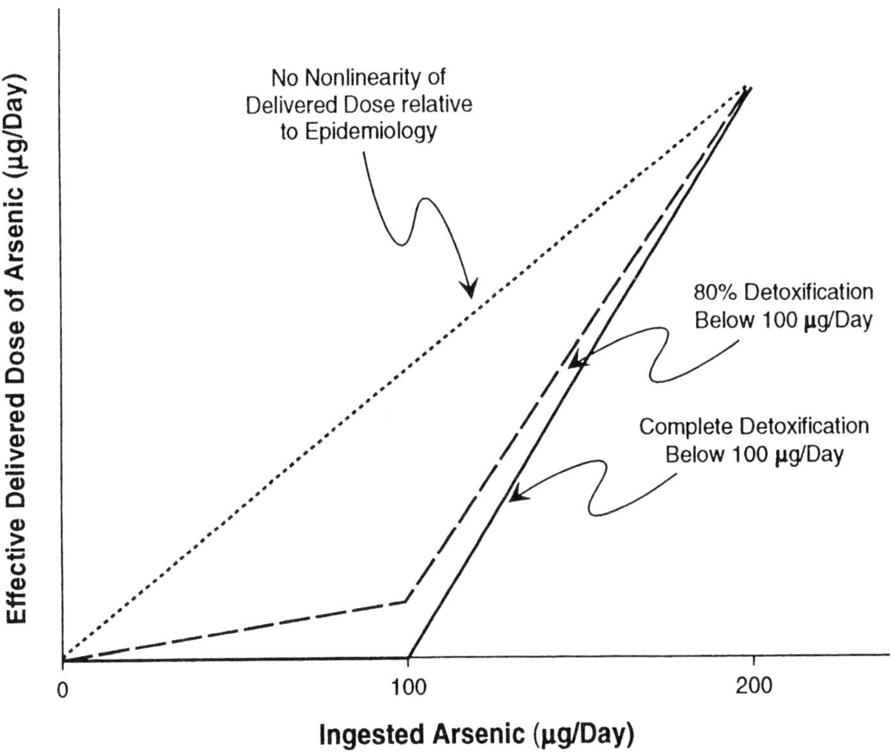

Figure 4 *Three possible relationships between ingested arsenic and effective dose*

an unresolved scientific issue regarding the health risks from ingesting arsenic. Uncertainty at each node is represented by two or three branches, each of which represents a possible outcome when the uncertainty is resolved. This is a simplification in that we represent all possible outcomes at each node with only two or three scenarios. This is analogous to approximating ("discretizing") a continuous probability distribution with a few values. While it is a limited representation of all possible outcomes, it can offer a good approximation while greatly reducing computations.

Our assignments of probabilities are consistent with our own reading of the scientific literature and our personal communication with experts. In a more fully implemented analysis a formal process of interviewing experts could be used to assign probabilities to each of the uncertain branches. This analysis should be viewed as illustrative since the results do not reflect an integrated assessment or a consensus among experts. We will now develop our assumptions for each uncertainty node.

Delivered dose

The first node represents uncertainty regarding the relationship between ingested arsenic and the effective dose delivered to target tissues.

The metabolism and excretion of arsenicals is complex. Further development of physiologically based pharmacokinetic (PBPK) models will be required to better understand the relation between exposures and the effective delivered dose of dangerous arsenicals. Until such models are developed and validated, the delivered doses of arsenicals remain uncertain. In the decision tree shown in Figure 3, this uncertainty is represented as a node with three branches. Each branch corresponds to an alternative delivered dose relationship which might best represent the actual relationship. The top branch represents the possibility that the relation between exposures and delivered dose is similar at the exposure levels occurring in the USA and exposures that were studied in Taiwan (*i.e.*, the proportion of ingested arsenicals affecting vulnerable tissues is the same). The middle branch represents the possibility that at doses below 100 μg day^{-1}, the delivered dose is 20% of the doses predicted from higher-exposure epidemiology studies. The bottom branch represents the possibility that detoxification is complete below 100 μg day^{-1}. For both the middle and bottom branches, we assume that the detoxification effect disappears when doses reach 200 μg day^{-1}. These three relationships between exposure and delivered dose are shown in Figure 4.

We assign a probability of 0.6 to the possibility that there is no nonlinearity, a probability of 0.3 to the possibility that there is a nonlinearity with an 80% risk reduction below 100 μg day^{-1}, and a 0.1 possibility that there is no risk below 100 μg day^{-1}. These probabilities are chosen by the authors as illustrative values consistent with our reading of the literature.

Low-dose model of potency

The second uncertainty node represents uncertainty regarding the relationship between the dose that actually reaches target tissue and the increased risk of cancer.

Arsenic dose-response data have been fit to multi-stage models assuming either

Table 2 *Drinking water exposure assumptions.*

Percent population	Number of people (millions)	Average arsenic concentration ($\mu g\ L^{-1}$)
Current standard, 50 $\mu g\ L^{-1}$:		
95.05%	237.6	2.0
4.4%	11.0	10.0
0.55%	1.4	30.0
Average value		2.51
New standard, 20 $\mu g\ L^{-1}$:		
95.05%	237.6	2.0
4.95%	12.4	10.0
Average value		2.40

low-dose linearity or low-dose quadratic and linear coefficients (US EPA 1988, Appendix B; Brown *et al.*, 1989). While other low-dose forms may prove more accurate, a plausible research goal is to determine whether the linear or the combined linear and quadratic low-dose relationship is more accurate. We consider the implications of a pure quadratic (*i.e.*, no linear component) dose-response model as a sensitivity case.

Table 1 summarizes estimates of the lifetime risk of cancer as a consequence of ingesting arsenic at a rate of 1 $\mu g\ kg^{-1}\ day^{-1}$ (1 μg of arsenic per kilogram of body mass every day for 76 years). (We have used 76 years as the median lifespan in the United States, following Brown *et al.*, (1989).)The values we use are the average of two of the studies shown: the risk values for males in the US EPA Risk Assessment Forum Report (US EPA, 1988) and for males in Brown *et al.*, (1989). The mean values are 0.004 risk per $\mu g\ kg^{-1}\ day^{-1}$ assuming a linear low-dose relation, and 0.0018 risk per $\mu g\ kg^{-1}\ day^{-1}$ assuming a linear and quadratic low-dose relation. (Note that we use a single linear-quadratic potency factor across all doses we consider, and are thus neglecting the curvature of the linear-quadratic model. This is a good approximation because (1) the quadratic term is very small at the 1 $\mu g\ kg^{-1}\ day^{-1}$ dose, so the curvature is very small; and (2) the doses we consider are close to 1 $\mu g\ kg^{-1}\ day^{-1}$, roughly ranging from 0-2 $\mu g\ kg^{-1}\ day^{-1}$.) For our illustrative analysis we assume each of the linear and linear-quadratic models has a 50% likelihood of more accurately reflecting the true relationship.

Internal organ cancers

The third uncertainty node in Figure 3 represents uncertainty regarding the risk of internal organ cancers caused by ingested arsenic. While epidemiological studies of arsenic ingestion have focused on skin cancer, several recent studies have suggested

Table 3 *Exposure assumptions for inorganic arsenic in food.*

Percent of the US population	Number of people (millions)	Daily exposure to inorganic arsenic in food ($\mu g\ day^{-1}$)
88%	220	7
10%	25	50
2%	5	100
Average food-borne arsenic exposure:		13.2

that internal organ cancers may also be a concern (Bates *et al.*, 1992; Chen and Chen, 1988; Gibb and Chen, 1989; Chen and Wang, 1990; Smith *et al.*, 1992). Arsenic has been implicated in cancer risk at a variety of internal organs, including the liver, lung, kidney, bladder, nasal cavity, prostate, and gastrointestinal tract.

If arsenic ingestion were found to increase the risk of internal organ cancers, the predicted number and severity of cancers associated with arsenic ingestion could significantly increase. At 50 $\mu g\ L^{-1}$ of arsenic in drinking water, estimates of risk for internal cancers associated with ingesting arsenic have included zero (implicit in analyses not considering internal cancers), about 7 per 1,000 among studied Taiwanese (Chen and Wang, 1990) and 13 per 1,000 for one liter of contaminated water per day^{-1} (Smith *et al.*, 1992). These estimates compare with skin cancer risk estimates (again with consumption of 50 $\mu g\ L^{-1}$ in water) of 2.6 per thousand using the linear-quadratic low-dose assumption and 5.7 per thousand using the linear low-dose assumption (calculated from Table 1: Risk per 1,000 = Risk ($\mu g\ kg^{-1}\ day^{-1}$) x 50 ($\mu g\ L^{-1}$) x 2 ($L\ day^{-1}$) x 1/70 $(kg)^{-1}$ x 1,000 people). Thus the risk of internal cancers associated with ingestion has been suggested to be in the range of zero to about five times the risk of skin cancer (although exposure assumptions have not been entirely comparable). We use a range of zero to 5 for the risk multipliers for the uncertainty node associated with internal cancers. These multipliers simply scale the total expected risk of cancers. The authors' probabilities are assigned to the branches as shown in Figure 3.

Including internal cancer risks requires extending the analysis beyond simply the *number* of cancer cases because skin cancer associated with arsenic exposure poses a much lower threat to health than internal cancer. If caught early, the skin cancer can be treated inexpensively and effectively with little risk of causing an earlier death. Internal cancers, however, may be difficult to treat and are likely to appreciably shorten life span. For this reason we evaluate skin and internal cancers differently when evaluating results.

Potential epidemiology biases

The fourth, and final, uncertainty node shown in Figure 3 represents uncertainty regarding how well the results of the Taiwanese studies will predict cancer incidence among the US populations. Two major sources of this uncertainty are: (1) the possibility that susceptibility to arsenic-induced cancer may differ between Taiwanese and US populations, and (2) aspects of the epidemiology studies that may detract from the accuracy of results.

Again, each branch of the node representing this uncertainty has an assigned "risk multiplier" that scales risk corresponding to an ultimate resolution of this uncertainty. The values of these multipliers were selected to span the authors' views of uncertainty. The top branch corresponds to the possible outcome that the US population experiences twice the risk suggested by the epidemiology studies. The middle branch corresponds to the possibility that the risk estimates from the epidemiology studies are accurate for the US population, so risks are the same for US citizens as would be predicted by the studies. Finally, the bottom branch corresponds to the US population having risks one fifth of those predicted by the studies. Probabilities are assigned to the branches as shown in Figure 3.

Exposure assumptions

The four uncertainty nodes provide a simplified, illustrative summary of uncertainties in risks associated with ingesting inorganic arsenic. The remaining information required to estimate risks is a set of projections for arsenic exposure among the US population. Since exposures vary considerably across the population, we will assign fractions of the population to groups with differing estimated exposures. For most people the dominant exposure sources of arsenic are in food and drinking water (US EPA, 1984). We consider each of these pathways. (For the full background and discussion of the values selected for the analysis see footnote 1 for reference.)

Drinking water exposure

For our analysis we need to estimate exposures to arsenic with the existing 50 $\mu g\ L^{-1}$ standard and then with a 20 $\mu g\ L^{-1}$ standard.

We have assumed three exposure groups, with the population size of each corresponding to US EPA estimates. We selected plausible estimates of the average exposure within each of these groups: for those drinking water over 20 $\mu g\ L^{-1}$, the average concentration is 30 $\mu g\ L^{-1}$, and for those drinking water between 5 and 20 μg L-^{1}, the average concentration is 10 $\mu g\ L^{-1}$ (Table 2). Finally, for the 95.05% of the population drinking water below 5 $\mu g\ L^{-1}$, we assume the average concentration is 2 $\mu g\ L^{-1}$. These averages yield a current expected exposure through drinking water of 2.51 $\mu g\ L^{-1}$, which is consistent with population average values estimated in several sources (US EPA, 1984; US EPA, 1988, recent communication with the US EPA and the Metropolitan Water District of Southern California).

The effect of imposing the lower standard is assumed to be that the 1.4 million people now drinking water over 20 $\mu g\ L^{-1}$ would have their water brought down to an average of 10 $\mu g\ L^{-1}$. We selected an average value of less than 20 $\mu g\ L^{-1}$ because some water supplies may cease using wells with high arsenic or use substitute sources that are well below the 20 μg L-^{1} level. Further research regarding compliance with the 1977 50 $\mu g\ L^{-1}$ standard would verify the degree to which this "compliance overshoot" occurs or is negated by non-compliance at other sites. Note

Table 4 *Annual costs and benefits of reducing the arsenic drinking water standard from 50 μg L^{-1} to 20 μg L^{-1}. (Based on illustrative data and assumptions.)*

Internal cancer reduction (expected cases):25.3	
Benefit @ $1.5 M per avoided internal cancer ($M):	$37.9
Skin cancer reduction (expected cases):18.0	
Benefit @ $25 K per avoided skin cancer ($M):	$0.4
Total annual benefits ($M):	$38.3
Annual costs ($M):[1]	$31.0
Annual net benefits ($M):	$7.3
Net present value of benefits over 30 years ($M):[2]	$112.0

[1] See discussion of cost estimates in text.
[2] Calculated using 5% discount rate net of inflation.

that while the 20 μg L^{-1} standard reduces high exposures, it only reduces *average* exposure by about 5%, to 2.4 μg L^{-1}. This is because most water supplies are already below the 20 μg L^{-1} level. This is discussed further with our results.

Food exposure

For most people the most important exposure route for both organic and inorganic arsenic is food. The 1984 US EPA study of arsenic exposure (US EPA, 1984) uses Food and Drug Administration estimates of about 45 μg day^{-1}, with regional averages ranging from about 25 to 80 μg day^{-1} (see also Pershagen, 1986). Most of this arsenic was believed to be bound in organic complexes that are readily excreted, although the organic fraction varies widely by food type. The 1988 US EPA Risk Assessment Forum report estimates daily arsenic ingestion in food to be about 50 μg day^{-1}, of which 13 μg day^{-1} was inorganic arsenic ((US EPA, 1988, pp. 101-104): total intake estimate of 17-18 μg day^{-1} minus water exposure of 5 μg day^{-1}). US EPA's 1992 Draft Drinking Water Criteria Document on Arsenic estimates adult exposure to arsenic at about 53 μg day^{-1}, with 11-14 μg day^{-1} in inorganic forms (US EPA, 1992). Total arsenic found in FDA's Total Diet Studies have ranged from about 30 to 60 μg day^{-1}, with values hovering around 35 μg day^{-1} for the past few years.

Regional and individual eating habits can have large impacts on inorganic arsenic exposure because certain foods are very high in arsenic, and because the fraction of total arsenic that is inorganic varies from 0-75% (US EPA, 1988). This suggests that a wide range of exposures are probably occurring within the US population. However, we are not aware of information characterizing the distribution of exposure. In the absence of detailed information, we arbitrarily selected exposure

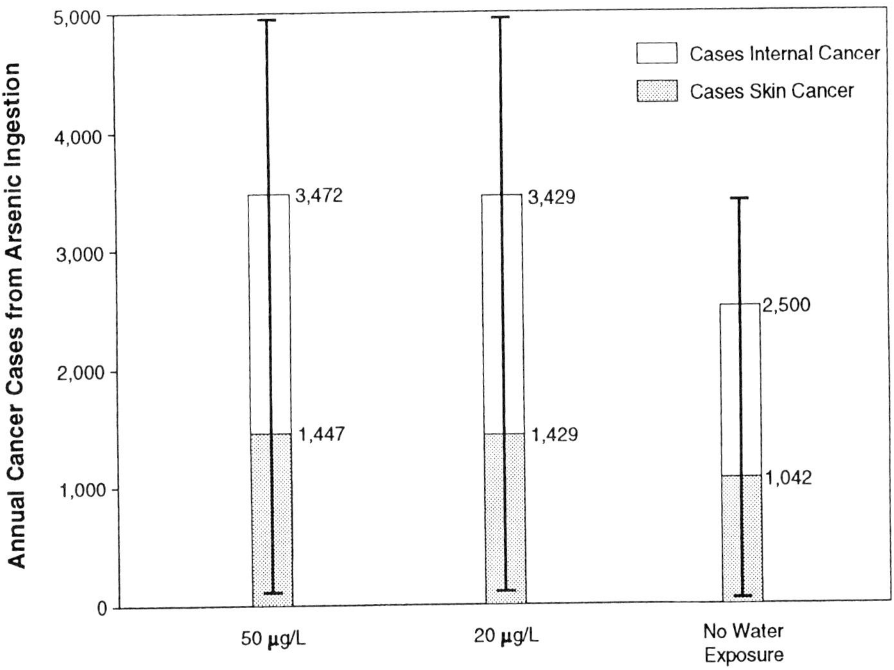

Figure 5 *Expected values of annual cancer cases for 2 drinking water standards (50 µg L^{-1} and 20 µg L^{-1}) and for the total elimination of drinking water exposure to arsenic. Based on illustrative data. (There is a 70% probability that the sum of skin and internal cancer cases will fall between the error bars shown).*

ranges and population estimates giving about an order of magnitude range in exposures and yielding a total average exposure to inorganic arsenic of roughly 13 µg day^{-1} (Table 3).

Cost of reducing drinking water standard to 20 µg L^{-1}

Based on personal communications with the US EPA and the American Waterworks Association (AWWA), we estimate capital costs at $140 million and annual operating costs at $22 million to bring the 600 water systems currently over 20 µg L^{-1} to under 20 µg L^{-1}. These estimates are preliminary in that the US EPA is expected to provide updated estimates with proposed regulations. In order to estimate a total annual cost we assume that the capital cost is financed at an interest rate of 5% above inflation and amortized (paid off) over 30 years. These assumptions lead to estimated annual capital plus operating costs of $31 million.

Corresponding annual cleanup costs per water system are about $50,000 ($31 million/600 systems) or $22 per person affected ($31 million/1.4 million people). However, costs per system and per household are expected to vary widely depending on the size and type of water system and the water characteristics.

Analysis

The first step in the value-of-information calculations is analyzing our situation *without* further information. For our example this entails calculating the costs and risks of the current 50 µg L^{-1} drinking water standard versus a 20 µg L^{-1} standard. The parameters from our decision tree were entered into a computerized spreadsheet to calculate the probabilities and risks of each possible outcome. The risks at each endpoint are calculated using the assumptions regarding exposure, delivered dose, dose-response, and risk multipliers for internal cancer risk and epidemiology uncertainty. The probability at each endpoint is the probability that the information in all branches leading to that point correctly represent the actual underlying risk. This is calculated as the product of probabilities on each branch leading to that endpoint. Once we have the risks and probabilities at each endpoint, we calculate the probability-weighted average, or "expected value," of cancers under each of the 2 water standards. To estimate the value of research alternatives, we then re-evaluate the expected value with the various uncertainties resolved. (For the purposes of our illustrative example we neglect concerns about the *distribution* of risk across society or societal risk aversion, allowing us to use the expected value as a basis for comparing the alternatives. Issues of risk distribution and aversion could be incorporated into the analysis without changing the overall approach, but would add complexity to the evaluation of the decision tree's endpoints.)

Results

Results of the decision tree calculations are shown in Figure 5. The values are expected values, *i.e.*, probability-weighted averages across all scenarios. There is a 15% chance that the total number of skin and internal cancer cases will exceed the upper error bound shown, and a 15% chance that the total will be less than the lower

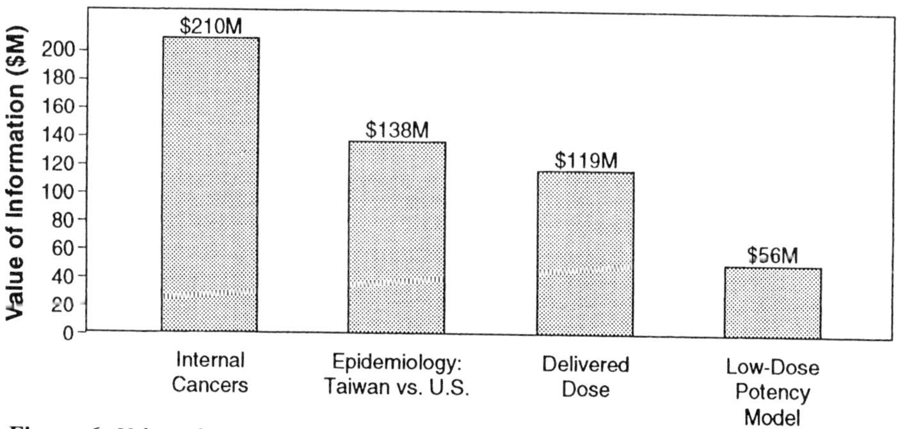

Figure 6 *Value of completely resolving uncertainty for each of 4 topics. These results are based on illustrative data and assumptions. Values are shown in mllions of dollars, and are expected net present values over 30 year horizon using 5% discount rate net of inflation.*

Table 5 *Effect of alternative prior probabilities on value of information. (Based on illustrative data and assumptions.)*

Internal cancer risk scenario:	Possible risk of internal cancer: 5 x skin cancer risk	Same as skin cancer risk	No internal cancer risk	Value of information ($M)
Base case	0.2	0.4	0.4	210
Low internal cancer risk	0.1	0.2	0.7	161
High internal cancer risk	0.3	0.4	0.3	165

bound shown. The expected number of skin cancer cases associated with arsenic ingestion (lower portion of each bar) at the 50 and 20 μg L^{-1} standards are similar to a recent estimate of 1,684 cases annually (US EPA, 1988, pp. 32). The reduction in cancer cases associated with the reduced drinking water standard is modest: 18 cases of skin cancer and 25 cases of internal organ cancer, for a total of 43, or 1.2% of the total of 3,472 cancer cases. This is because (1) about 75% of exposure to inorganic arsenic is associated with food consumption, and (2) most water systems are already below the 20 μg L^{-1} standard. For comparison, Figure 5 also shows the risk neglecting all water exposure. This illustrates that in our analysis about 3/4 of the total risk from arsenic-induced cancer is due to food-borne arsenic.

In order to decide between the 50 μg L^{-1} and 20 μg L^{-1} standards, we must evaluate the benefits of reduced cancer cases versus the costs of implementing the lower standard. To do this we assign monetary values to avoiding a case of skin cancer or an internal organ cancer. Although such valuation is difficult, methods that do not explicitly value health effects must do so implicitly. There has been considerable discussion regarding the use of such valuations in setting public policy (Travis *et al.*, 1987; Miller, 1990).

For skin cancer the risk of mortality is low with early treatment. However, the risk of mortality is significant with internal cancer. For this reason the value assigned to an avoided internal cancers should be considerably higher than the value assigned to an avoided skin cancer. For the purposes of this analysis we initially assign a value of $25,000 to avoiding one case of skin cancer and $1.5 million to avoiding one case of an internal cancer.

Given these assumptions, the costs and benefits of reducing drinking water standards from 50 μg L^{-1} to 20 μg L^{-1} are shown in Table 4. As shown, reducing the drinking water standard from 50 μg L^{-1} to 20 μg L^{-1} appears cost-effective, offering $7.3 million annually in net benefits. Using a 5% real interest rate and a 30 year horizon, this corresponds to a net present value of $112 million in benefits.

The analysis of expected costs and benefits described above is useful for

Table 6 *Reducing uncertainty regarding one issue (e.g., internal cancer) affects the value of information from other research. (Based on illustrative data and assumptions.)*

Remaining uncertainty	Value of information before resolution of internal cancer uncertainty ($M)	Value of information if internal cancer risk uncertainty resolved, and risk found to be:		
		5 x Skin ($M)	Same as Skin ($M)	None ($M)
Nonlinearity of delivered dose	119	40	84	0
Low-dose potency model	56	0	52	0
Epidemiology applicability to USA	138	1	97	0

evaluating the best action, but our primary goal is to consider the value of arsenic research options. Using the decision tree and the methods described above, the values of fully resolving the uncertainty at each of the four uncertainty nodes were calculated. The results, as present values over 30 years, are shown in Figure 6. The values range from $56 million, for resolving uncertainty regarding the low-dose potency model, to $210 million for resolving uncertainty regarding the importance of internal cancers. Resolving uncertainty regarding either the nonlinearity in delivered dose or the applicability of epidemiology studies to the US have values of about $120-140 million.

An US EPA spokesman recently estimated the cost of undertaking research on these topics to be about $2.6 million (Fowle, 1991). Although our results are highly approximate, they suggest that the value of research for any one of the four topics considered here are far higher than this cost. Thus it appears that biologically-based research regarding health risks associated with arsenic ingestion may be a good public investment.

Discussion

High stakes and close decisions lead to high values of information

To improve our intuition regarding value of information, let us consider two important factors that contributed to the high values of information. First, substantial risks and resources are at stake: the present value of the $31 million annual costs to achieve the 20 $\mu g\ L^{-1}$ standard over 30 years is $478 million, and the present value of the annual total benefits of $38.3 million is $590 million. Consequently, hundreds of millions of dollars in potential costs and benefits are at stake. It is interesting to note that costs may be about 10 times higher to achieve a 5 $\mu g\ L^{-1}$ standard and fifty times higher to achieve a 2 $\mu g\ L^{-1}$ standard (personal communication with US EPA

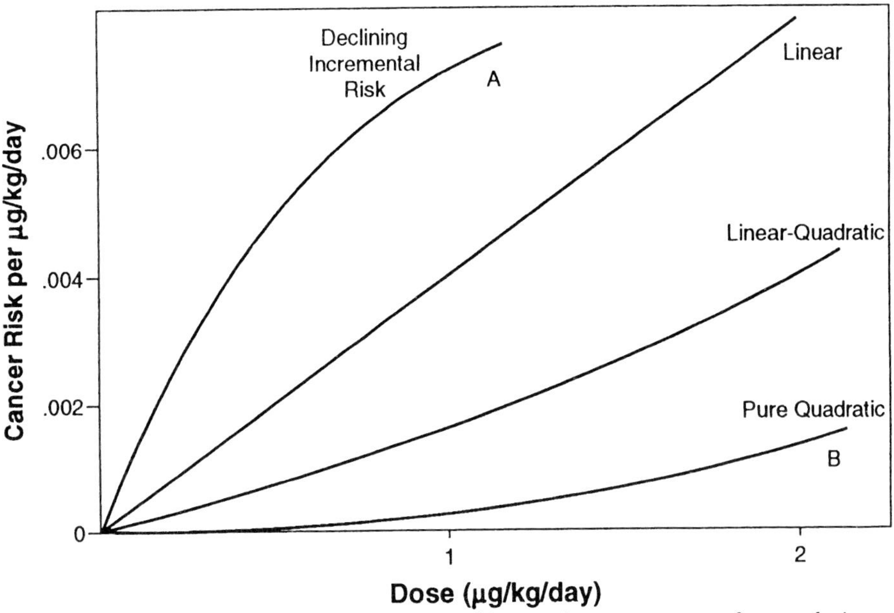

Figure 7 *Range of expected annual cases of internal organ cancers for resolutions of each uncertainty. (Values shown for each assume other uncertainties are not resolved.) Results are based on illustrative data and assumptions.*

and American Water Works Association). Thus the stakes, and the estimated values of information, would tend to rise considerably for more stringent standards.

Second, costs and benefits are similar and uncertainties are large. In our example, benefits are only about 25% greater than costs, and uncertainties affect risks by factors of 2 or greater. As a result, the resolution of an uncertainty can readily change the balance between costs and benefits, and can therefore change the decision. In this situation the value of information tends to be high. Conversely, when benefits are far larger than costs, or *vice versa*, the favored action is clear and the decision tends to be robust. In such one-sided cases, the value of resolving uncertainty is lower since further information is less likely to change the decision.

We can illustrate both the effect of having higher stakes and having a larger difference in the evaluation of the decision alternative by considering whether to adopt a 5µg L^{-1} standard or keep the 50 µg L^{-1} standard. In order to evaluate this decision we: (1) increased the control cost estimate from $31 million annually to $295 annually (capital cost of $1,320 million and annual cost of $210 million), again based on recent personal communication with the US EPA and the American Water Works Association, and (2) assumed that under the 5 µg L^{-1} standard all water supplies currently exceeding 5 µg L^{-1} are brought down to an average of 4 µg L^{-1}. The effect of this reduction is to reduce total cancers by an additional 1.5% (50 annual cases) below the cancer cases under the 20 μg L^{-1} standard, for a total reduction of 2.7% (93 annual cases) relative to the 50 μg L^{-1} standard.

Valuing avoided cancers as above, the costs of the 5 μg L^{-1} standard exceed the benefits by about $154 million annually, or by a present value of $2,400 million over 30 years. Thus, with these assumptions, additional regulation to achieve the 5 μg L^{-1} does not appear cost-effective. Further, the decision is not particularly close, with benefits being only about 50% of costs. It is not surprising that with the $1.5 million valuation for avoided internal cancer cases the 5 μg L^{-1} standard appears less cost-effective than the 20 μg L^{-1} standard, since the cost is roughly 10 times higher but the number of cancer cases avoided only increases by a factor of 2. If the value of avoided internal cancers is increased to $3 million, then the costs and benefits of the 5 μg L^{-1} standard are about equal.

The values of information are also changed in the new decision context of the 5 μg L^{-1} standard. Using the $1.5 million value for internal cancer cases, the value of research regarding internal cancers increases by about a factor of 3 to $622 million. This reflects the significantly higher stakes of an action that costs much more but also may avoid more cancer. However, for two of the four uncertainties the value of information is actually reduced to zero. This is because the choice is sufficiently one-sided (costs far greater than benefits) that the decision would not be changed regardless of research outcomes on these two topics. The value of research regarding the applicability of epidemiology results to the USA remains substantial at $75 million, but is less than its value of $138 million when the 20 μg L^{-1} standard was being considered, again because the choice is now more one-sided. If the value of avoided cancers is raised to $3 million, then all of the uncertainties can easily swing the decision, and the increased stakes lead to very high values of information for all four uncertainties. Present values range from $695 million for uncertainties in the dose-response model to $2,145 million for uncertainties regarding internal cancer risk.

Greater uncertainty increases the value of information

A wider range of possible outcomes regarding an uncertainty tends to increase the value of resolving the uncertainty. To explore the effect of increasing uncertainty, we revisit the range of uncertainty regarding the dose-response model. In the baseline analysis we considered a dose-response model that was purely linear or a model including both linear and quadratic terms. However, models implying either higher or lower risks at low exposures cannot be eliminated using currently available arsenic data. This is because the Taiwanese data have considerable uncertainty in exposure estimates. Exposures, for which elevated cancer risks have been observed in Taiwan, are 5 to 50 times higher than exposures of nearly all US citizens. Thus it is necessary to extrapolate to lower doses, where the nature of the dose-response relationship may be different.

The actual low-dose potency may be found to be either higher or lower than the potency predicted by extrapolation of the linear or linear-quadratic models. For example, some mechanisms of action could suggest decreasing incremental risk as exposure increases. Such a model could predict risk following a curve similar to curve "A" shown in Figure 7. Mechanisms leading to such a relationship could arise, for example, if a metabolic pathway producing a carcinogenic metabolite became saturated at high dose levels (Bailar *et al.*, 1988).

Conversely, mechanisms may be found that suggest very low risks at low doses, with risks increasing more rapidly at moderate and high doses. For example, a mechanism could suggest that the dose-response model is a pure quadratic at low doses, as shown by Curve "B" in Figure 7. For example, such a curve may approximate the potency if risk follows second-order kinetics with respect to arsenic concentration. If chromosome alterations at two or more points were required (as in sister chromatid exchanges), such a relationship could hold. If arsenic consumes a cellular resource important to DNA repair, such as methyl donors, cancer risk may remain low until dose reaches a level that begins to exhaust the resource (Mass, 1992).

To demonstrate the effect of increasing the uncertainty of the dose-response model we keep the linear model as one option, with a lifetime cancer risk of 4.0×10^{-3} for 1 $\mu g\ kg^{-1}\ day^{-1}$ of arsenic, and select a pure quadratic low-dose model as the other alternative. Our least-square fit of a pure quadratic function to data for either 69 year-old Taiwanese males or females yields a risk of 1.1×10^{-4} for 1 $\mu g\ kg^{-1}\ day^{-1}$. Fitting a pure quadratic through the lowest dose data for 69 year old Taiwanese males or females yields risks of 1.8×10^{-4} and 4×10^{-4}, respectively, for 1 $\mu g\ kg^{-1}\ day^{-1}$. For our example we select an average value of 2×10^{-4} at 1 $\mu g\ kg^{-1}\ day^{-1}$ as the risk implied at low doses using a pure quadratic low-dose model. (We emphasize that this very simple derivation is only intended to provide a plausible value for the purpose of illustration.) Thus we replaced the linear-quadratic low-dose potency of 1.8×10^{-3} per $\mu g\ kg^{-1}\ day^{-1}$ with our pure-quadratic estimate of 2×10^{-4} at 1 $\mu g\ kg^{-1}\ day^{-1}$ and re-calculated the value of information assuming that the linear and pure quadratic relationships are equally likely. The result is that the value of information for resolving uncertainty regarding the low-dose potency model rises from \$56 million to \$168 million. With the current standard of 50 $\mu g\ L^{-1}$ the expected total number of cancer cases also falls by about 27%, to 2,514. If the pure quadratic dose response relation were found to be correct, the expected number of cancer cases would fall to 239, or about 10% of the expected value with the uncertainty in the dose-response relationship included.

Probabilities of research outcomes affect value of information

Probabilities assigned to outcomes also affect the value of information, as was demonstrated in the example of the hypothetical carcinogen. Table 5 shows the dependence of the value of information regarding internal cancers from ingesting arsenic on the probabilities assigned to internal cancer risk scenarios. Two sets of estimated probabilities are shown in addition to the base case set. One of the new cases has a lower expected risk of internal cancers than the base case and the other has a higher expected risk. The value of information is about 20% lower in both cases than it is in the base case. This is because the base case probabilities in the situation where expected benefits and costs of the 20 $\mu g\ L^{-1}$ standard were closest. Both of the other sets of probabilities lead to decisions in which the comparison of expected costs and benefits is not as close, with a resulting reduction in the value of information. However, in this example (1) the value of information is not very sensitive to probability assignments; and (2) the value of information remains high relative to research costs.

Value of resolving uncertainty depends on other uncertainties
Resolving one uncertainty usually changes the value of resolving another. For example, Table 6 shows how resolving the uncertainty regarding internal cancer will affect the value of information regarding the other uncertainties. The values of resolving each remaining uncertainty is shown for each of the possible resolutions of internal organ cancer risk.

In each case the value of information is reduced by knowledge about the importance of internal cancer. This is consistent with the general (but not universal) trend that as more is known about a problem (*i.e.*, uncertainties are resolved) the value of resolving remaining uncertainties falls. This reflects a sort of declining marginal value of information in that, as we learn more about a situation, the chances are reduced that new information will change our decision-making. Since we have not included all possible sources of uncertainty, we will generally underestimate total uncertainty. This result suggests that our values of information may underestimate the true value of resolving uncertainties.

The value of all information falls to zero if internal cancers are not found to be a problem. This is because if internal cancers are not associated with arsenic ingestion, then reducing the standard to 20 $\mu g\ L^{-1}$ is not cost-effective, regardless of the outcome of additional research. The values of information are also lower if internal cancers are found to occur with high frequency relative to skin cancer. This is because the reduced standard is likely to be cost-effective under these circumstances, regardless of additional findings.

Conclusions

Over the medium and long term, research is our principal means for improving the cost-effectiveness of risk reduction measures. For this reason, and because research can be expensive and have long lead times, the selection and prioritization of research should be a major component of policies aimed at reducing cancer risks. Value-of-information calculations offer one means to estimate the potential value of alternative research options. These calculations involve:

(1) Identifying the important regulatory decisions,
(2) Identifying the important uncertainties, and possible resolution of these uncertainties, that affect our ability to regulate cost-effectively, and
(3) Obtaining expert assessments of the likelihood that various uncertainties will be resolved in various ways.

While numerical results are the primary goal, these steps may also add to the understanding of the decisions and uncertainties in regulating a potential carcinogen.

The arsenic analysis uses illustrative data, although we believe the estimates are realistic based on our review of the literature. The following conclusions and comments should be considered in the context of our assumptions. First, our calculations suggest that biologically-based arsenic research may be highly cost-effective. Under most scenarios we estimate values of resolving uncertainties regarding arsenic potency and mechanism of action to be in the \$100 million to \$1,000 million range. This suggests, for example, a research program costing \$10 million over 5 years is more likely to be an under-investment than an over-investment. The values of information shown in this analysis for arsenic research are high because: (1) widespread exposure and the potency of arsenic may lead to many cancer cases; (2) costs of reducing exposure through drinking water are high, and (3) uncertainties are great.

Information resolving uncertainty about the risk of internal organ cancers may have the highest value. In the base-case analysis, resolving this uncertainty has a value of over \$200 million. If the alternative regulatory level is 5 $\mu g\ L^{-1}$ or less, and if avoided cases of cancer are assigned a higher value, then (at least for the \$3 million/case), then the value of information could exceed \$1,000 million. This information has high value because internal organ cancers are much more dangerous than skin cancer and there is considerable uncertainty about the magnitude of the risk of internal organ cancer.

Our illustrative analysis suggests that some reduction in drinking water standards may be cost-effective, but it also highlights the limitations of lower standards. First, most people have low exposure to arsenic in drinking water. As a result, reducing the drinking water standard from 50 $\mu g\ L^{-1}$ to 20 $\mu g\ L^{-1}$ only reduces the population-wide risk associated with drinking water by about 5%, and a reduction to 5 $\mu g\ L^{-1}$ reduces drinking water risk by just another 10%. This is because widespread exposure to low doses in water creates most of the population-wide risk. Second, about 3/4 of expected exposure to inorganic arsenic may be from food. This, combined with the widespread low concentration exposure to water-borne arsenic mentioned above, may limit the cancer reductions from lower drinking water standards to only about 1-5% of total arsenic related cancers. As shown in our example, lowered drinking water standards may still be cost-effective, depending on the assumptions used. In addition, there may be important public health issues, and opportunities to reduce risks, associated with arsenic ingested in food.

References

Bailar, J.C. III *et al.*, 1988. One-hit models of carcinogenesis, conservative or not? *Risk Analysis*, **8**(4), 485-497.

Bates, M.N., Smith, A.H. and Hopenhayn-Rich, C. 1992. Arsenic ingestion and internal cancers: a review. *American Journal of Epidemiology*, **135**(5), 462-476.

Brown, K.G., Boyle, K.E., Chen, C.W. and Gibb, H.J. 1989. A dose-response analysis of skin cancer from inorganic arsenic in drinking water. *Risk Analysis*, **9**(4), 519-528.

Chen, C.W. and Chen, C-J. 1988. Integrated quantitative cancer risk assessment of inorganic arsenic. *Proceedings of a Symposium on Health Risk Assessment of Environmental, Occupational and Life Style Hazards,* 20-22 December 1988. Taipei, Taiwan.

Chen, C-J. and Wang, C-J. 1990. Ecological correlation between arsenic level in well water and age-adjusted mortality from malignant neoplasms. *Cancer Research*, **50,** 5470-5474.

Clemen, R.T. 1991. *Making Hard Decisions: An Introduction to Decision Analysis.* PWS-Kent Publishing Company, Boston, Massachusetts.

Fowle, J.R. III. 1991. Health effects of inorganic arsenic in drinking water: research needs - a case study. *Presentation to the 25th Conference on Trace Substances in the Environment and Annual Meeting of the Society for Environmental Geochemistry and Health, 21 May 1991. Columbia, Missouri.*

Fowle, J.R. III. 1992. Health effects of arsenic in drinking water: research needs. *Env. Geochem. Health*, **14,** 63-68.

Gibb, H. and Chen, C. 1989. Is inhaled arsenic carcinogenic for sites other than the lung? In: U. Mohr (ed.), *Assessments of Inhalation Hazards.* Springer-Verlag Berlin Heidelberg.

Howard, R.A. and Matheson, J.E. (eds). 1983. *The Principles and Applications of Decision Analysis,* Strategic Decisions Group, Menlo Park, California.

Mass, M.J. 1992. Human carcinogenesis by arsenic. *Env. Geochem. Health*, **14**, 49-54.

Miller, T.R. 1990. Plausible range for the value of life: red herring among the mackerel. *J. Forensic Economics*, **3,** 17-40.

Pershagen, G. 1986. Sources of exposure and biological effects of arsenic. In: Fishbein, L. (ed.), *Environmental Carcinogens Selected Methods of Analysis, Some Metals: As, Be, Cd, Cr, Ni, Pb, Se, Zn*, pp. 45-61. International Agency for Research on Cancer (IARC). Lyon, France.

Raiffa, H. 1968. *Decision Analysis: Introductory Lectures on Choices Under Uncertainty.* Addison-Wesley Publishing Company, Reading Massachusetts.

Smith, A.H., Hopenhayn-Rich, C., Bates, M.N., Goeden, H.M., Hertz-Picciotto, I., Duggan, H., Wood, R., Kosnett, M. and Smith, M.T. 1992. Cancer risks from arsenic in drinking water. *Environ. Health Perspect.*, **97,** 259-267.

Travis, C.C., Crouch, E.A.C, Wilson, R. and Klema, E.D. 1987. Cancer risk management: a review of 133 Federal regulatory decisions. *Env. Sci. Tech.*, **21**, 415-420.

US EPA. September 1984. *Arsenic, Occurrence in Drinking Water, Food, and Air.* Science and Technology Branch, Criteria and Standards Division, Office of Drinking Water. US EPA Contract No. 68-01-6388. Work Assignment 29. US EPA Task Manager, William Coniglio.

US EPA. July 1988. *Special Report on Ingested Inorganic Arsenic: Skin Cancer; Nutritional Essentiality*. Risk Assessment Forum. US EPA/625/3-87/013.

US EPA Science Advisory Board. September 1989. *Science Advisory Board's Review of the Issues Relating to Arsenic Contained in the Phase II Proposed Regulations from the Office of Drinking Water.* Letter report to the Administrator.

US EPA Science Advisory Board. May 1992. *Review of Arsenic Research Recommendations: Review by the Drinking Water Committee of the Office of Research and Development's Arsenic Research Recommendations.* EPA-SAB-DWC-92-018.

US EPA Office of Water. 1992. *Second draft for the drinking water criteria document on arsenic.* Prepared Under ICAIR Program No. 1524, for EPA Contract 68-C8-0033, ERG Subcontract No. LSI-8700, ERG Work Assignment No. 2-19, Life Systems, Inc. Work Assignment No. 391524.

2 Human oral exposure to inorganic arsenic

Denis R. Borum and Charles O. Abernathy*

Human Risk Assessment Branch (4304), Office of Water, US EPA, Washington, D.C. 20460-0001, USA

Abstract

The US EPA is in the process of regulating inorganic arsenic (As) in drinking water. To accomplish this task, it is important to develop scenarios that accurately reflect human exposure patterns. This report will focus on exposures from food and water and only briefly mention air and soil exposure. It is not intended as a comprehensive assessment of exposure variations; rather, it describes common sources and levels of As that may be typically ingested. In industrial settings, or other occupational situations, people may experience exposures to inorganic As far greater than the average non-occupationally exposed individual. For the non-occupationally exposed person, drinking water and food represent nearly all of the As intake. Available studies indicate that As uptake after dermal exposures from soil or water are very low. Similarly, inhalation exposures are normally low (approximately 2% of total exposure at levels of about 0.5 μg day^{-1}) compared to the ingestion exposures. Drinking water intakes of 20 μg As day^{-1} represent approximately 60% of total intake, while dietary intake represents approximately 35 to 40% (estimated intake range of 11 to 14 μg day^{-1}).

Introduction

Arsenic (As), is widely distributed, yet constitutes only 0.0005% of the earth's crust. It occurs most often in complexes; there are over 100 arsenic-containing minerals. Arsenic and its compounds are used in industry. The trioxide is commercially obtained as a smelter product of various arsenic-containing ores. Arsenical pesticides were used extensively, however, these uses sharply declined after the 1960s. Arsenicals are also used in pharmaceuticals for both humans and animals. Other

---

The opinions expressed in this manuscript are those of the authors and do not necessarily reflect the opinions or policies of the US Environmental Protection Agency.

*To whom correspondence should be addressed.

uses include wood preservatives, feed additives, dye stuffs, hide preservatives, glass and enamel production, various metallurgy applications, and the manufacture of transistors, lasers and semiconductor devices (US EPA, 1984; Wickstrom, 1972; Suta, 1980).

Arsenic is released to the environment from a variety of natural and anthropogenic sources. Weathering, dissolution and erosion of As into natural waters, dustfall from meteoritic and volcanic activity, and animal and plant decay are among the major natural sources (NRCC, 1978; Walsh *et al.*, 1979). The major industrial sources of release include the metal and metal alloy manufacturing industries, petroleum refining, burning of fossil fuels (predominantly coal), various agricultural, wood and waste incinerations, and manufacturing (Walsh *et al.*, 1979; Suta, 1980; US EPA, 1980; Berry and Wallace, 1974).

Total As is comprised of inorganic and organic forms. The inorganic form is the most toxic; organic forms exhibit comparatively little or no toxicity. The primary sources of human exposure to As are ingestion, dermal contact and inhalation. Since As is found in both food and water, these two sources will be considered as separate sources of exposure. Available information on the levels found in drinking water and in foods are summarized and used to make estimates of the As intake by a non-occupationally exposed individual. Brief sections describing dermal and inhalation sources are also presented.

Exposure

Dermal exposure

Dermal exposure occurs after contact with water and/or soil. In soil, As is immobilized in numerous mineral complexes, as well as by soils high in organic content. Some highly volatile arsines may be released into the atmosphere. However, they are also readily oxidized and reenter water and soil via dry fallout or precipitation (Lisk, 1972; Suta, 1980). Soil concentrations of As vary markedly. For example, Wickstrom (1972, in Lisk, 1972) found As in soil ranging from 1 to 40 mg kg^{-1}. These levels are similar to those found in Southwest England, which ranged from 0.1 to 40 mg kg^{-1} (Colburn *et al.*, 1975). A recent report using human skin from cadavers estimated 24-hour percutaneous inorganic As absorption from soil to be 0.8% (Wester *et al.*, 1993). The same study indicated percutaneous As absorption from water was 1.9%. Additional work is needed, but these data indicate that dermal absorption of As from either soil or water is low.

Air exposure

The atmosphere is a significant channel for the recycling of As to water and soil via fallout or precipitation (US EPA, 1984). Although As may be recycled back into the atmosphere, it is usually fixed in soil and sediments. Since photolysis is not an important process, As concentrations in the ambient air are usually low. However, concentrations up to 1 μg m^{-3} may be found near power plants burning fossil fuel with high As content or near ore smelters (IPCS, 1981)) and have been reported as high as 25 μg m^{-3} in New York City (Leaderer *et al.*, 1978). In such situations, local As exposure from the air could be significant.

The most recent information available for As occurrence in air is from the Aerometric Information Retrieval System (AIRS), the air monitoring data base of the EPA Office of Air Quality Planning and Standards (OAQPS). Data from 8 states (Arizona, Colorado, Illinois, Kansas, Minnesota, Montana, Tennessee and Texas) were reviewed for airborne As levels for the years 1980 to 1991. Arsenic was detected at 118 of 257 sites sampled, based on a reporting limit of 0.01 μg m^{-3} (0.25 μg m^{-3} for the state of Texas). Average concentrations from the sites ranged from 0.01 to 0.454 μg m^{-3}. Due to the possible presence of As at concentrations below the detection limit, the overall mean was calculated in three ways: substituting values at or below the detection limit with zero, the detection limit, and one-half the detection limit. Based on these three calculations, low, high and median estimates of the mean were calculated to be 0.0205, 0.0285 and 0.0254 μg m^{-3} (OAQPS, 1992).

Data from the EPA studies in the 1980s described above are generally higher than EPA and state studies from the 1970s. EPA's National Air Sampling Network (the predecessor of AIRS) results for composite sampling data from 1974 show a range of annual average concentrations of not detected to 0.083 μg m^{-3}, with an overall annual average of 0.003 μg m^{-3}. A few available state sampling studies (from New Jersey, New York, California, Texas and Washington) from the 1970s also show comparably lower levels than the AIRS data (US EPA, 1993). It is not clear from these studies whether the more recent higher data are due to greater industrial releases or other factors. It is likely that analytical measurement capabilities have improved; however, more research is needed to understand any trends and the implication of higher recent levels of As in air.

As part of the 1980 Total Exposure Assessment Methodology (TEAM) prepilot study, the As exposure of nine individuals from Elizabeth and Bayonne, New Jersey and three individuals from Research Triangle Park, North Carolina were measured. Of the nine individuals from New Jersey, approximately 5 were classified as occupationally exposed to As while all of the North Carolina and the remaining New Jersey participants represented low to moderate air exposures. In New Jersey, a total of 18 ambient air samples were collected from areas near the breathing zones of the nine individuals. Concentrations reportedly ranged from below the detection limit to 0.057 μg m^{-3}. In North Carolina, As was not detected in any of the nine samples collected (Pellizzari *et al.*, 1980; Sparacino *et al.*, 1980). Detection limits were not reported.

Oral exposure

(a) Drinking water exposure: The focus of this section is on studies which are national in scope, but brief comparison is made for available state studies. It is not an exhaustive review, but does provide a reasonable basis for estimating average As exposure. Laboratories that provide analyses on drinking water samples typically report the results as total inorganic As. In water, As in the +5 valence state is the predominant form followed by +3, with the +3 state generally found only under anaerobic conditions. US drinking water data are available from four national surveys of public drinking water supplies. The national surveys were designed such that data from water supplies throughout the US would comprise the ranges and averages indicated. While these are, no doubt, somewhat representative of the country, they are also rather limited in that the number of supplies sampled is obviously far less than the vast number of supplies throughout the country.

In the 1978 Community Water Supply Survey (1978 CWSS), the US EPA surveyed 500 community water supplies for organic and inorganic compounds. Both raw and finished water were collected from each system (one to five samples per system), with finished water samples collected at the plant and/or in the distribution system. The detection limit varied with each sample, ranging from 2.5 to 5 $\mu g\ L^{-1}$. Due to analytical problems, results from only 259 ground-water and 94 surface water supplies are available. Arsenic was detected in 49 ground-water supplies (19%) and 3 surface water supplies (3%), with concentrations ranging from 2.5 to 28 $\mu g\ L^{-1}$ and 2.5 to 10.7 $\mu g\ L^{-1}$, (respectively US EPA, 1983).

The Rural Water Survey (RWS) was conducted by the EPA between 1978 and 1980 to assess the quality of drinking water in rural America. Samples were collected from 2,654 households (648 public water supplies) across the U.S. and analysed for physical, radiological, and chemical constituents. Of the supplies surveyed, only 71 ground-water supplies and 21 surface water supplies were analysed for As, a specified subset of the RWS scheduled for inorganics sampling (detection limit of 2 to 5 $\mu g\ L^{-1}$). Arsenic was detected in 23 ground-water supplies (32%), with a range of 2 to 82 $\mu g\ L^{-1}$. Two surface water supplies (10%) had detectable levels of As, with concentrations of 3 $\mu g\ L^{-1}$ and 5 $\mu g\ L^{-1}$ (US EPA, 1989).

As part of the National Organics Monitoring Survey (NOMS) conducted in 1976 and 1977, levels of 27 trace elements were sampled in the distribution systems of 113 supplies. Results were available for As in 88 surface and 18 ground-water supplies (detection limit was 5 $\mu g\ L^{-1}$). A total of 21 surface water supplies and 7 ground-water supplies contained detectable As levels. Concentrations ranged from 5 to 23 $\mu g\ L^{-1}$ for surface water supplies (mean of 9.2 $\mu g\ L^{-1}$ and median of <5 $\mu g\ L^{-1}$) and a range of 5 to 18 $\mu g\ L^{-1}$ for ground-water supplies (mean concentration of 9.3 $\mu g\ L^{-1}$ and median of <5 $\mu g\ L^{-1}$; (US EPA, 1989).

The National Inorganics and Radionuclides Survey (NIRS) was conducted in the mid-1980s to provide data on the occurrence of inorganics and selected radionuclides in drinking water from ground-water sources. A total of 1,000 community water supplies were selected from the Federal Reporting Data System (FRDS) data base. To enhance the representativeness of the sampling group, the community ground-water supplies in the United States (47,700 based on FRDS inventory for fiscal year 1985) were stratified into four categories based on population served (25-500; 501-3,300; 3,301-10,000; and >10,000). Random samples were then selected from each size category in proportion to the number of supplies in those size groups nationally. The resulting sample represented about 2% of the nation's community ground-water supplies for each group. The minimum reporting level for

As was 5 μg L^{-1}. Results indicated that 7% (64 of 982 samples) had detectable levels of As. Values ranged from 5 to 48 μg L^{-1} (mean of positives of 13 μg L^{-1} and median of 9 μg L^{-1}; (US EPA, 1988).

National exposure to As from public water supplies for ground-water systems was estimated using the NIRS data. For surface water systems, the estimates were made based on the combined results of the 1978 CWSS, NOMS, and RWS. The distributions provide a basis for estimating the likelihood that As will be present in a public water supply. A more detailed explanation of this analysis is described in US EPA (1993). The total estimated cumulative population exposures for community and non-transient non-community (NTNC) ground-water and surface water supplies indicate that about 15% of the population is exposed to As levels greater than 2 μg L^{-1}. With the exception of the NTNC surface water supplies (which show lower concentrations), about 5% of the population is estimated to be exposed to levels greater than 5 μg L^{-1}. Approximately 2% of the total population is estimated to be exposed to levels greater than 10 μg L^{-1}.

Results from additional surveys and other data sources indicate similar levels of As in drinking water (US EPA, 1993). A large nationwide survey by Greathouse and Craun (1978) found detectable As levels in 67% of 3,834 drinking water samples. The overall mean concentration was about 2.4 μg L^{-1} (detection limit = 0.1 μg L^{-1}). However, numerous studies from areas associated with elevated geologic concentrations of As show much greater concentrations in drinking water. Studies from California and Nevada indicate mean values up to 80 μg L^{-1} (Fontaine, 1991). Studies from California, Oregon, Utah and Alaska show maximum As concentrations of 1,400, 2,150, 270 and 2,450 μg L^{-1}, respectively (US EPA, 1993). Under these conditions, drinking water exposure could potentially pose a significant health risk. According to the Federal Reporting Data System, public water supply violations of the current As standard of 50 μg L^{-1} are most often seen in smaller systems (*i.e.*, those serving fewer than 500 people).

(b) Dietary exposure: Since 1961, the US Food and Drug Administration (FDA) has periodically conducted the Total Diet Study, also known as the Market Basket Study. This survey is used to estimate the dietary intake of pesticides, industrial chemicals, toxic chemicals, radionuclides, and essential minerals.

In 1982, the FDA abandoned the analysis of composite food samples in favor of individual food analysis and revised the diet to reflect updated consumption and nutritional data. The individual food analysis approach allowed the studies to represent the US dietary intake with a relatively small number of food items; subsequently, the FDA was able to provide dietary intakes for a greater number of agesex groups.

The FDA has analysed foods for As content since the early 1960s. The most recent available results from the FDA consisted of separate food analyses conducted from April 1982 to April 1988 and June 1988 to April 1990 in each of which a total of 234 foods were analyzed for As content. To calculate total dietary intakes, results from all samplings are combined by averaging together residues from duplicate foods. Results from each separate food are then categorized into one of 11 food

Table 1 *Percentage of total arsenic present as inorganic arsenic in each food group:a preliminary analysis.*[a]

Food	% of inorganic arsenic
Milk and dairy products	75
Meat– beef and pork	75
Poultry	65
Fish– saltwater	0
– fresh water	10
Cereals	65
Rice	35
Vegetables	0 5
Potatoes	10
Fruits	10

[a]Speciation of the arsenic content of basic food groups based on preliminary data from the Ontario Research Foundation and other sources.

Source: Weiler, 1987 in US EPA, 1988.

groups and averaged. Total dietary intake for three different age groups is then calculated from the residue averages using average daily consumption data for each food group. Data are reported by the FDA as total As (inorganic and organic forms) content in food and was measured as elemental arsenic. Based on these calculations, the average infant (6 months old), toddler (2 years old), and adult were found to have total As intakes of 21.5, 27.6 and 52.6 μg day^{-1}, respectively. In the adult diet, meat, fish, and poultry accounted for the largest contribution to the total intake with a contribution of 31 μg day^{-1} or 60% of the total intake (with fish and seafood comprising the majority at 27.3 μg day^{-1}), followed by grain at 4.4 μg day^{-1} (8.4%) and vegetables at 3.3 μg day^{-1} (5.6%). For infants, milk and milk products contributed 13.4 μg day^{-1} or 63% of the total intake. In the toddler diet, both meat, fish, and poultry; and milk and milk products were the largest contributors with respective contributions of 8.3 μg day^{-1} (30%) and 8.5 μg day^{-1} (31%) (FDA, 1992).

FDA results from the early 1960s for adult dietary intake were approximately 75.7 μg day^{-1} for total As; however, from 1973 to 1975, dietary intake was reportedly reduced to only 12.7 μg day^{-1} due to the decrease in As-containing pesticides (U.S. EPA, 1993). These levels, however, are believed to be an underestimation of the As content resulting from the use of less accurate analytical methods (improved analytical methodology was introduced in 1976). From 1976 to 1981, the total As intake in adult diets ranged from 44.8 to 54.2 μg day^{-1}. As was true with the most recent study results, As was contributed mainly by meat, fish, and poultry. In the infant and toddler surveys, first implemented in 1975, total As intake concentrations fluctuated between 1 to 4 μg day^{-1} for infants and 8 to 18 μg day^{-1} for toddlers. As with the adult studies, analytical methods employed in the earlier studies lacked sensitivity to determine low As levels; therefore, dietary intake may be underestimated.

As stated earlier, the toxicity concerns for As are primarily with its inorganic forms. Although FDA does not report on specific organic and inorganic forms of As in the various foods, other studies have characterized As species in foods. Foods sampled by the Canadian Ministry of the Environment indicate that inorganic As comprises most of the As found in meats (75%), poultry (65%), dairy products (75%), and cereals (65%) (Weiler, 1987 cited in US EPA, 1988). Organic forms predominate in fruits, vegetables, and fish/seafood, with inorganic As contributing 10%, 5%, and 0 to 10%, respectively. About 20% of daily intake of dietary As is inorganic.

Using the most recent FDA Total Diet Study data (FDA, 1992), and the inorganic As percentages from the Canadian study (Table 1), the range of inorganic As from an adult diet is calculated to be between 10.9 and 14.0 μg day^{-1}. The range is due to uncertainty in the percentages of inorganic As from eggs, legumes/nuts/seeds, condiments/fats/oils, sugars/adjuncts, and soft drink beverages, as these were not characteried by the Canadian study. These food categories, however, represent only 11.9% (6.27 μg day^{-1}) of the estimated total dietary As intake for an adult. Using the assumption that none of the As in these categories are inorganic As, the resulting calculation is 10.9 μg day^{-1}. With the assumption that 50% of these categories are made up of inorganic As, the resulting estimate is 14.0 μg day^{-1}, or approximately 20 to 25% of total dietary intake.

Relative source contribution analysis

For most US citizens, food is the major source of total As intake, with inorganic As expected to be about 20% of the dietary total. Air is usually a small source of As intake and soil intake is also likely to be a minor exposure pathway.

The drinking water sampling survey data show As is usually present at levels below 10 μg L^{-1} (an approximation of the mean of positive samples from Federal survey data) when accounting for non-detect samples. Thus, for a majority of persons in the United States using public water supplies, exposure is likely to be less than 10 μg L^{-1}. If As is present at 10 μg L^{-1} in drinking water, an adult consuming 2 L day^{-1} of water would be exposed to 20 μg of As day^{-1}. Using the estimated adult dietary intake range of 10.9 to 14.0 μg As day^{-1}, drinking water would contribute between 58 and 63% of total intake for an adult. Estimating inhalation exposure is much more difficult as the data are not as representative as the data for food and drinking water. Using either the median or high estimates of the mean concentration of As in air (from the AIRS data base), ambient air exposure is relatively insignificant at about 0.5 to 0.6 μg day^{-1}.

Assuming the more common As concentration in drinking water of 2-3 μg L^{-1}, drinking water would contribute less to overall exposure. In these cases, the contribution would be about 22 to 35% of total exposure. On the other hand, in regions of the country (*e.g.*, the western US) with elevated As levels, drinking water not only contributes over 99% of total exposure, but also represents levels posing a potentially significant health risk.

References

Berry, W.L. and Wallace, A. 1974. *Trace Elements in the Environment - Their Role and Potential Toxicity as Related to Fossil Fuels - A Preliminary Study.* Prepared by the University of Southern California, Los Angeles for the Atomic Energy Commission under Contract No. AT(04-1) GEN 12; January.

Colburn, P., Alloway,B.J. and Thornton, I. 1975. Arsenic and heavy metals in soils associated with regional geochemical anomalies in South-west England. *Sci. Total Environ.*, **4**, 359-363.

FDA. 1992. US Food and Drug Administration. *Printout of FDA Market Basket Studies, 1982-1990.*

Fontaine, J.A. 1991. Supervisor, Public Health Engineering, Consumer Health Protection Services, Department of Human Resources, State of Nevada, Letter to Denis Borum, U.S. Environmental Protection Agency, February 19.

Greathouse, D.G. and Craun, G.F. 1978. Cardiovascular disease study-Occurrence of inorganics in household tap water and relationships to cardiovascular mortality rates. In: Hemphill, D.D. (ed.), *Trace Substances in Environmental Health-XII; A Symposium.* University of Missouri, Columbia, MO.

Hinman, K., Schwartz, D. and Soffer, E. 1986. *Analysis of Risks from Exposure to Air Toxics (Santa Clara Valley Integrated Environmental Management Project.* US Environmental Protection Agency, Policy Planning and Evaluation Revised Stage I Report, May 30. pp. 3-1 (122).

IPCS. 1981. International Programme on Chemical Safety. *Environmental Health Criteria.* **18**. *Arsenic.* World Health Organisation, Geneva.D.M.

Leaderer, B.P., Bernstein, D.M., Daisey, J.M., Kleinman, M.T., Kneip, T.J., Knutson, E.O., Lippman, M., Lioy, P.J., Rahn, K.A., Sinclair, D., Tanner R.L. and Wolff, G.T. 1978. Summary of the New York summer aerosol study (NYSAS). *J. Air Pollut. Contr. Assoc.*, **28**(4), 321-327.

Lisk, D.J. 1972. Trace metals in soils, plants and animals. *Adv. Agron.*, **24**, 267-325.

Merck Index. 1989. Budavari, S., (ed.), 11th edn., Merck & Co., Inc., Rahway, NJ, p. 126.

NRCC. 1978. National Research Council of Canada. *Effects of Arsenic in the Canadian Environment.* Associate Committee of Scientific Criteria for Environmental Quality.

OAQPS. 1992. *Printout of OAQPS Air Monitoring Data, 1980-1991.* Office of Air Quality Planning and Standards, US Environmental Protection Agency.

Pellizzari, E.D., Hartwell,T., Zelon,H., Leninger,C., Erickson, M. and Sparacino, C. 1980 (approximately). *Total Exposure Assessment Methodology (TEAM): Prepilot Study-Northern New Jersey.* Research Triangle Institute, NC. Prepared for US Environmental Protection Agency, Washington, D.C. under Contract No. 68-01-3849.

Sparacino, C., Leninger, C., Zelon, H., Hartwell, T., Erickson, M. and Pellizzari, E. 1980 (approximately). *Sampling and Analysis for the Total Exposure Assessment Methodology (TEAM): Prepilot Study _ Research Triangle Park.* Research Triangle

Institute, NC, Prepared for U.S. Environmental Protection Agency, Washington, D.C. under Contract No. 68-02-2688.
Suta, B.E. 1980. *Human Exposure to Atmospheric Arsenic, Final Report.* SRI International, Menlo Park, CA, prepared for U.S. Environmental Protection Agency under Contract No. 68-01-4314 and 6802-2835.
US EPA. 1993. US Environmental Protection Agency. *Drinking Water Criteria Document for Arsenic*, 2nd draft. Office of Water, Washington, D.C.
US EPA. 1989. *Estimated National Occurrence and Exposure to Arsenic in Public Drinking Water Supplies.* Office of Drinking Water, Washington, D.C., prepared by SAIC, Incorporated, Arlington, VA and Wade Miller Associates, Arlington, VA, June 29.
US EPA. 1988. *Special Report on Ingested Inorganic Arsenic. Skin Cancer; Nutritional Essentiality.* Risk Assessment Forum, U.S. EPA, Washington, D.C. EPA Report No. 625/3-87/013.
US EPA. 1984. *Health Assessment Document for Inorganic Arsenic.* Office of Health and Environmental Assessment, Washington, D.C. March, pp. 1-1 to 3-22, 9-1 to 9-4.
US EPA. 1983. *Data on Inorganics from the Community Water Supply Survey.* Provided by E.M. Glick, Drinking Water Quality Assessment Branch, Technical Support Division, Office of Drinking Water, Cincinnati, Ohio.
US EPA. 1980. *Treatability Manual, Volume I: Treatability Data.* Office of Research and Development. EPA600/8/80/042, July.
Walsh, P.R., Duce, R.A. and Fasching,J.L. 1979. Considerations of the enrichment, sources, and flux of arsenic in the troposphere. *J. Geophys. Res.*, **84**(C4), 1719-1726.
Weiler, R.R. 1987. Unpublished data, Ministry of the Environment, Report No. 87-48-45000-057, Toronto, Ontario. Cited in US EPA, 1988.
Wester, R.C., Maibach, H.I., Sedik, L., Melendres, J. and Wade, M. 1993. *In vivo* and *in vitro* percutaneous absorption and skin decontamination of arsenic from water and soil. *Fundamental and Applied Toxicology*, **20**, 336-340.
Wickstrom, G. 1972. Arsenic in the ecosystem of man. *Work Environ. Health*, **9**, 2-8

3 Nationwide and California arsenic occurrence studies

Marshall K. Davis[1*], Kenneth D. Reich[2] and Maria W. Tikkanen[3]

[1]*Metropolitan Water District of Southern California, 700 Moreno Ave., La Verne, CA 91750, USA*
[2]*Central and West Basin Municipal Water Districts, 17140 So. Avalon Blvd., Suite 210,Carson, CA 90746-1218, USA*
[3]*Association of California Water Agencies, 910 K St., Sacramento, CA 95814-3577, USA*

Abstract

In 1994/95, the US Environmental Protection Agency (US EPA) will promulgate a more stringent maximum contaminant level (MCL) for arsenic in drinking water: between 0.5 and 20 µg L^{-1}. In order to evaluate the impact of this more stringent drinking water regulation, three low-level (detection limit, 0.5 µg L^{-1}) arsenic occurrence studies have recently been conducted, one nationwide, one statewide (in California), and one along the coastal plain of southern California.

In the nationwide survey, samples were collected and analysed for 140 utilities across the continental USA from December 1992 to May 1993. The surveyed population represented 36 percent of the US population. The nationwide survey results showed that approximately 56 percent of the utilities surveyed had arsenic levels greater than 0.5 µg L^{-1}, 16 percent had values greater than 2 µg L^{-1}, and 5 percent had values greater than 5 µg L^{-1}. This survey also indicated that higher arsenic levels were located in the western and southwestern USA as compared to the eastern states. The California statewide survey, scheduled to be completed by fall 1993, will detail arsenic levels in over 1,000 representative ground- and surface-water sources. The southern California survey, completed in October 1992, sampled 262 wells, indicating that 90 percent of the wells had detectable arsenic levels of ≥ 0.5 µg L^{-1} and 60 percent of the wells had levels of ≥ 2 µg.L^{-1} Based on the results of these three surveys, the western USA would be severely impacted by a low-level revised arsenic regulatory limit, and the state of California – particularly central and southern California – would be hit especially hard.

*To whom correspondence should be addressed.

Introduction

In 1986, Congress, through the Safe Drinking Water Act Amendments, mandated that the US EPA finalize an interim MCL for arsenic by 1989. The US EPA is now under a court order to propose a revised MCL by fall 1994.

Recent scientific health-effects studies indicate that very low levels of arsenic in drinking water may cause cancer of the internal organs in humans (Bates *et al.*, 1992; Brown *et al.*, 1989; Chen *et al.*, 1992; Smith *et al.*, 1992). The US EPA has already indicated that the new health-risk data will probably result in a much more stringent MCL for arsenic.

The US EPA typically sets MCLs at levels at which the associated risk of excess cancer in the exposed population is considered to be between one in 10,000 and one in 1,000,000. Preliminary estimates of the arsenic levels that correlate with these health risks are 0.25 $\mu g\ L^{-1}$ and 0.0025 $\mu g\ L^{-1}$, respectively (Smith *et al.*, 1992). The current MCL for arsenic is 50 $\mu g\ L^{-1}$. The new MCL range suggested by the US EPA will be 0.5 to 20 $\mu g\ L^{-1}$ or, more likely, 0.5 to 5 $\mu g\ L^{-1}$.

The regulatory time frame will allow the water supply community to provide input into the arsenic rule-making process until March 1994. Intensive research efforts are still urgently needed to provide the US EPA with the best scientific data available. Such research is being planned or is now being conducted by a number of professional and governmental organizations, *e.g.*, the American Water Works Association (AWWA), the AWWA Research Foundation (AWWARF), and the US EPA. As an initial step, this paper will provide arsenic occurrence information for the United States, for the state of California, and for southern California.

Nationwide arsenic occurrence study (Metropolitan Water District of Southern California)

The Metropolitan Water District of Southern California (Metropolitan) is a public, municipal, wholesale water agency that delivers approximately 2.5 million acre-feet of water to 15 million people in a 5,200-square-mile, six-county area of southern California. In light of recent toxicological information and the US EPA's mandate to reconsider the arsenic MCL, Metropolitan decided to conduct a nationwide, low-level arsenic occurrence study. Previous studies--such as the 1978 Community Water Supply Survey (US EPA, 1983), the Rural Water Survey (Brower, 1983), the National Organic Monitoring Survey (US EPA, 1980), and the National Inorganics and Radionuclides Survey (US EPA, 1986)--have included arsenic occurrence information. However, the analytical detection limit for arsenic in these previous studies ranged from 2.5 to 5 $\mu g\ L^{-1}$, which is 5 to 10 times higher than the minimum suggested MCL. Consequently, occurrence data at a lower analytical detection limit were required.

Nationwide study protocol

The nationwide arsenic occurrence study used the AWWA Water Industry Database to identify 160 utilities within the 10 US EPA regulatory regions. Of the 160 utilities in the continental USA that were requested to participate, 147 collected samples for arsenic analyses and 140 allowed their arsenic data to be reported. For the utilities surveyed, 53 (38 percent) had groundwater sources, 78 (56 percent) used surface-water sources, and nine (6 percent) used a blend of both surface and groundwaters. Eighty-eight percent of the utilities surveyed had populations greater than 50,000, with a total surveyed population of 81.6 million people (which represents approximately 36 percent of the US population).

Sampling program and analytical method

Each participating utility received two 500 mL, acid-washed, polyethylene containers to sample for arsenic. Utilities providing treatment (*i.e.*, flocculation, coagulation, sedimentation, and filtration) were requested to sample their treatment-plant influent and effluent. The other utilities (those providing no treatment, or providing only disinfection) were requested to sample two locations, representing different water sources, within their distribution system. After collection, the samples were shipped by overnight express to Metropolitan's Water Quality Laboratory (WQL) in La Verne, California. The samples were acidified immediately upon receipt at the WQL. Each sample was then digested using sample and standard preparation method 3114 B.(4.*d*) from *Standard Methods for the Examination of Water and Wastewater* (APHA, 1992) and was analysed using continuous hydride generation/atomic absorption spectrometric method 3114 C. of *Standard Methods* (APHA, 1992).

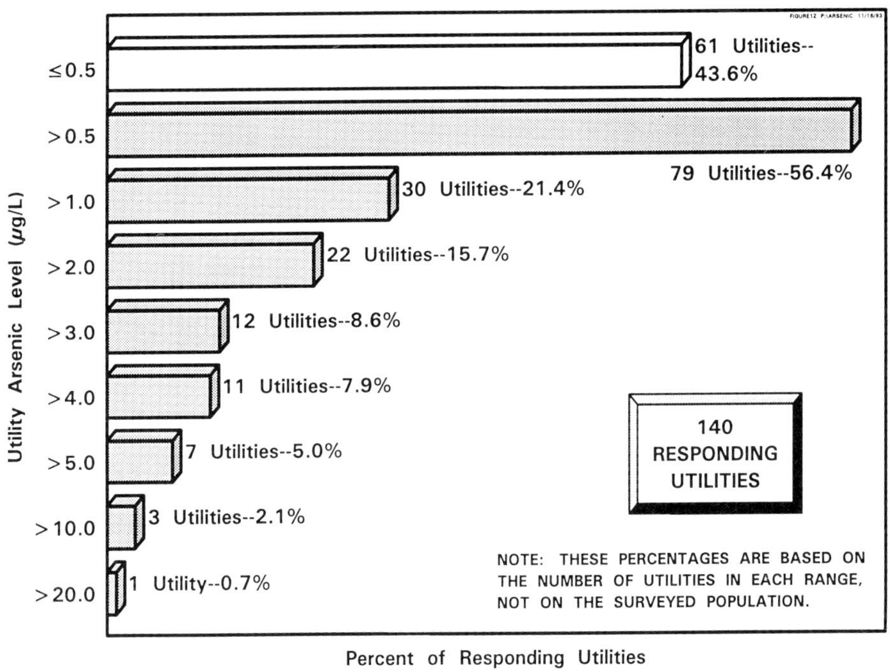

Figure 1 *Percent of responding utilities with arsenic levels greater than selected values (effluents).*

Quality control (QC) consisted of 34 duplicates and 34 spikes. All QC samples were within precision and accuracy control limits established by the WQL.

Nationwide study results

Effluent arsenic levels for utilities using groundwater averaged 2.4 μg L^{-1} and ranged from <0.5 to 38.6 μg L^{-1}. Surface-water systems had a much lower average arsenic value of 0.6 μg L^{-1} and a narrower range of arsenic concentrations (<0.5 to 4.6 μg L^{-1}). Systems using a blend of ground and surface water had an average arsenic concentration of 1.5 μg L^{-1} and ranged from <0.5 to 4.2 μg L^{-1}. This information indicates that groundwater sources may have higher arsenic levels than surface-water sources.

The distribution of arsenic levels in treated water for all participating utilities is shown in Figure 1. Approximately 44 percent of the utilities surveyed had treated-water arsenic levels of less than 0.5 μg L^{-1}, and 56 percent had levels greater than 0.5 μg L^{-1}. If the future arsenic regulation were set at 2 μg L^{-1}, 16 percent of the utilities surveyed would have to provide additional arsenic removal to comply with the new standard. Similarly, 5 percent of the utilities would have to provide arsenic removal if the revised standard were set at 5 μg L^{-1}.

Figure 2 shows the distribution of population by exposure to different arsenic levels. Approximately 36 percent of the population was exposed to arsenic levels of

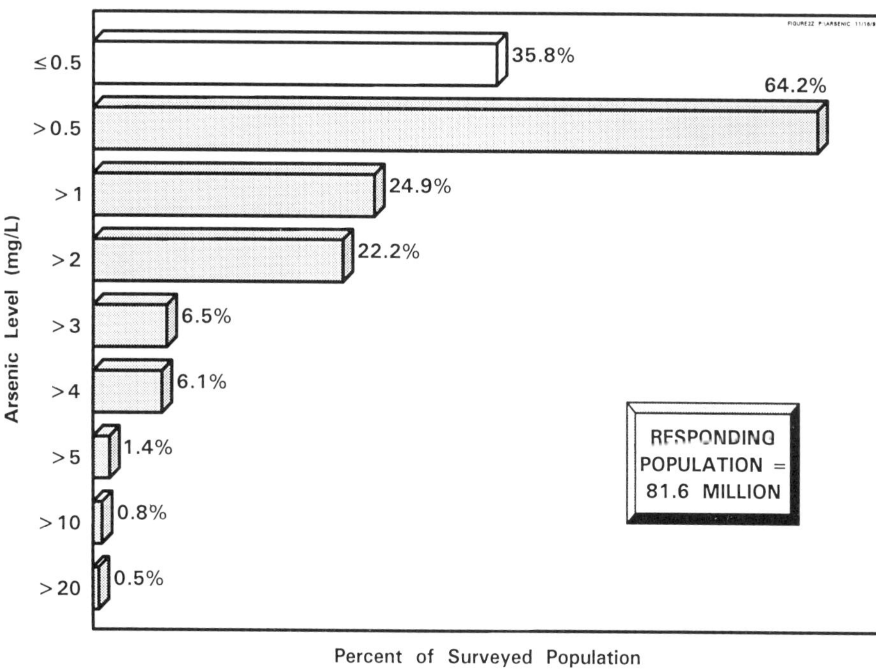

Figure 2 *Percent of surveyed population with arsenic levels greater than selected values (effluents).*

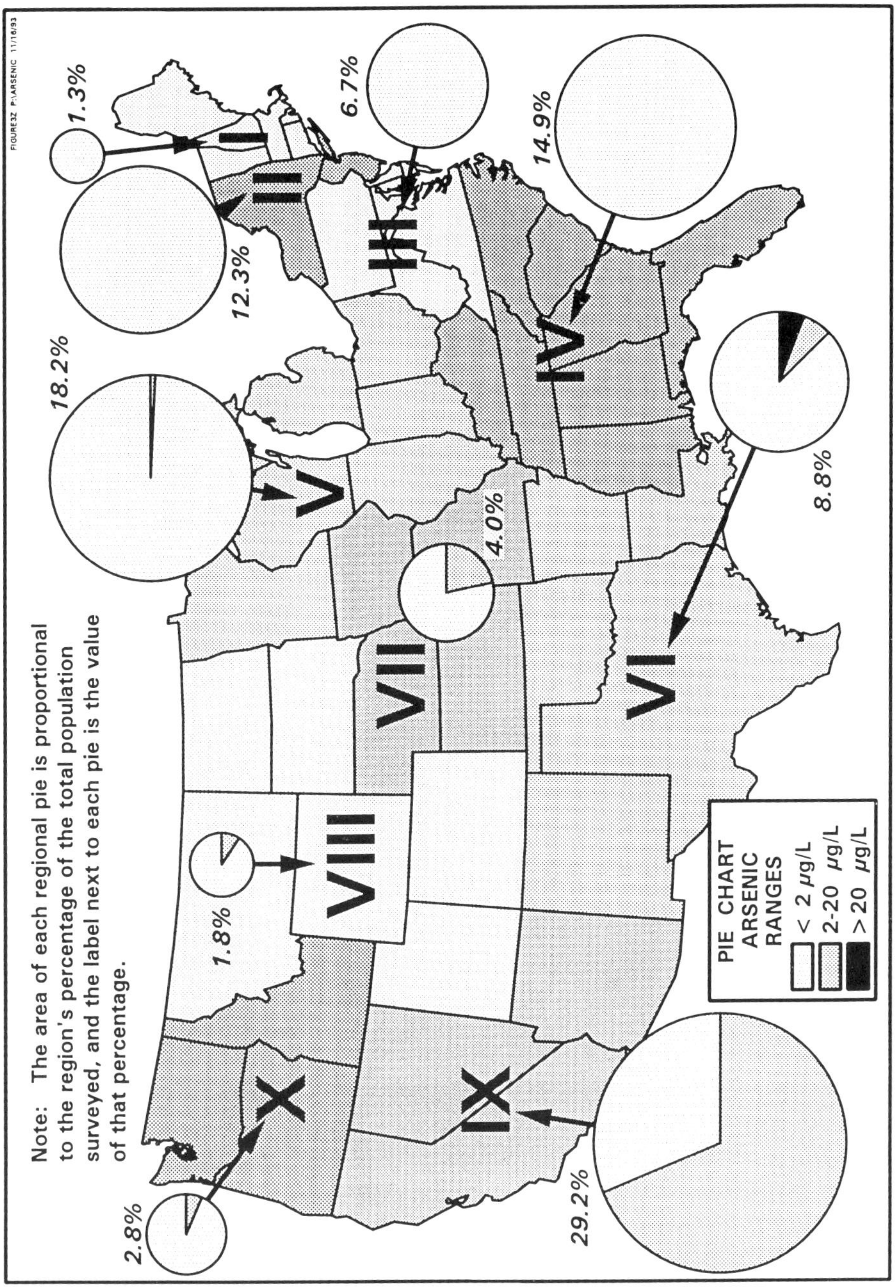

Figure 3 *Arsenic distribution in finished-water effluent for each US EPA region (survey population = 81.6 million).*

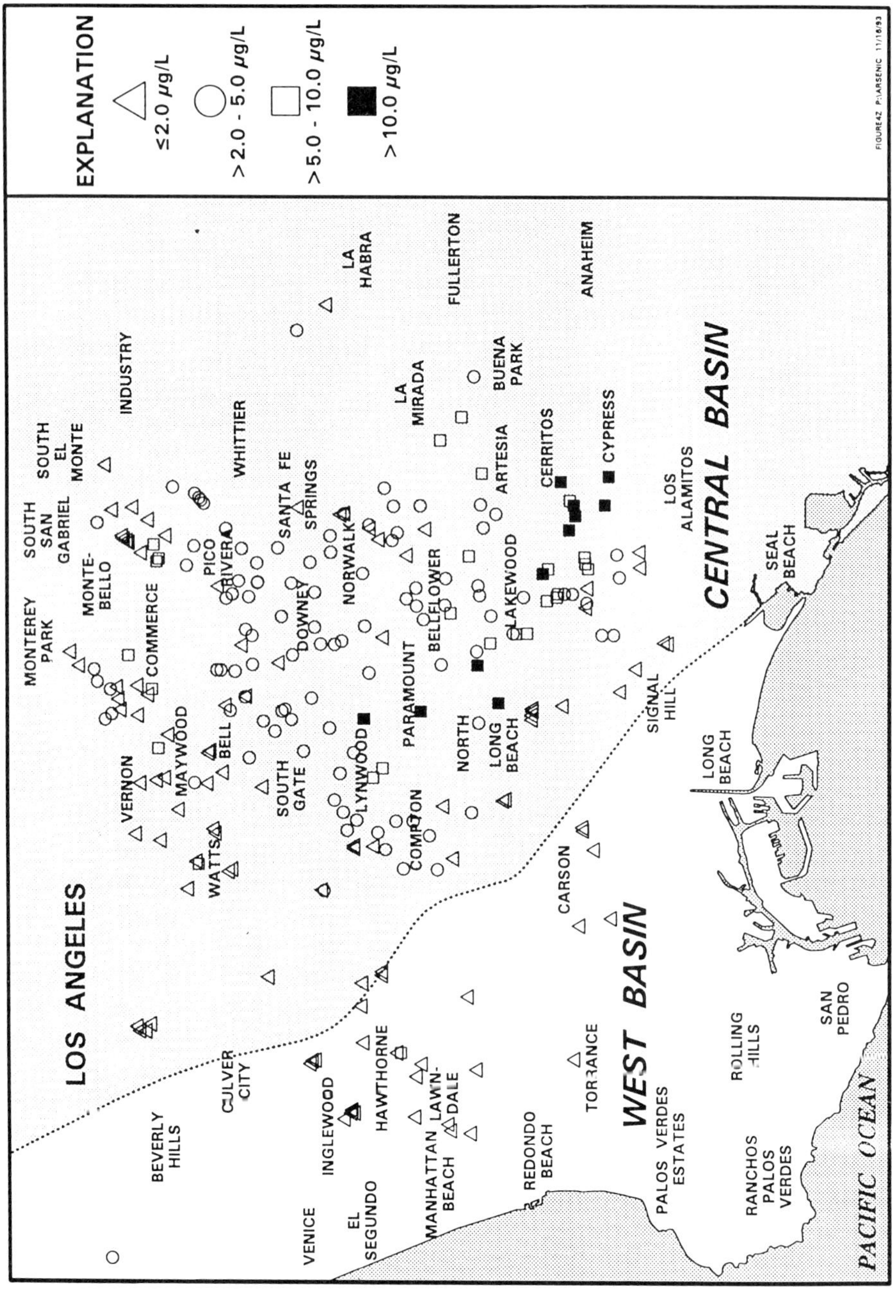

Figure 4 *Distribution of arsenic in Central and West Basin groundwater.*

less than 0.5 μg L^{-1}, with 64 percent exposed to levels greater than 0.5 μg L^{-1}. Twenty-two percent of the population surveyed was exposed to levels greater than 2 μg L^{-1}. By comparison, only 1.4 percent of the population was exposed to arsenic levels greater than 5 μg L^{-1}.

Figure 3 presents arsenic occurrence by geographic region and provides two particularly important pieces of information: First, each numerical value outside a circle represents the percentage of the total surveyed population within that specific US EPA region. The East Coast regions (I through IV) had approximately 35 percent of the total surveyed population, the West Coast regions (IX and X) had 32 percent, and the Central states had approximately 33 percent. Thus, if the US were split into three areas (East Coast, West Coast, and Central states), the surveyed population would be considered evenly distributed.

The second important piece of information in Figure 3 is provided by the pie charts, which indicate the arsenic exposure level within each US EPA region. It is clear that the surveyed population on the East Coast had arsenic levels of less than 2 μg L^{-1}. In fact, only seven of the 45 East Coast utilities surveyed had arsenic levels greater than 0.5 μg L^{-1}. Regions VI through IX show substantially higher population exposure than the other US EPA regions, and a majority of the population surveyed in Region IX had arsenic levels between 2 and 20 μg L^{-1}. One of the utilities in Region VI had an arsenic level greater than 20 μg L^{-1}. This suggests that exposure to arsenic increases west of the Mississippi River, and a majority of the West Coast population surveyed is currently drinking water with arsenic levels greater than 2 μg L^{-1}. The arsenic survey data collected by Metropolitan have been provided to the AWWA in order to provide input from the water supply community into the regulatory process with the US EPA.

Statewide arsenic occurrence studies (Association of California Water Agencies)

The Association of California Water Agencies (ACWA) is an organization of 410 public water agencies throughout the state. ACWA is the largest statewide coalition of public water utilities in the USA. The association's member agencies deliver approximately 90 percent of California's domestic and agricultural water. Recognising that a lower arsenic standard, particularly in the range of 0.5 to 5 μg L^{-1} could be extremely costly to its membership, ACWA initiated the collection of existing arsenic information in 1992. It became obvious that the 1992 survey information had some limitations because the typical reporting detection limit was 10 μg L^{-1}; lower detection limits, if available, ranged from 2 to 5 μg L^{-1}. Thus, information on arsenic occurrence in some water supplies, particularly groundwater supplies, could not be used to evaluate the impact of a low-level arsenic standard in the range of 0.5 to 5.0 μg L^{-1}. ACWA determined from the 1992 survey results that a statewide low-level arsenic occurrence study was needed.

In 1993, ACWA initiated a statewide survey that queried more than 100 domestic water systems serving groundwater, surface water, and blended water. More than

1,000 domestic water supplies were sampled and analysed using platform-stabilised graphite furnace atomic absorption spectrophotometry method 200.9 (US EPA, 1991). Information on this survey is currently being collected and correlated by ACWA and should be available in the fall of 1993. This information will be used by ACWA and the AWWA to represent the California coalition of water agencies during the regulatory process.

Southern California arsenic occurrence study (Central and West Basin Municipal Water Districts)

The Central and West Basin Municipal Water Districts (C&WBMWDs) are major water wholesalers and member agencies of Metropolitan, supplying water to 2.4 million people in 40 cities in the southern coastal plain of Los Angeles County. The southern California arsenic survey was conducted as part of the Cooperative Basin-Wide Title 22 Groundwater Quality Monitoring Program administered by these agencies (C&WBMWDs, 1993).

Southern California survey criteria
In September 1992, the C&WBMWDs conducted a comprehensive monitoring survey of arsenic levels in domestic groundwater sources located in the Central and West Coast groundwater basins of Los Angeles County. The agencies sampled 227 active and standby wells in the Central Basin and 35 wells in the West Basin during September and October 1992. The samples were collected in polyethylene bottles preserved with nitric acid and then analysed by inductively coupled plasma-mass spectrometry (ICPMS). The analytical detection limit for the ICPMS method was 0.5 $\mu g\ L^{-1}$.

Field QC samples included 22 duplicates (5 percent) and 32 duplicate matrix spikes (7 percent). All QC was within the established control limits.

Southern California survey results
Tables 1 and 2 present statistical summaries of the monitoring survey data. Twenty-four of the 31 Central Basin groundwater pumping agencies in the monitoring survey had average arsenic levels greater than 2.0 $\mu g\ L^{-1}$ in their wells. Seven of these agencies had average arsenic levels greater than 5.0 $\mu g\ L^{-1}$. The basinwide average arsenic level in Central Basin wells was 3.8 $\mu g\ L^{-1}$; 50 percent of the Central Basin wells tested had arsenic levels greater than 2.6 $\mu g\ L^{-1}$; and the maximum concentration was 40.5 $\mu g\ L^{-1}$ (Table 1). The distribution of wells in various concentration ranges, as shown in Table 2, indicates that 69 percent of the Central Basin wells tested above 2.0 $\mu g\ L^{-1}$ and 16 percent above 5.0 $\mu g\ L^{-1}$. The significance of these data is apparent when one takes into account the general consensus that the US EPA is likely to propose an MCL at least as low as 5 $\mu g\ L^{-1}$ and possibly as low as 0.5 $\mu g\ L^{-1}$.

The West Basin wells had much lower levels of arsenic than the Central Basin wells. In the West Basin, the overall average was 0.8 $g\ L^{-1}$ and the median level was 0.6 $\mu g\ L^{-1}$. None of the participating West Basin agencies had average arsenic levels above 2.0 μg L^{-1} in their wells. In fact, only two West Basin wells averaged greater than 2.0 μg L^{-1}.

The geographic distribution of arsenic in the Central and West Basin groundwater wells is shown in Figure 4. Evaluating the US EPA option of setting the MCL lower than 5 μg L^{-1}, Figure 4 clearly demonstrates that nearly the entire geographic area of the Central Basin could be impacted by the revised regulation, especially if the MCL is set as low as 2 μg L^{-1}. West Basin wells would be less likely to be affected unless the MCL is set at less than 1 μg L^{-1}.

Conclusions

The nationwide, low-level arsenic occurrence study results described in this paper indicate that the East Coast water systems surveyed do not normally have elevated arsenic levels. However, major compliance problems will be experienced by systems in the western and southwestern states if a low-level arsenic standard is promulgated. The southern California survey (1) indicates that a majority of the wells tested had arsenic levels greater than 2 μg L^{-1} and (2) identifies areas where costly arsenic removal treatment alternatives would have to be implemented, particularly if the new standard for arsenic is in the range of 0.5 to 5 μg L^{-1}. Of particular concern are the systems using groundwater, many of which would have to provide individual wellhead treatment systems that could become very costly. Water systems treating surface water may be able to optimise their existing coagulation treatment processes to comply with a low-level arsenic standard, depending on the concentration of arsenic in their source water.

Information collected in previous surveys does not reflect sufficiently low analytical detection limits to evaluate a low-level arsenic regulation. Thus, analytical methods for arsenic determination need to be refined to allow for lower detection limits. Furthermore, "rugged" analytical methods need to be developed to identify the various arsenic species. Much of this work is currently being performed under contract with the AWWA and AWWARF.

The survey information presented in this paper indicates that groundwater sources are more likely to have higher arsenic levels than surface-water sources. Because the nationwide survey consisted primarily of large water systems, this survey information could severely underestimate the effects of a more stringent arsenic regulation on small- and medium-sized systems. Therefore, another nationwide survey should be conducted to project the cost impact on small- and medium-sized systems. The California statewide survey currently being conducted by the ACWA will be very beneficial in providing this much-needed information.

Acknowledgements

The authors gratefully acknowledge the editorial assistance of Peggy Kimball, graphics preparation by Robert La Londe, the clerical support of Patti Arlt, and technical support by Jeffry Dennis and Paul Evans. Of particular importance was the assistance provided by staff from the utilities that participated in these surveys, without which the survey information would not be available for use during the regulatory process.

References

American Public Health Association (APHA). 1992. *Standard Methods for the Examination of Water and Wastewater,* 18th edn. APHA Publication Office, Washington, D.C.

Bates, M.N., Smith, A.H. and Hopenhayn-Rich, C. 1992. Arsenic ingestion and internal cancers: A review. *Amer. J. Epidemiol.*, **135,** 462-476.

Brower, B. 1983. *Rural Water Survey.* Cornell University Department of Rural Sociology, Ithaca, New York.

Brown, K.G., Boyle, K.E., Chen, C.W. and Gibb, H.J. 1989. A dose response analysis of skin cancer from inorganic arsenic in drinking water. *Risk Anal.*, **9,** 519-528.

Chen, C.J., Chen, C.W., Wu, M.M. and Kuo, T.L. 1992. Cancer potential in liver, bladder, and kidney due to ingested inorganic arsenic in drinking water. *Brit. J. Cancer,* **66,** 888.

C&WBMWDs. 1993. *Cooperative Basin-Wide Title 22 Groundwater Monitoring Program: 1992 Annual Water Quality Report.* In-house report, C&WBMWDs, Carson, California.

Smith, A.H., Hopenhayn-Rich, C., Bates, M.N., Goeden, H.M., Hertz-Picciotto, I., Duggan, H.M., Wood, R., Kosnett, M.J. and Smith, M.T. 1992. Cancer risks from arsenic in drinking water. *Envir. Health Perspect.*, **97,** 259-267.

US EPA. 1980. *National Organics Monitoring Survey.* Technical Support Division, Office of Drinking Water, Cincinnati, Ohio.

US EPA. 1983. *Community Water Supply Survey.* Drinking Water Assessment Branch, Technical Support Division, Office of Drinking Water, Cincinnati, Ohio.

US EPA. 1986. *National Inorganics and Radiological Survey (NIRS).* Water Supply Technology Branch, Technical Support Division, Office of Drinking Water, Cincinnati, Ohio.

US EPA. 1991. *Methods for the Determination of Metals in Environmental Samples.* EPA/600/4-91/010, Office of Research and Development, Washington, D.C.

4 Dietary intake and hazards of arsenic

Michael A. Adams*, P. Michael Bolger[2] and Ellis L. Gunderson[2]

[1]Chemistry Review Branch (HFS-247) and [2]Contaminants Standards Monitoring and Programs Branch (HFS-308), U S Food and Drug Administration, Center for Food Safety and Applied Nutrition, 200 C Street, S.W., Washington, D.C. 20204, USA.

Abstract

The predominant route of arsenic exposure for the general population occurs through the food supply, with the exception of those exposed in the workplace and those exposed at isolated sites with high environmental arsenic burdens (e.g., chemical disposal sites, smelting operations). It is estimated that in the United States the mean total arsenic intake from all food, excluding shellfish, is 30 μg/person/day. For many individuals, however, seafood is a significant source of dietary arsenic. Arsenic is present in seafood in an organic form that has very low toxicity.

Much of the inorganic and organic arsenic that is ingested is absorbed from the gastrointestinal tract. The organic forms (as found in seafood, for example) are excreted unchanged in the urine within a short period of time (days). Inorganic arsenic is converted to a much less toxic form by the liver, cleared by the kidneys, and excreted in the urine. The available evidence suggests that this metabolic capacity must be exceeded before arsenic-induced toxicity can be elicited.

In laboratory animal studies, inorganic arsenic has elicited effects such as gastrointestinal irritation, anemia, neuropathy, dermal and vascular lesions, and hepatic and renal injury. High oral dose levels are associated in some species with teratogenic and fetotoxic effects, and death at the highest dose levels.

Studies of human populations exposed to inorganic arsenic in their drinking water have reported an increased occurrence of certain dermal lesions (e.g., hyperkeratosis, melanosis), cardiovascular lesions (e.g., blackfoot disease), and basal cell carcinoma.

Introduction

Compounds of arsenic are widely distributed in nature. The arsenic in these compounds exists primarily in two oxidation states, arsenite (+3) or arsenate (+5).

---

* To whom correspondence should be addressed.

Arsenic trioxide is produced as a by-product of copper, lead, and nickel smelting. Dimethylated arsenic compounds arise from the methylation of arsenic by microorganisms. Arsenic compounds are used as defoliants in cotton fields, as fungicides, herbicides, insecticides, algicides, and wood preservatives, and for hardening and corrosion resistance in alloys of copper, lead, and bronze. Arsenic compounds are also used in semiconductor manufacture, as medicinals, and as components of antifouling paints.

Arsenic exposure may occur by several different pathways. These pathways are summarized in Figure 1 (left), which shows the sources of arsenic exposure for the general population. From the most recent Food and Drug Administration (FDA) Total Diet Study (TDS) information, food accounts for about 93% of the total arsenic exposure, with water, air, and soil making additional contributions (FDA, 1993). Examining the food category in more detail (Figure 1, right) shows that fish and other seafood account for 90% of the total food arsenic exposure, with all other foods accounting for the remaining 10%.

Arsenic in the diet

Total Diet Study - design and capabilities

The TDS is a tool used by the FDA to monitor and estimate arsenic exposure in the US diet (Pennington, 1983). The primary purposes of the TDS are to determine the dietary intakes of pesticide residues, heavy metals, industrial chemicals, essential minerals, and radionuclides, and to compare the intakes of these substances with Acceptable Daily Intakes, Recommended Dietary Allowances, or Estimated Safe and Adequate Daily Dietary Intakes (Pennington and Gunderson, 1987). In addition, the TDS allows identification of trends; it may identify isolated contamination sources, and it serves as a final check on the effectiveness of pertinent US regulations and initiatives.

Out of a total of approximately 5,000 foods routinely consumed in the United States (as determined by the US Department of Agriculture Nationwide Food

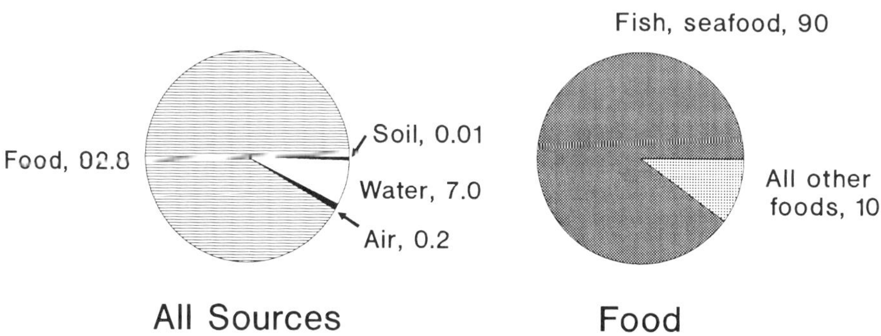

Figure 1 *Sources of arsenic exposure shown as percentages of total exposure. Left, sources of exposure for the general population. Right, arsenic from food.*

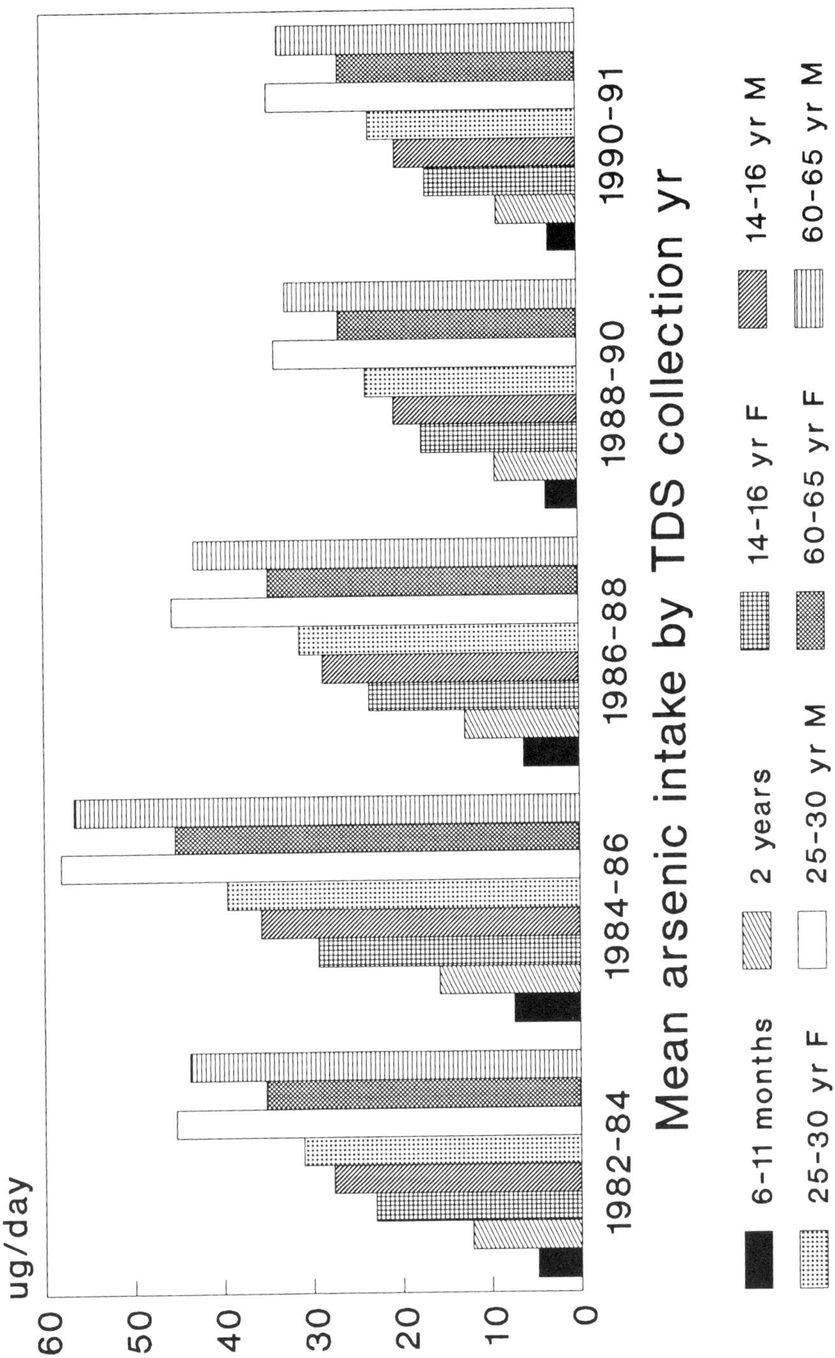

Figure 2 *Arsenic levels in the diets of eight age-sex groups, as determined by the FDA Total Diet Study over a 10-year period.*

Table 1 *TDS foods having arsenic concentrations above 50 ppb (1990-91 market baskets).*

Fish	Meat	Vegetable
Tuna in oil	Turkey breast	Rice, cooked
Fish sticks		Rice cereal
Haddock, fried		Mushrooms, raw
Shrimp, boiled		Olive/safflower oil
Tuna noodle casserole		
Fish sandwich		
Clam chowder		

Table 2 *Mean daily arsenic intakes (μg day^{-1}) as determined by FDA's Total Diet Study for market baskets collected from April, 1982 through April, 1991.*

	Years of collection (number of collections)				
Population group	1982-84 (8)[a]	1984-86 (8)[b]	1986-88 (8)[b]	1988-90 (8)[b]	1990-91 (5)[b]
6-11 months	4.9	7.4	6.2	3.6	3.2
2 yr	12.2	15.8	12.8	9.3	9.0
14-16 yr, F	23.0	29.3	23.5	17.5	16.9
14-16 yr, M	27.7	35.7	28.7	20.6	20.3
25-30 yr, F	31.0	39.5	31.3	23.7	23.3
25-30 yr, M	45.3	58.1	45.6	34.0	34.6
60-65 yr, F	35.2	45.3	34.8	26.7	26.6
60-65 yr, M	43.8	56.6	43.1	32.7	33.4

[a]In calculating intakes, "trace" concentrations were assigned a value of one-half the limit of quantitation; the "trace" concentration assigned to arsenic during these years was 0.01 mg kg^{-1}.
[b]In calculating intakes for this time period, "trace" concentrations were estimated by the analyst.

Consumption Survey), 234 foods were selected as representative of the total (Pennington, 1983). Most of the individual foods selected represent a group of foods similar in type and nutrient content. From these foods, diets were constructed for eight age-sex groups, from infants to mature adults, including males and females. Diets were designed to be excessive in caloric content; for example, the caloric content of the diet of a young male is 3,900 kcal day^{-1}.

Total Diet Study - arsenic levels

The TDS has been very useful in monitoring the overall level of total arsenic (speciation of arsenic in foods is not reported) in the US diet over a period of years. Results of a decade of tracking arsenic levels in the diet are presented in Figure 2, for the eight age/sex groups routinely monitored. For the 1990-91 time period, mean arsenic intake is shown to be substantially lower than the peak intakes of the 10-year span.

Very few foods surveyed for the TDS have been found to have arsenic concentrations above 50 parts per billion (ppb). Table 1 shows the foods from the 1990-91 sample with arsenic levels above 50 ppb.

The results of 10 years of TDS collections and analyses for arsenic are presented in Table 2, for several age and sex groups.

Introduction of arsenic into the food supply

The use of arsenical insecticides has left many agricultural soils contaminated with residual levels of arsenic. Arsenic from these pesticide residues, as well as from deposited fly ash, sewage sludge, and manure used as fertilizer, may be transferred from contaminated soil to plants (Nriagu and Azcue, 1990). This is seldom worrisome because plants with an affinity for arsenic may be poisoned before accumulating amounts of arsenic that might be toxic to humans. Also, uptake of arsenic by plants is in competition with uptake of phosphorus. Because there is usually far more phosphorus than arsenic in soil, arsenic uptake is restricted.

In contrast to terrestrial plants, arsenic uptake by marine organisms is more facile. Bioconcentration of arsenic occurs in aquatic organisms, primarily in algae and lower invertebrates. Organoarsenic compounds such as dimethylarsenosugars, methanoarsonate, dimethylarsinate, and arsenolipids account for 60-99% of the arsenic in marine algae. Arsenocholine and arsenobetaine account for 80-90% of the arsenic in ocean-caught fish. Between 80 and 90% of the arsenic found in freshwater fish is in the form of an unidentified, methanol-soluble material (Nriagu and Azcue, 1990).

Animal feed is a source of arsenic in meat and meat products. An analysis of nonmedicated animal feed showed an average arsenic content of 0.4 $\mu g\ g^{-1}$ (Anderson, 1983). Capar and coworkers have reported that a typical medically unsupplemented feedlot diet for cattle contained 0.1 μg arsenic g^{-1} (Capar *et al.*, 1978). In addition, certain organic and inorganic arsenicals are approved by FDA for use as medicating agents for animal feed. Arsenic supplements have been found to increase weight and feed conversion efficiency in poultry and swine. They have also proved useful for controlling swine dysentery, preventing liver and intestinal disease (blackhead) in turkeys, and increasing the pigmentation of chicken and turkeys. Arsenicals have proven safe for these uses because arsenic is not bioaccumulated to any great extent. Once the medicated feed is withdrawn, tissue arsenic levels quickly decline. The rate of depuration depends on several factors, including the type of animal, the amounts of food and water consumed, and the health of the animal. If medicated feed is not withdrawn early enough, arsenic residues may be present in

Table 3 *Arsenicals approved by FDA for use as animal feed supplements.*

Arsenical	Listing in Code of Federal Regulations	
Arsanilic acid, sodium arsanilate	21 CFR	558.60 558.62
Roxarsone (3-nitro-4-hydroxyphenyl arsonic acid or sodium salt)	21 CFR	520.2087 520.2088 558.62
Nitarsone (4-nitrophenylarsonic acid)	21 CFR	558.369
Carbarsone (p-ureidobenzene arsonic acid)	21 CFR	558.120

tissues at the time of slaughter, thus contributing to the arsenic burden of the human diet (Calvert, 1975).

Four arsenicals are now approved for use as medicating agents for animal drinking water or feeds. Table 3 lists the compounds and their regulatory citation in the Code of Federal Regulations (CFR).

Arsenic-based additives must be withdrawn from the animal drinking water or feed several days before slaughter to allow residual arsenic to clear from the tissues. In addition to this clearance time, there exist tolerances for arsenic in the tissues of animals intended for human consumption. These tolerances are found in 21 CFR 556.60. In the edible tissues and eggs of chicken and turkeys only 0.5 mg arsenic kg^{-1} is allowed in uncooked muscle tissue, 2 mg arsenic kg^{-1} in uncooked edible by-products, and 0.5 mg arsenic kg^{-1} in eggs. In the edible tissue of swine, 2 mg arsenic kg^{-1} is allowed in uncooked liver and kidney, whereas only 0.5 mg arsenic kg^{-1} is permitted in uncooked muscle tissue and by-products other than liver and kidney.

Arsenic in foods - hazards and risk

Other than exposure in the workplace and in isolated sites with high environmental burdens of arsenic (*e.g.*, waste chemical disposal sites, smelting operations), the major route of exposure to arsenic for the general population occurs through the diet - food and drinking water. For many individuals, seafood is a significant source of dietary arsenic, which occurs primarily in an organic form either as arsenobetaine or arsenocholine (ATSDR, 1991; WHO, 1981; WHO, 1989).

Arsenic forms a variety of inorganic and organic compounds that vary widely in toxicity. As a general rule, soluble inorganic compounds of trivalent arsenic are the most toxic, whereas the organic forms demonstrate a fairly low order of toxicity. The soluble forms of pentavalent arsenic demonstrate intermediate toxicity, which may be due, in part, to the *in vivo* reduction to the trivalent form (Arnold, 1988; ATSDR, 1991; WHO, 1989).

Most of the inorganic and organic forms of arsenic that are ingested are absorbed from the gastrointestinal tract. Once absorbed, soluble trivalent and pentavalent arsenic is widely distributed throughout the body. Pentavalent arsenic undergoes reductive conversion to the trivalent form, which in turn is methylated in the liver, primarily to dimethylarsenic acid. The methylated arsenicals are much less toxic than the parent compound and are readily excreted in the urine. This methylation of the inorganic forms of arsenic is efficient at low to moderate exposure levels, but becomes less efficient as the dose levels become greater. This suggests that the metabolic capacity is saturable and at highest exposure levels will result in the deposition of inorganic forms of arsenic in tissues. Indeed, until recently the available evidence suggested that this metabolic capacity must be exceeded before arsenic-induced toxicity can be elicited (ATSDR, 1991; Ishinishi *et al.*, 1986). However, a recent analysis of data from studies of urinary arsenic levels in humans suggests a wide range of methylation capacity. In addition, it was observed that 20-25% of inorganic arsenic remains unmethylated regardless of the level of exposure. This analysis of epidemiological and experimental human data argues against the saturable methylation hypothesis (Hopenhayn-Rich *et al.*, 1993). The organic forms of arsenic found in foods are readily excreted unchanged in the urine within a short period of time (days) (Luten *et al.*, 1982; Tam *et al.*, 1982). Studies with one form of organic arsenic found in seafood, arsenobetaine, have shown that it is excreted unchanged in thc urine within days of ingestion (Cannon *et al.*, 1983).

In laboratory animals, the acute to chronic oral exposure to inorganic arsenicals, over a range of dose levels, results in a variety of toxic effects including gastrointestinal irritation, anemia, neuropathy, dermal and vascular lesions, and hepatic renal injury. High oral dose levels in several species are associated with teratogenic and fetotoxic effects and eventually death at the highest doses. Studies of human populations exposed to inorganic arsenic in their drinking water have reported an increased occurrence and risk of certain dermal (*e. g.*, hyperkeratosis, melanosis) and cardiovascular lesions (*e.g.*, blackfoot disease), basal cell carcinoma and certain internal cancers, such as hepatic angiosarcomas, renal adenocarcinomas, and cancers of the mammary and lymphatic glands (Arnold, 1988; ATSDR, 1991; EPA, 1988; WHO, 1989).

In contrast, studies of organic arsenicals in animals have demonstrated a fairly low order of toxicity. This lack of toxicity has been further demonstrated in humans. There are no reports of adverse effects in those populations who consume large quantities of seafood which can result in daily exposures of organic arsenic of as much as 50 μg. This information indicates that in contrast to the inorganic forms of arsenic, the organic forms found in food, particularly fish, present little or no hazard to human health (ATSDR, 1991; Coulsen *et al.*, 1935; Jongen *et al.*, 1985; Welch and Landau, 1942; Welch and Welch, 1938).

References

ATSDR (Agency for Toxic Substances and Disease Research). 1991. *Toxicological Profile for Arsenic - Update 1991,* pp. 7-95. ATSDR, Atlanta, GA.

Anderson, C. E. 1983. Arsenicals as feed additives for poultry and swine. In: W. H. Lederer and R. J. Fensterheim (eds.), *Arsenic: Industrial, Biomedical, Environmental Perspectives,* pp. 89-97. Van Nostrand Reinhold, New York.

Arnold, W. 1988. Arsenic . In: *Handbook on Toxicity of Inorganic Compounds.* Chap. 8, pp. 79-93. Marcel Dekker, Inc., New York.

Calvert, C. C. 1975. Arsenicals in animal feeds and wastes. In: E. A. Woolson (ed.), *Arsenical Pesticides,* pp. 70-80. American Chemical Society Symposium Series, Washington, DC.

Cannon, J., Saunders, J. and Toia, R. 1983. Isolation and preliminary toxicological evaluation of arsenobetaine - the water soluble arsenical constituent of the hepatopancreas of the Western rock lobster. *Sci. Total Environ.,* **31**, 181-185.

Capar, S. G., Tanner, J. T., Friedman, M. H., and Boyer, K. W. 1978. Multielemental analysis of animal feed, animal wastes, and sewage sludge. *Environ. Sci. Technol.,* **12**, 785-790.

Coulsen, E., Remington, R. and Lynch, K. 1935. Metabolism in the rat of the naturally occurring arsenic of shrimp as compared with arsenic trioxide. *J. Nutr.,* **10**, 255-270.

EPA (Environmental Protection Agency). 1988. *Special Report on Ingested Inorganic Arsenic: Skin Cancer; Nutritional Essentiality,* pp. 11-38. Environmental Protection Agency, Washington, DC.

FDA (Food and Drug Administration). 1993. *Guidance Document for Arsenic in Shellfish,* pp. 25-27. US Food and Drug Administration, Washington, DC.

Hopenhayn-Rich, C., Smith, A. H. and Goeden, H. M. 1993. Human studies do not support the methylation threshold hypothesis for the toxicity of inorganic arsenic. *Environ. Res.,* **60**, 161-177.

Ishinishi, N., Tschiya, K., Vahter, M. and Fowler, B. 1986. Arsenic (Chapter 3). In: *Handbook on the Toxicology of Metals,* Vol. II, pp. 43-79. Elsevier Science Publishing Co., Inc., New York.

Jongen, W., Cardinaals, J., Bos, P. and Hagel, P. 1985. Genotoxicity testing of arseno-betaine, the predominant form of arsenic in marine fishery products. *Food Chem. Toxicol.,* **23**, 669-673.

Luten, J., Riekwel-Booy, G. and Rauchbar, A. 1982. Occurrence of arsenic in plaice (*P. platessa*), nature of the organoarsenical compound present and its excretion by man. *Environ. Health Perspect.,* **45**, 165-170.

Nriagu, J. O. and Azcue, J. M. 1990. Food contamination with arsenic in the environment. In: *Food Contamination from Environmental Sources,* pp. 121-143. Wiley Interscience, New York.

Pennington, J. A. T. 1983. Revision of the Total Diet Study food list and diet. *J. Am. Diet. Assoc.,* **82**, 166-173.

Pennington, J. A. T. and Gunderson, E. L. 1987. History of the Food and Drug Administration's Total Diet Study - 1961 to 1987. *J. Assoc. Off. Anal. Chem.,* **70**, 772-782.

Tam, G., Charbonneau, S., Bryce, F. and Sandi, E. 1982. Excretion of a single oral dose of fish-arsenic in man. *Bull. Environ. Contamin. Toxicol.,* **28**, 669-673.

Welch, A. and Landau, R. 1942. The arsenic analogue of choline as a component of lecithin in rats fed arsenocholine chloride. *J. Biol. Chem.,* **144**, 581-588.

Welch, A. and Welch, M. 1938. Lipotopic action of certain compounds related to choline chloride. *Proc. Soc. Exp. Biol. Med.,* **39**, 7-9.

WHO (World Health Organization). 1981. International program chemical safety (IPCS). *Environmental Health Criteria,* **18**, 33-48. Geneva, Switzerland.

WHO (World Health Organization). 1989. *33rd Report of the Joint WHO/FAO Expert Committee on Food Additives.* Technical Report Series 776, pp. 27-28. Geneva, Switzerland.

5 Determination of total arsenic and arsenic compounds in drinking water

Kurt J. Irgolic

Institut für Analytische Chemie der Karl-Franzens-Universität Graz, Universitätsplatz 1, A8010 Graz, Austria

Abstract

Normally, drinking water contains arsenic as arsenate. Arsenite may be found, when the water is anaerobic. Methylarsonic acid and dimethylarsinic acid will be present in water supplies only rarely. These organic arsenic compounds are either formed from inorganic arsenic through microbial activity or are introduced through the application as herbicides. In most cases the determination of total arsenic will provide the information needed for risk assessment. The silver diethydithiocarbamate method, the hydride generation technique, flame atomic absorption spectrometry, graphite furnace atomic absorption spectrometry, electrochemical methods, and atomic emission spectroscopic methods are presented and evaluated for their usefulness for the determination of arsenic in drinking water under the assumptions that the maximal allowable concentration of arsenic remains at 50 μg L^{-1} or will be lowered to 20, 2, or 0.002 μg L^{-1} Lowering the limit to 20 μg L^{-1} will have only a small effect on the monitoring of the arsenic concentrations. At the lower limits serious problems will arise. The rising costs of the required instruments and of the skilled personnel will not allow the water utilities to perform the analyses without considerable investments in analytical equipment and personnel trained in analytical chemistry.

Introduction

Under normal circumstances women, men, and children must eat, drink, and breathe regularly. These indispensable activities provide the body with organic compounds needed for the synthesis of tissues and for the generation of energy; with water to serve as a reagent and solvent for anabolic and metabolic reactions; and with oxygen required for the intracellular, energy-generating combustion of organic compounds. Through food and water the body receives also inorganic ions (for instance, Na^+, Ca^{2+}, Cl^-, phospate, a series of trace elements) employed as building materials for bones, as electrolytes in body fluids, and as components of enzymes. A well balanced diet contains all the nutrients and essential elements. Unfortunately, food items and water may also contain compounds of elements not needed by the body, provide a daily dose of essential elements much in excess of the demand, and in extreme situations deliver toxic amounts of substances. Generally, acute cases of intoxications are easily recognized and can be appropriately controlled. The long-term, low-level chronic exposure to a toxic substance with insidious and not easily recognized effects may become a public health issue. Compounds of arsenic, particularly the inorganic, water-soluble compounds with arsenite or arsenate as anionic components, belong to this class of long-term, low-level agents. For the evaluation of the exposure, for the correlation of the exposure to clinical symptoms, for the promulgation of regulatory exposure limits, and for the routine control of the limits accurate, precise, robust and inexpensive analytical methods should be available that do not require highly trained (and thus expensive) operators. Whether or not these requirements can be fulfilled, will depend on the state of the art of analytical methodology and instrumentation, on the elements and their compounds to be determined, and on the concentrations of the analytes in the samples of interest.

In the following sections of this paper these issues will be discussed as they apply to arsenic and arsenic compounds in drinking water. An attempt will be made to point out the problems that increase rapidly as the allowable limits for arsenic in drinking water are set at ever lower concentrations.

Arsenic compounds in drinking water

In the absence of man-made pollution and biological activity water will contain only inorganic arsenic compounds. In anaerobic water masses the trivalent arsenic compound - customarily called arsenite - should be the only arsenic compound present. Arsenous acid, H_3AsO_3, is the weak acid (pK = 9.2), from which - through replacement of one, two, or three protons by suitable cations arsenites are formed. However, in aqueous solutions with pH values less than 8 arsenous acid will be present as the neutral molecule H_3AsO_3. In aerobic systems arsenate is the thermodynamically stable arsenic compound. The term "arsenate" stands for the anions derived from the rather strong arsenic acid, H_3AsO_4 ($pK_1 = 2.2$, $pK_2 = 6.9$). Because surface and groundwater generally is neutral to slightly acidic, "arsenate" will be found as a mixture of the mononegative dihydrogen arsenate anion ($H_2AsO_4^-$) and the dinegative hydrogen arsenate anion ($HAsO_4^{2-}$). The concentration ratio of these two anions will be determined by the pH value of the water that may be governed by a buffer system. The concentration of these arsenic species in drinking water rarely exceeds 50 µg As L^{-1}($6.6x10^{-7}$ mole L^{-1}). At such low concentrations the ion products of arsenates (arsenites) and common cations present in water (Ca^{2+}, Mg^{2+}) are smaller than the solubility products (approximately 10^{-20}). Therefore, precipitates will not form. The high affinity of arsenate and arsenite for aluminum and iron(III) hydroxides must be mentioned. Should such hydroxides form in water (for instance, upon exposure of anaerobic water containing Fe^{2+} to air) arsenate and arsenite might coprecipitate.

The organic arsenic compounds methylarsonic acid ($CH_3AsO_3H_2$) and

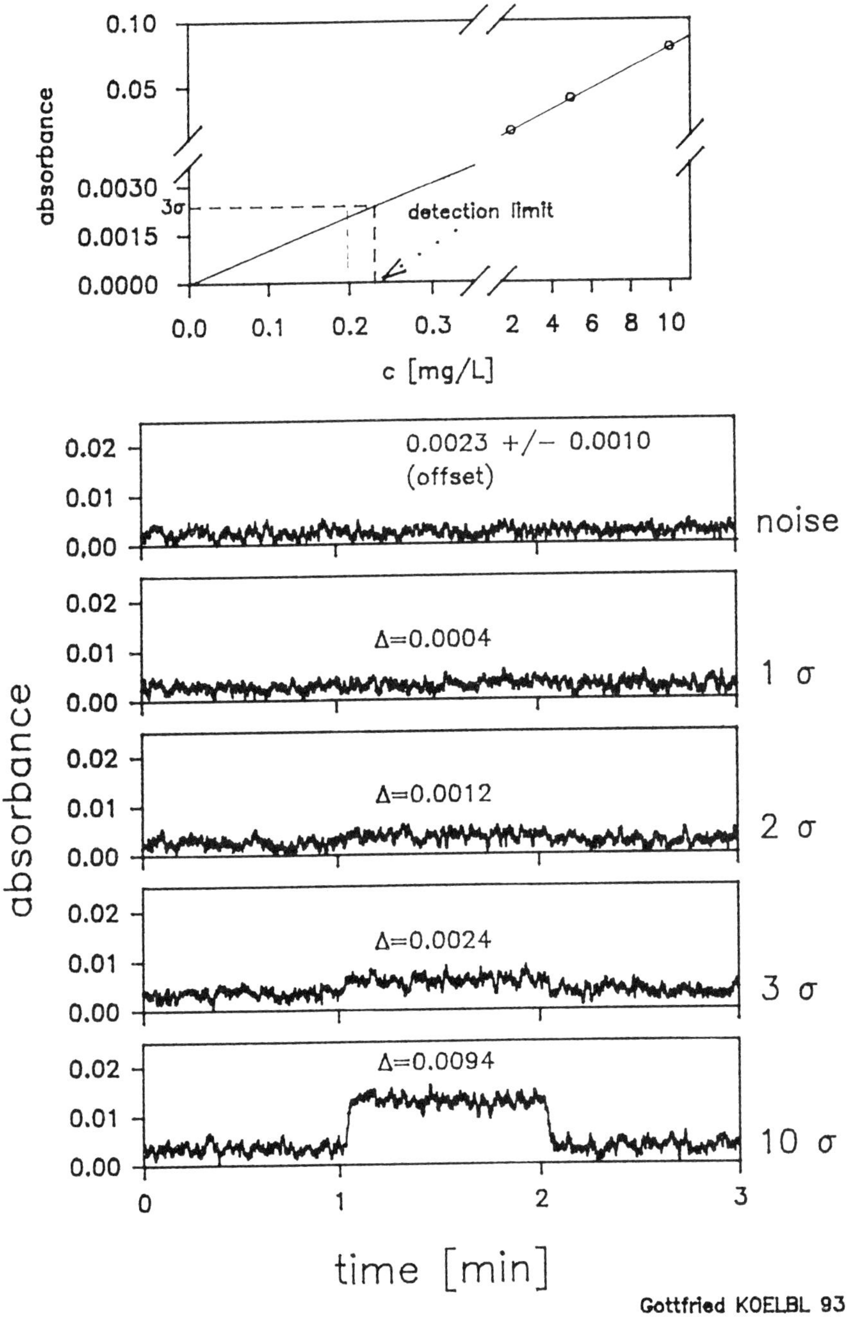

Figure 1 *Determination of detection limits for an element by flame atomic absorption spectrometry.*

dimethylarsinic acid [$(CH_3)_2AsO_2H$] could enter water systems through biological conversion of inorganic arsenic to these methylated arsenic derivatives and through leaching from soil treated with these herbicidal compounds. Many microorganisms are capable of converting inorganic arsenic compounds in a multistep reaction to methylarsonic and dimethylarsinic acid (Cullen and Reimer 1989). Although these methylated arsenic compounds were found in surface waters, the concentrations were in the low nanomolar range (1 nanomole L^{-1} = 75 ng As L^{-1} = 0.075 μg L^{-1}) (Anderson and Bruland 1991). Methylated arsenic compounds were not present at concentrations above 1 μg L^{-1}in the arsenic-rich well waters from Taiwan (Irgolic 1982). The very low concentrations of methylarsonic acid and dimethylarsinic acid in fresh waters and the much lower toxicity of these compounds in comparison to arsenite and arsenate make the determinations of these methylated arsenic compounds in the context of risk evaluation unnecessary. Among the more complex organic arsenic compounds (trimethylarsine oxide, tetramethylarsonium salts, arsenocholine, arsenobetaine, arsenic-containing riboses and lipids) only trimethylarsine oxide was detected at low concentrations in fresh water (Braman 1975). Therefore, these compounds are also of no importance in connection of arsenic in drinking water. However, all of these organic compounds of arsenic must be considered, when the intake of arsenic through food is evaluated.

Determination of total arsenic in drinking water

Because arsenic occurs in drinking water predominantly and in most cases exclusively as arsenate, the determination of total arsenic without consideration of the chemical form is sufficient to evaluate exposure and risk. Recent reviews summarize these methods (Irgolic 1983 1992). Here, only the most widely used methods will be presented without details on instrumentation and on the chemical bases of the procedures. Detection limits, skills required of the operator, and suitability for routine operation will be emphasized.

Detection limits

Detection limits provide important information about the methods available for the determination of arsenic. However, these limits should not be interpreted as the concentrations or absolute amounts of arsenic that can be quantitatively determined in a water sample. Generally, detection limits are obtained from the standard deviation (σ) of the baseline signals (the signals produced, when distilled water or a solution of the same composition as the sample but without arsenic) is analysed. The detection limit is then the concentration of arsenic, that produces a signal twice (2σ) or three times (3σ) as intense as the standard deviation of the baseline signals. These 2σ and 3σ detection limits are taken from a calibration curve constructed over a suitable concentration range (Figure 1). At these concentrations an accurate and precise quantification of arsenic is not possible. Figure 1 shows the signals generated with a flame atomic absorption spectrometer for the blank (distilled water without analyte) and for solutions containing the analyte at concentrations of 1, 2, 3, and 10σ.

At concentrations of 1 and 2σ the analyte signal is almost impossible to detect. At 3σ the signal becomes recognizable but not usable for quantification. At 10σ a quantifiable signal is present. Under the assumption that the standard deviations of these signals remain the same at σ the relative standard deviation of the 1σ signal [100σ /(average baseline signal +σ,] is approximately 40%. Results with such a huge relative standard deviation cannot be called quantitative. The 10σ signal has a relative standard deviation of 10%. Quantitative determinations begin at this concentration. As a rule of thumb quantitative information may be obtained with a method at an analyte concentration ten-times the detection limit.

When "real" samples - samples that are not solutions of the analyte in distilled water but contain other substances subsumed under the term "matrix" - are analyzed, the "distilled-water" detection limits do not apply. The sample matrix often increases the detection limits and thus the quantification limit. Instruments and methods are sold on the basis of the low "distilled-water" detection limits. An evaluation of the suitability of a method for an analytical task must take this situation into account and carefully consider the effect of the sample matrix on the detection and quantification limits. In the case of drinking water, the concentrations of the substances forming the matrix are rather low. Therefore, the "distilled-water" detection limits may be used as good approximations for the detection limits achievable with drinking water samples.

The detection limits for analytical methods are expressed as absolute limits (for instance, 1 ng As) or as concentration limits (1 mg As L^{-1}). The requirements of a method with respect to the sample introduction govern, in which manner the detection limits are reported. When the experimental conditions are reported in sufficient detail, the two types of detection limits can be converted into each other, although such a conversion might not be easy in all cases. In the following discussion of the methods both detection limits are given whenever possible.

Silver diethyldithiocarbamate method

Arsenite and arsenate are reduced in aqueous solution at pH 1 with sodium borohydride to arsine (AsH_3), which is transferred in a nitrogen stream into a solution of silver diethyldithiocarbamate in chloroform containing morpholine. The intensity of the red color that develops is measured at 520 nm with a spectrophotometer. Methylarsonic acid, dimethylarsinic acid, antimony, and certain cations interfere with the determination. The concentrations of these potential interferants in drinking water are so low that problems from them will be encountered only rarely. This method is practiced widely in the water industry (Greenberg *et al.*, 1991).

The detection limit for this method is approximately 1 microgram arsenic. The relative standard deviation of the results obtained with solutions of arsenite/arsenate containing 10 μg arsenic are 10%. When 100 mL of solution are analysed, the concentration detection limit is 10 μg As L^{-1} (10 ppb). The quantification limit is approximately ten-times higher. When larger volumes of sample are used the concentration detection limit will be lower and can reach the EPA limit of 50 μg As L^{-1}.

Hydride generation methods

Arsenite and arsenate are reduced by sodium borohydride at pH 1 to arsine. The arsine is removed in an inert gas stream from the solution and thus separated from the matrix, collected in a trap cooled with liquid nitrogen, and volatilized from the trap into a detector. The arsine can also be transferred directly into the detector without prior trapping. Collection of the arsine in the trap makes the detection independent of the kinetics of the reduction process. A variety of detectors can be used. Flame atomic absorption spectrometers, a quartz-tube atomizer in conjunction with an atomic absorption spectrometer, an atomic fluorescence spectrometer, a helium glow discharge (atomic emission), direct-current or inductively coupled argon plasma emission spectrometers, microwave-plasma emission spectrometers, inductively coupled plasma mass spectrometers, and mass spectrometers were connected to hydride generators. Detectors employed in gas chromatography may also be used. The achievable detection limits will depend on the detector employed and can be as low as 0.01 ng (0.1 ng As L^{-1} = 0.1 ppt, when 100 mL of sample are used). Although a hydride generation system consisting of a reduction vessel, traps, a helium plasma, a monochromator, a photomultiplier tube, an amplifier, and a strip chart recorder are not too costly, the operation requires analytical skills.

Flame atomic absorption spectrometry

Flame atomic absorption spectrometry has detection limits for arsenic rarely better but often worse than 1 mg L^{-1}. The reason for these high detection limits are light absorption by the flame gases at the wavelengths (189.0 193.7 197.2 nm) used for the determination of arsenic and chemical interferences. Pre-concentration of arsenic via extraction into an organic solvent followed by re-extraction into water and aspiration of the thus obtained aqueous solution into the flame made arsenic detectable at 20 µg L^{-1} of original sample. This detection limit is still too high to be useful for the determination of arsenic in drinking water samples that do not exceed the 50 µg L^{-1} limit. The extraction/back-extraction sequence is too time-consuming for routine analyses.

Graphite furnace atomic absorption spectrometry

Graphite furnace or electrothermal atomic absorption spectrometry is widely used for the determination of arsenic in water samples. The order-of-magnitude absolute detection limits for this method are 10 picogram of arsenic. This impressive absolute detection limit turns into a less imposing concentration detection limit of 0.1 to 0.5 µg L^{-1} (ppb) when the volumes of sample (100 20 µL) that are normally introduced into the furnace are considered. These detection limits are achievable only in the absence of interferences and with a spectrometer operating with background correction. In "Standard Methods for the Examination of Water and Wastewater" (Clesceri 1989) the estimated detection limit is given as 1 µg L^{-1} with an optimal concentration range for quantitative determinations from 5 to 100 µg L^{-1}. This method must be carried out by skilled analysts to obtain reliable results. Interlaboratory determinations of total arsenic in a drinking water sample with 9.78 µg As L^{-1} gave a single analyst precision and an interlaboratory precision of approximately 25% (relative standard deviation) (Clesceri 1989). This "imprecision"

Table 1 *Methods for the determination of total arsenic in drinking water*

Method	Detection limit		Instrument cost	Operator	Applicable at As limit of			
	Absolute	Concentration	\$ x 10^{-3}	skill level	50 μg L^{-1}	20 μg L^{-1}	2 μg L^{-1}	2 ng L^{-1}
Ag diethyldithiocarbamate	1 μg	10 μg L^{-1} (100 mL)	5	low	yes	no	no	no
Hydride generation								
Thermal conductivity		10 μg L^{-1}	40	high	yes	yes	yes	no
Atomic fluorescence		20 ng L^{-1}	20	high	yes	yes	yes	no
Graphite furnace AAS	0.2 ng	2ng (100 mL)	60	high	yes	yes	yes	no
Quartz-tube AAS	0.5ng	32 ng L^{-1} (15 mL)	60	high	yes	yes	yes	no
Helium discharge	0.1ng	1 ng L^{-1} (100mL)	20	high	yes	yes	yes	no
Microwave plasma	20pg	1.4 ng L^{-1} (15mL)	20	high	yes	yes	yes	no
DC-plasma	0.4 μg	4 μg L^{-1} (100 mL)	60	very high	yes	yes	no	no
ICP emission	2 ng	20 ng L^{-1} (100 mL)	80	very high	yes	yes	yes	no
ICP-MS	10 pg	0.1 ng L^{-1} (100 mL)	200	very high	yes	yes	yes	no
Differential pulse polarography	20 ng	1 μg L^{-1} (20mL)	20	high	yes	yes	no	no
Anodic stripping voltammetry	6 pg	0.3 ng L^{-1} (20 mL)	30	high	yes	yes	yes	yes
Flame atomic absorption		1 mg L^{-1}	20	low	no	no	no	no
Graphite furnace AAS		1 μg L^{-1}	80	high	yes	yes	no	no
DCP or ICP AES		10 μg L^{-1}	80	very high	yes	yes	no	no
ICP-MS		0.05 μg L^{-1}	200	very high	yes	yes	yes	no

indicates, that the determination of arsenic by graphite furnace atomic absorption spectrometry is not a routine method in the concentration range including the 50 μg L^{-1} limit. Graphite furnace atomic absorption spectrometers are not inexpensive to buy and maintain.

Electrochemical methods
Several electrochemical methods were developed for the determination of arsenite in aqueous media such as fresh water and seawater (Greschonig 1992). The most widely used method is differential pulse polarography employing the dropping mercury electrode. In acidic solution arsenite is reduced at approximately 0.4 V against the saturated calomel electrode. Detection limits of a few microgram arsenic per liter and working ranges of 1 to 300 μg L^{-1} are found in the literature. The use of catalytic waves for the polarographic determination of arsenite extends the working range to 0.1 μg L^{-1} . Preconcentration of arsenic on the electrode via anodic stripping allows arsenite to be determined at concentrations as low as 0.3 ng L^{-1}

Arsenate is not electroactive under these conditions and must be reduced to arsenite, for instance with hydriodic acid, sodium hydrogen sulfite, or hydrazine. Arsenate becomes electrochemically reducible in acidic solution in the presence of polyhydroxyalkanes such as mannitol. These electrochemical methods can be used for the quantification of total arsenic in drinking water with arsenic concentrations not lower than a few micrograms per liter.

The electrochemical methods have not found wide application in spite of the relatively low cost of the required equipment. The rapid development in the field of electrochemically based sensors, the success in producing special electrodes specific to certain ions, and the integration of such disposible electrodes into easy to handle measuring devices raise hopes that a system suitable for determining arsenic in drinking water can be found. The less than ideal electrochemical properties of arsenite and especially of arsenate should not deter from attempts to develop such a system.

Atomic emission methods
Flames are not hot enough to excite arsenic atoms in sufficient numbers to make flame atomic emission spectrometry a viable method for the determination of arsenic in drinking water directly aspirated into the flame. Emission spectrometry becomes useful for arsenic determinations when the much hotter plasmas are used for the excitation. Direct-current (DC) and inductively coupled argon plasmas (ICP) as light sources for sequential or simultaneous spectrometers have detection limits in the range of 10 μg–50 μg As L^{-1} with approximately 10 mL as the minimal volume of sample required for aspiration. The DC and ICP emission spectrometers have high acquisition, operating, and maintenance costs and require highly skilled operators. The detection limits of these instruments are low enough to monitor drinking water supplies as long as the allowable limit is not set much lower than 50 μg L^{-1}. Inductively coupled argon plasma mass spectrometers have detection limits of approximately 0.05 μg L^{-1}, but cannot be used for routine monitoring of water supplies because of the high cost of the instruments and the high cost of the required skilled operators.

Methods for the determination of arsenic and concentration limits

When a limit is set for total arsenic in drinking water, precise and accurate methods must be available for the determination of arsenic with a working range that includes the limit and has the limit not too close to either end of the range. If such a method does not exist, a suitable method has to be developed. Such a method must be suitable for routine application in water treatment plants and should not require highly skilled personnel. Equipment and reagents should not be too costly. Table 1 summarises the currently available methods suitable for the determination of arsenic in drinking water, their detection limits, the cost of equipment, and the required skill level of the operator.

Flame atomic absorption spectrometry cannot be used even at the current limit of 50 μg L^{-1} limit. The silver diethyldithiocarbamate colorimetric method, the least expensive with respect to equipment and personnel, ceases to be applicable, when the limit is lowered to 20 μgL^{-1}. At the 20 μg L^{-1} limit the hydride generation technique with any of the detectors listed in Table 1, electrochemical methods, graphite furnace atomic absorption spectrometry, and plasma emission spectrometry can be used for the determination of arsenic. At the 2 μg L^{-1} limit only the hydride generation technique with costly detectors, anodic stripping voltammetry, and inductively coupled argon plasma mass spectrometry remain. Finally at the 2 ng L^{-1} limit the only choices are hydride generation with the inductively coupled argon plasma mass spectrometric detector and anodic stripping voltammetry. Table 1 clearly points out, that a limit of 20 μg L^{-1} can be supported by the available analytical methodology, but will bring higher costs for equipment and personnel to the utilities. The 2 μg L^{-1} limit, although analytically manageable in experienced laboratories, will cause great difficulties to the utilities. The 2 ng L^{-1} limit can neither be supported in a reasonable way by the available analytical methods, nor financed by the utilities.

Determination of arsenic compounds

Most drinking water supplies will contain only arsenate, in some arsenite may also be present, and in a few methylated arsenic compounds may be found. The hydride generation technique can be used to identify and quantify arsenite, arsenate, methylarsonic acid, and dimethylarsinic acid. Only arsenite is reduced by sodium borohydride at pH 5 and detected as arsine. At pH 1 arsenate is converted to arsine, methylarsonic acid to methylarsine, and dimethylarsinic acid to dimethylarsine. These arsines are collected in a liquid nitrogen-cooled trap, and then volatilized in sequence of increasing boiling temperature into the detector. A gas chromatograph may be used to separate the arsines. The detection limits dependent on the detector used are approximately 0.1 ng for arsine (arsenite and arsenate) and 10 to 100 ng for the methylarsines.

The arsenic compounds in the water samples can also be separated by anion exchange chromatography or ion-pairing chromatography and detected and quantified with element-specific detectors. Generally, the detection limits for the liquid chromatography systems are not as good as the limits for the hydride generation method. Details about these methods can be obtained from the literature (Gailer 1993; Irgolic 1992). These methods are not for routine use by personnel with low analytical skills. Fortunately, unless special circumstances exist (pollution of water supplies by arsenical herbicides, high biological activity), the determination of arsenic compounds in drinking water supplies is not necessary.

Acknowledgement

Appreciation is expressed to Mag. G. Kölbl for the experimental work required for and the preparation of Figure 1.

References

Anderson, L.C.D. and Bruland, K.W. 1991. Biogeochemistry of arsenic in natural waters: the importance of methylated species. *Environ. Sci. Technol.*, **25**, 420-427.

Braman, R.S. 1975. Arsenic in the environment. *ACS Symp. Ser.*, **7**, 108-123.

Clesceri, L.S., Greenberg, A.E. and Trussel, R.R. (eds.) 1989. *Standard Methods for the Examination of Water and Wastewater*, 17th edn., pp. 3-35 to 3-43. APHA, AWWA, WPCF, Washington D.C.

Cullen, W.R. and Reimer, K.J. 1989. Arsenic speciation in the environment. *Chem. Rev.*, **89**, 713-764.

Gailer, J. and Irgolic, K.J. 1994. The Ion-Chromatographic Behavior of Arsenite, Arsenate, Methylarsonic Acid, and Dimethylarsinic Acid on the Hamilton RPR-X100 Anion Exchange Column. *Appl. Organomet. Chem.*,**8,** 129-140.

Greenberg, A.E., Eaton, A.D.and Clesceri, L.S. (eds.) 1991. *Standard Methods for the Examination of Water and Wastewater*, Suppl. to the 17th edn., pp.37-40. APHA, AWWA, WPCF, Washington, D.C.

Greschonig, H. and Irgolic, K.J. 1992. Electrochemical methods for the determination of total arsenic and arsenic compounds. *Appl. Organomet. Chem.*, **6**, 565-577.

Irgolic, K.J. 1982. *Speciation of arsenic compounds in water supplies.* Report 1982, US EPA/600/182/010, Washington, DC; Order No. PB82257817, NTIS, Springfield IL, 125 pp.

Irgolic, K.J., Stockton, R.A. and Chakraborti, D. 1983. Determination of arsenic and arsenic compounds in water supplies. In: W.H. Lederer and R.J. Fensterheim (eds.), *Arsenic: Industrial, Biomedical, Environmental Perspectives*, pp. 282-308. Van Nostrand Reinhold, New York.

Irgolic, K.J. 1992. Arsenic. In: M. Stoeppler (ed.), *Hazardous Metals in the Environment*, pp. 287-350. Elsevier, Amsterdam.

6 Sources and pathways of arsenic in south-west England: health implications

Iain Thornton

Imperial College Centre for Environmental Technology, Environmental Geochemistry Research, Royal School of Mines, Prince Consort Road, London SW7 2BP, UK

Abstract

Arsenic, together with copper and tin, has been mined and smelted in south-west England for centuries, either directly, as arsenopyrite, or indirectly, as an undesirable constitutent of copper and tin ores. Consequently, quantities of arsenic-rich materials were dumped alongside the large numbers of mines in Devon and Cornwall, or spread over wider areas by plume fall-out from the roasting of ores. With the decline of the mining industry, the reclamation of sites and the increased urbanisation of some areas, arsenic-rich materials have been further dispersed.

There have been several studies concerning arsenic and associated metals in stream sediments, soils, waters, and forage and crop plants. The Wolfson Geochemical Atlas of England and Wales clearly delineates the major arsenic province in Devon and Cornwall. Within this area over 700 km^2 of agricultural and urban land are contaminated with arsenic, ranging up to 2,500 $\mu g\ g^{-1}$ in soils and up to 460 $\mu g\ g^{-1}$ in household dusts. Although little soil arsenic is taken up into agricultural, horticultural and home-grown crops, direct ingestion of soil arsenic by grazing cattle and of dust arsenic by young children comprise important exposure pathways.

There are as yet no proven health effects associated with arsenic in local populations, though it may be concluded that there is a need for careful assessment of exposure, particularly in young children. Epidemiological studies are also necessary in exposed populations within the heavily contaminated towns and villages.

Introduction

In Britain, the oldest and most extensive sources of metal pollution are related to the history of metalliferous mining which commenced in Roman and earlier times, thrived to the end of the nineteenth century and then declined rapidly. Historically, south-west England was the premier metalliferous mining centre of the world. In Cornwall, more than six hundred mines have operated in the past. The density of the

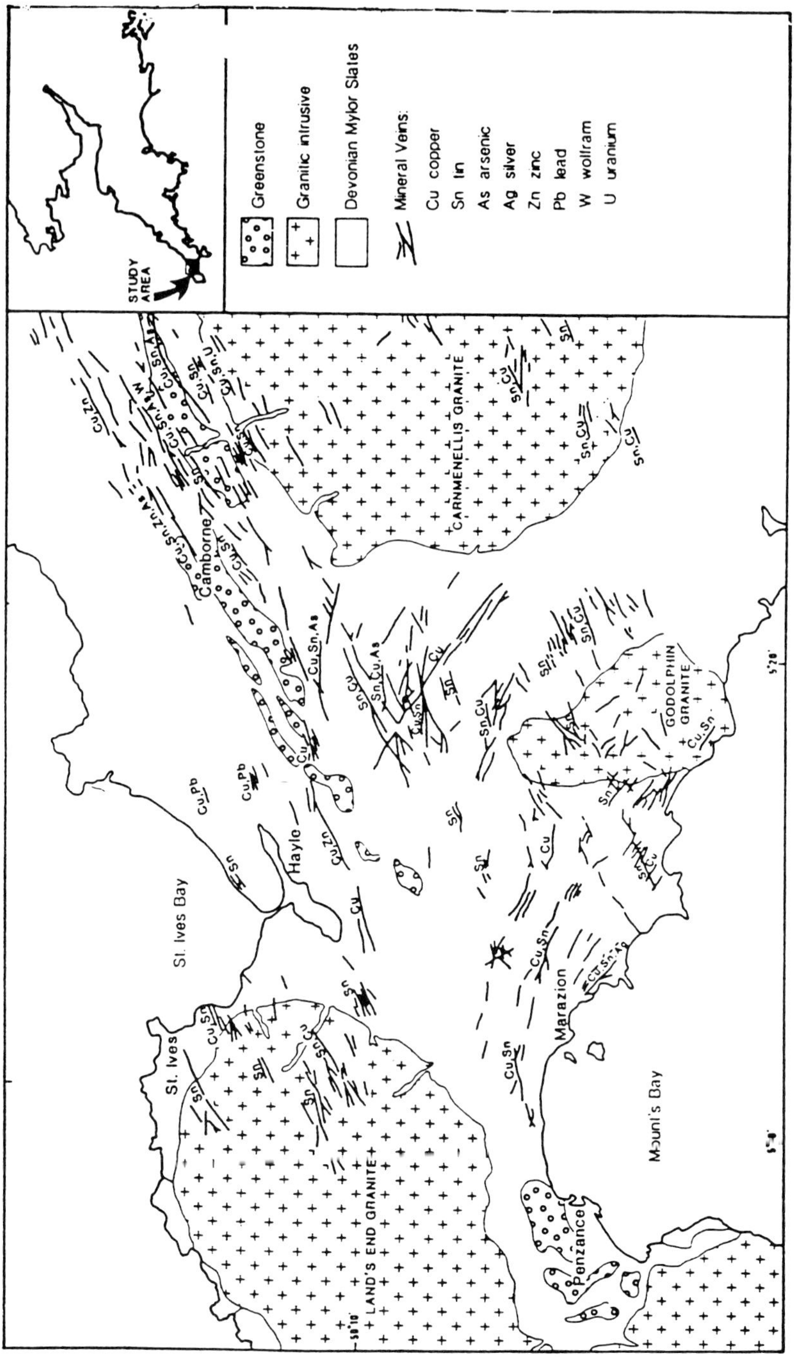

17

Figure 1 *Geology and mineral occurrences in Western Cornwall, England*

Table 1 *Mean contribution of soil to total daily intake of arsenic in 11 herds of cattle in Cornwall, England*

Farm type (based on soil As)	Mean % soil ingested	Soil As $\mu g\ g^{-1}$	Faeces As $\mu g\ g^{-1}$DM	Total daily intake of As $\mu g\ day^{-1}$	% As ingested as soil
20-40 $\mu g\ g^{-1}$ As					
$n = 3$	1.4	24	1.3	7	67
55-140 $\mu g\ g^{-1}$ As					
$n = 3$	1.4	85	3.1	21	76
160-250 $\mu g\ g^{-1}$ As					
$n = 5$	1.1	202	8.1	52	58

mineral veins worked is illustrated in Figure 1. Even before its commercial use, arsenic's association with tin ores in this region was of prime concern to the mining industry, as the arsenic impurity produced a relatively hard, less ductile tin product (Barton, 1967). At that time arsenic-rich materials were dumped alongside the large numbers of mines, or spread over wider areas by plume fall out from the roasting of ores.

Once a demand for arsenic arose with its use as an insecticide in the late 1880s, its production became an important by-product of many tin mines. In Cornwall, arsenic was mined as the ore arsenopyrite. This was crushed and calcined, and the vapour passed through chambers and flues where arsenic oxide accumulated as a grey powder known as "arsenic soot" or "crude arsenic" (Dines, 1956). From about 1860 to 1900, Cornwall was the world's major producer of arsenic and at the height of the industry also imported arsenical ores for smelting.

These mining and smelting activities have left behind a legacy of dereliction and pollution with arsenic-rich mine tailings and other wastes abundant. Further extensive areas of land were contaminated with fallout from the smelting process.

This paper summarises the results of several studies into the extent and degree of contamination of the surface environment in south-west England, research into the geochemical pathways influencing human and animal exposure to arsenic, and possible health implications to local communities.

Geochemical mapping

A map showing the regional distribution of arsenic in stream sediments was published as part of the Wolfson Geochemical Atlas of England and Wales (Webb *et al.*, 1978) (Figure 2) and clearly shows the extensive nature of the arsenic province in south-west England. Subsequent studies comparing the distribution patterns of arsenic in stream sediments and soils have confirmed extensive soil contamination throughout south-west England, with in excess of 700 km^2 effected (Colbourne *et al,*

1975; Abrahams and Thornton, 1987). With the exception of some derelict areas, the majority of the contaminated area is under agricultural use for arable and vegetable crop production and for livestock farming. A number of villages and towns are located within the mineralised area and urban development has sometimes taken place on contaminated land.

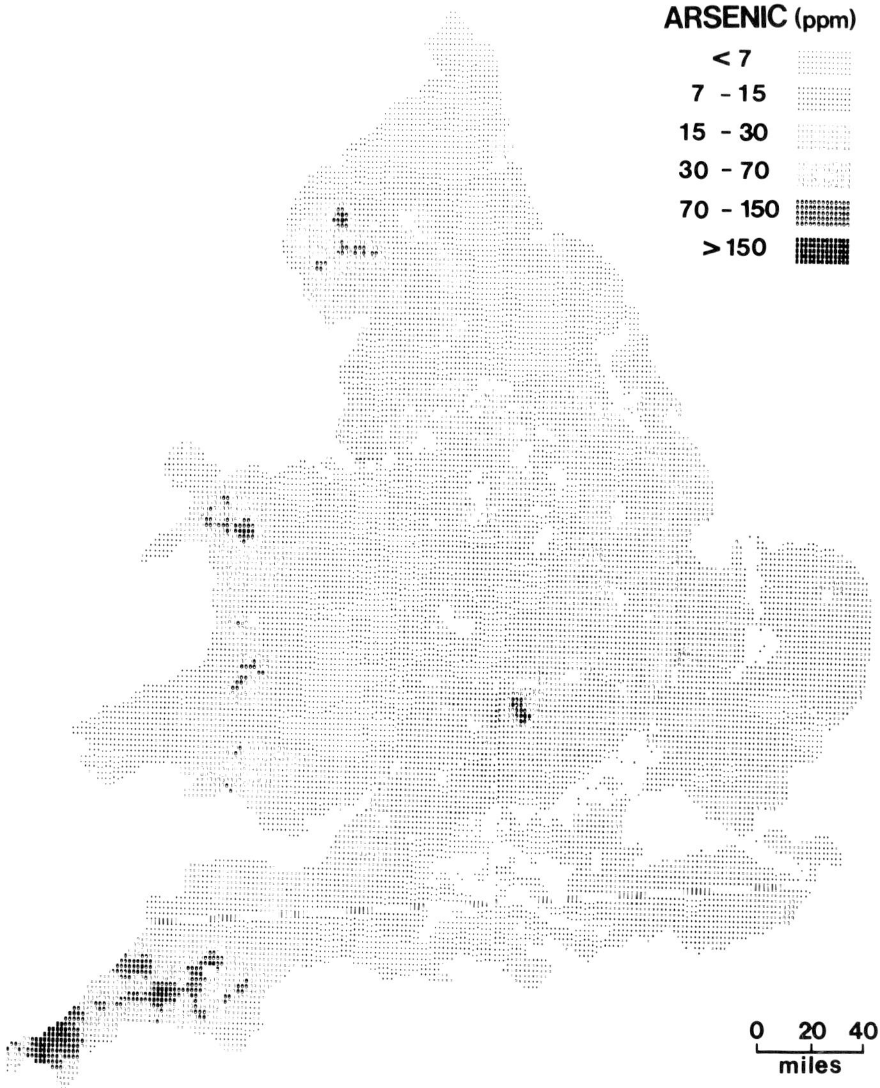

Figure 2 *Map showing the distribution of arsenic in stream sediments in England and Wales.*

Table 2 *Concentrations of arsenic in garden soils and vegetables in mineralised areas of Cornwall, England*

Total As in soil ($\mu g\ g^{-1}$)	Lettuce	Onion	Beetroot	Carrot	Pea	Bean
		$\mu g\ g^{-1}$ dry matter				
Range						
144 - 892	0.15 - 3.88	0.10 - 0.49	0.02 - 0.93	0.10 - 0.93	0.01 - 0.11	0.02 - 0.09
Geometric mean						
322	0.85	0.20	0.17	0.21	0.04	0.04
No. of samples						
32	28	23	23	19	19	7

Contamination of agricultural land

Using the Wolfson Geochemical Atlas as a basis for delineating areas within which surface contamination could be expected, soils were sampled on traverse lines across the anomalous geochemical patterns and at additional alluvial sites below mines. Concentrations of arsenic ranged from 60 - 2,500 $\mu g\ g^{-1}$ in soils in the vicinity of upland mining areas and 90 -900 $\mu g\ g^{-1}$ on contaminated alluvium (Colbourne *et al*, 1975). More systematic studies in that part of Cornwall shown in Figure 1 showed arsenic in agricultural soils (0 - 15 cm) ranging up to 700 $\mu g\ g^{-1}$ As (Abrahams, 1983; Abrahams and Thornton, 1987). A 'normal' range for arsenic concentrations in agricultural soils of England and Wales has been reported as 2 - 53 $\mu g\ g^{1}$, with a median value of 10 $\mu g\ g^{-1}$ (Archer and Hodgson, 1987). A statistical evaluation of the data, using a method for assessing soil contamination proposed by Davies (1971), has led to soils with arsenic concentrations ranging from 110 - 190 $\mu g\ g^{-1}$ being defined as moderately contaminated (exceeding the 70th percentile) and those exceeding 190 $\mu g\ g^{-1}$ as highly contaminated (exceeding the 90th percentile). Of the total area of 9,000 km^2 surveyed in this study, 1.3% has been designated as highly contaminated and 6.6% as moderately contaminated (Abrahams and Thornton, 1987). In the vicinity of roasting ovens and smelter stacks, arsenic may range as high as 0.1 to 1.0% or more in surface soils on which livestock are grazing at this present time.

A study of arsenic uptake into pasture herbage showed a maximum concentration 9.6 $\mu g\ g^{-1}$ As dry matter in washed grass growing on a soil containing in excess of 1,000 $\mu g\ g^{-1}$ As (Thoresby and Thornton, 1979). Highest concentrations were found in early spring and late autumn and may possibly reflect some degree of soil contamination. A later study confirmed large contamination of herbage by soil in early spring and showed that washing grass in deionised water could reduce the arsenic content of herbage dry matter from as much as 58 to as little as 1 $\mu g\ g^{-1}$ (Abrahams, 1983).

Soil contamination of herbage plays a major role in the intake of arsenic by grazing cattle. Livestock may involuntary ingest soil along with grass by as much as 1 to over 10% of their dry matter intake (Healey, 1968). The contribution of soil to the daily intake of arsenic in 11 herds of cattle in south-west England is summarised in Table 1 and in general terms ranges from around 50 –80% of the total arsenic ingested (Thornton and Abrahams, 1983).

Arsenic uptake into barley grain has also been studied. The concentrations measured tended to increase with those in soil, though actual uptake was very small, not exceeding 0.4 $\mu g\ g^{-1}$ As DM (Thoresby and Thornton, 1979). Arsenic also increased in both lettuce and strawberries with soil content but concentrations were again small, peaking at 1.4 $\mu g\ g^{-1}$ and 0.7 $\mu g\ g^{-1}$ DM respectively.

Arsenic in home gardens and home grown vegetable crops

An examination of 32 home gardens in the Hayle-Camborne area (see Figure 1), showed arsenic concentrations to range widely from around 150 to nearly 900 $\mu g\ g^{-1}$ As (Xu and Thornton, 1985). Water-soluble arsenic constituted 0.1 - 0.5% of the total and acid-fluoride extractable arsenic less than 5% of the total. Both were significantly correlated with the total ($R = 0.80$ and 0.75 respectively, $p < 0.01$). Arsenic was determined in six garden crops: lettuce, onion, beetroot, carrot, pea and bean. Total concentrations in these did not exceed 1 $\mu g\ g^{-1}$ As, dry matter, with the exception of lettuce (Table 2). Arsenic uptake into these vegetables is species dependent with maximum concentrations in lettuce (geometric mean 0.86 $\mu g\ g^{-1}$ dry matter) and lowest in peas and beans (0.04 $\mu g\ g^{-1}$). Arsenic in vegetables increased with both total and extractable arsenic in soils. Although soil arsenic would seem to be the main factor controlling arsenic levels in these vegetables, the effects of soil iron, aluminium, calcium and phosphorous and pH were tested using ridge regression analysis to construct models for the prediction of arsenic uptake. An example is shown for lettuce:

$$\text{As (lettuce, } \mu g\ g^{-1}) = 0.381 \times 10^{-2}\ \text{As (soil total, } \mu g\ g^{-1}) +$$

$$0.3096 \times 10^{-3}\ \text{P (soil total, } \mu g\ g^{-1}) - 0.109\ \text{Fe (soil, total, \%)} - 0.4773$$

This illustrates the effect of soil iron on arsenic uptake, probably due to the low solubility product constant of iron arsenate and adsorption of arsenic on iron hydrous oxides.

In the UK, the statutory limit for arsenic in most foods offered for sale is 1 mg kg^{-1} ($\mu g\ g^{-1}$) fresh weight (Arsenic in Food Regulations, 1959). All of the vegetables sampled were well below this permitted level, in spite of the large amounts of arsenic present in these soils. It has been suggested that in this "geochemical hotspot" situation in Cornwall plants are acting as "geochemical barriers" in the environment and are only making a small contribution to man's exposure (Xu and Thornton, 1983). Speciation and bioavailability of arsenic in these soils has yet to be determined.

Arsenic in drinking water

Using data from the Wolfson Geochemical Atlas of England and Wales, referred to above, comparisons were made between arsenic concentrations in stream sediments and in waters in three river catchments. Eighty two per cent of the sediments ranged from 20 - 200 ppm As and 13% from 200 to in excess of 500 ppm. This compares with the normal or background abundance in England and Wales of 5 - 15 ppm As. On one of the rivers studied concentrations of 200 - 1500 ppm in sediments were associated with waters containing from 10 - 50 µg L^{-1} As. In another contaminated tributary, highest water arsenic concentrations exceeded 250 µg L^{-1} (Aston *et al.*, 1975). Leaching experiments have shown that the dissolution of arsenic from spoil tip material may be much less significant than that from condensed smelter fumes and arsenical ores (Aston *et al.*, 1975). However the dissolved arsenic is then very easily adsorbed by hydrated iron oxides and other detritus such as clays and organic aggregates (Ferguson and Jarvis, 1972).

On the basis of this exploratory study, it has been shown that within south-west England the distribution of arsenic in river waters, both on a regional basis and on a more localised catchment basis can be related to the arsenic content of stream sediments. However, abstraction of surface waters for processing and distribution avoids whenever possible waters contaminated by past mining activities, and arsenic in Cornish drinking water rarely exceeds 10 µg L^{-1} (MAFF, 1982)

Arsenic exposure in local populations

Some attempts have been made to determine arsenic exposure of young children from dusts and soils in and around the home environment. In a study of some seventy households in the Camborne/Hayle area of Cornwall, concentrations of arsenic in surface garden soils ranged from 120 - 1,100 µg g^{-1} (median 340 µg g^{-1}) and in housedusts ranged from 9 - 330 µg g^{-1} (median 56 g g^{-1}) (Johnson, 1983). The author assumed that a young child could ingest as much as 42 µg As day^{-1} in this contaminated locality.

Harding (1993) examined garden soils and house dusts in twenty-three households in villages associated with past arsenic mining in Cornwall. Hand wipe samples were also taken from twenty-seven children aged 1 - 3 years. Garden soils ranged up to 770 µg g^{-1} As and housedusts to 460 µg g^{-1} As. Quantities of arsenic on children's hands in no cases exceeded 3.5 µg. Assuming hand to mouth activity occurring on average ten times per day and the total arsenic removed on each occasion, it is estimated that arsenic intake via this route can range up to 35µg day^{-1}. Alternatively, if one assumes that the two year old child ingests around 100 mg dust day^{-1}, at a maximum dust arsenic concentration of 460 µg g^{-1}, the child will ingest 46 µg day^{-1}. Both these approaches are at best semi-quantitative and more detailed studies are now required. However it would seem possible that the young child in this heavily contaminated environment may well have an arsenic intake approaching or even exceeding the WHO Provisional Tolerable Daily Intake for inorganic arsenic of 2 μg kg^{-1} body weight (WHO, 1983).

Analysis of inorganic arsenic and its methylated metabolites in urine from adults and children from this Cornish mining area showed in general only slightly elevated urinary arsenic levels (Johnson and Farmer, 1989). First-void samples from 9 children ranged from 2.5 - 39.9 μg L^{-1} As and for 29 adults 1.7 - 23.7 μg L^{-1} As, showing no appreciable difference to values found within the city of Glasgow (which is well removed from any arsenic mineralisation).

Discussion

There is no dispute that concentrations of arsenic in agricultural and urban soils in parts of south-west Britain associated with copper-tin-arsenic mineralisation and the mining and smelting of arsenopyrite are extremely large ranging up to as much as 1,000 μg g^{-1} or more. A recent study has shown locations immediately adjacent to smelters with as much as 2 per cent arsenic in surface soil and in smelter slag (Frizzell, 1993). However there is little evidence of significant arsenic uptake into locally grown farm, horticultural and home garden crops. It would seem that arsenic in these soils is present in forms that are relatively insoluble and thus of low bioavailability. Of that portion of arsenic taken into plant roots very little is translocated to the aerial parts of the plant. Although food plants do respond incrementally to increasing arsenic concentrations in the soil, uptake is too small to constitute a significant exposure route to local populations. In all cases studied, concentrations of arsenic in crops have fallen below the UK statutory limit of 1 μg g^{-1} fresh weight.

Similarly uptake from agricultural soils into pasture herbage is again small, though direct ingestion of arsenic as soil may be as great as 136 mg day^{-1} in grazing cattle (Thornton and Abrahams, 1983). There are however as yet no studies of arsenic in locally produced milk and other agricultural products.

Evidence of arsenic exposure in local populations is extremely limited and the few data for urine arsenic available do not seem to indicate any appreciable uptake. However intake by young children through the hand to mouth exposure route may possibly be important and needs more careful and detailed investigation.

There are at present no clearly recognised health effects associated with this unique high arsenic province in Cornwall, although maps of cancer mortality show high standardised mortality ratios (SMR > 125) for melanoma of the skin in both men and women for south-west England (Gardner *et al.*, 1983). Various relationships between the incidence of cancers and multiple sclerosis and toxic elements including arsenic have in the past been proposed but are not supported by acceptable statistical data (Shaper, 1979). For example, it has been suggested that environmentally high exposure to arsenic played a contributory role in the raised incidence of malignant melanoma in Cornwall (Clough, 1980). Again this is not supported by sound epidemiological evidence. There is obviously a need for a multi-disciplinary, in-depth study concerning arsenic sources, dispersion, environmental pathways, exposure and possible health effects in this region.

Guidance on the assessment of contaminated land scheduled for redevelopment is at present based on tentative "threshold trigger concentrations" for a range of metallic substances (ICRCL, 1987). The current values listed for arsenic are 10 $\mu g\ g^{-1}$ for land to be used for urban development and for the growth of crops and 40 $\mu g\ g^{-1}$ for land to be used as parkland or for amenity use. These figures relate to total concentrations of arsenic in soil. The author is unaware of any relationships between these levels and recognised human health effects, nor would such modest levels seem appropriate for the regulatory process. It is perhaps worth concluding that whatever future regulations for arsenic and soil may be proposed and indeed implemented, it is highly likely that the concentrations reported in this paper in south-west Britain will greatly exceed these limits. It is thus seen that the measurement of total concentrations of arsenic (and indeed of metalloids and metals in general) is inappropriate for the regulatory process and that more specific measurements of arsenic species, their solubility and their potential bioavailability are a requisite for future regulatory processes.

References

Abrahams, P. 1983. Distribution, dispersion and agricultural significance of metals in soils of the mining regions of southwest England. PhD Thesis, University of London.

Abrahams, P. W. and Thornton, I. 1987. Distribution and extent of land contaminated by arsenic and associated metals in mining regions of southwest England. *Trans. Instn. Min. Metall. (Sect. B: Appl. earth sci.).* **96,** 131-138 The Institute of Mining and Metallurgy.

Archer, F. C. and Hodgson, I. H. 1987. Total and extractable trace element contents of soils in England and Wales. *Journal of Soil Science*, **38**, 421 - 431.

Arsenic in Food Regulations. 1959. S.I. 831.

Aston, S. R., Thornton, I. and Webb, J. S. 1975. Arsenic in stream sediments and waters of south west England. *The Science of the Total Environment*, **4**, 347 - 358.

Barton, D. B. 1967. *A history of tin mining and smelting in Cornwall.* Truro, Cornwall: D. Bradford Barton Ltd.

Clough, P. 1980. Incidence of malignant melanoma of the skin in England and Wales. *British Medical Journal*, **1**, 112.

Colbourne, P., Alloway, B. J. and Thornton, I. 1975. Arsenic and heavy metals in soils associated with regional geochemical anomalies in south-west England. *Sci. Total Env.*, **4**, 359 -363.

Davies, B. E. 1971. Trace element content of soils affected by base metal mining in the west of England. *Oikos*, **22**, 366 -72.

Dines, H. G. 1956. *The metalliferous mining region of south-west England, Vol. 1. Memoirs of the Geological Survey of Great Britain.* HMSO, London.

Ferguson, J. F. and Garvis, J. 1972. *Water Res.*, **6**, 1259.

Frizzell, P. M. 1993. Verification of a preliminary site inspection guidance manual for the identification of potentially contaminated land. MSc. Dissertation, University of London.

Gardner, M. J., Winter, P. D., Taylor, C. P. and Acheson, E. D. 1983. *Atlas of Cancer Mortality in England and Wales 1968 -1978.* John Wiley and Sons, Chichester and New York.

Harding, E. R. 1993. Arsenic contamination of garden soils and housedusts in West Cornwall. BSc. Dissertation, Imperial College of Science, Technology and Medicine.

Healy, W. B. 1968. Ingestion of soil by dairy cows. *New Zealand J. of Agric. Res.*, **11**, 487 - 499.

ICRCL 59/83. 1987. *Guidance on the assessment and redevelopment of contaminated land, 2nd edn.*, Department of the Environment, London.

Johnson, L. R. 1983. A study of arsenic in housedusts and garden soils with relation to geochemistry and health. MSc. Thesis, University of London.

Johnson, L. R. and Farmer, J. G. 1989. Urinary arsenic concentrations and speciation in Cornwall residents. *Environmental Geochemistry and Health*, **11,** 39 - 44.

Ministry of Agriculture, Fisheries and Food. 1982. Survey of Arsenic in Food. *Food Surveillance Paper No. 8.* HMSO, London. 35p.

Shaper, A. G. 1979. Epidemiology for geochemists. *Phil. Trans. R. Soc. Lond. B* **288**, 127 - 136.

Thoresby, P. and Thornton, I. 1979. Heavy metals and arsenic in soil, pasture herbage and barley in some mineralised areas in Britain: Significance to animal and human health. In: Hemphill, D. D. (ed.), *Trace Substances in Environmental Health*, Vol. 13.

Thornton, I. and Abrahams, P. 1983. Soil ingestion - A major pathway of heavy metals into livestock grazing contaminated land. *The Science of the Total Environment*, **28**, 287 - 294.

Webb, J. S., Thornton, I., Thompson, M., Howarth, R. J. and Lowenstein, P. L. 1978. *The Wolfson Geochemical Atlas of England and Wales.* Oxford University Press, Oxford.

World Health Organisation. 1983. 27th Report of the Joint FAO/WHO Expert Committee on Food Additives, p. 29. *Technical Report Series 696.* World Health Organisation, Geneva.

Xu, J. and Thornton, I. 1985. Arsenic in garden soils and vegetable crops in Cornwall, England: implications for human health. *Environmental Geochemistry and Health*, **7** (4), 131 -133.

7 Exposure assessment for arsenic concentrations from electric power stations

Lawrence B. Gratt [1] and Leonard Levin[2]

[1] *IWG Corp,.1940 Fifth Avenue, Suite 200, San Diego, CA 92101, USA*
[2]*Electric Power Research Institute, 3412 Hillview Avenue, P.O. Box 10412, Palo Alto, CA 94303, USA*

Abstract
Arsenic is the key contributor to health risks due to stack emissions from fossil-fueled electric power generating stations according to many current regulatory public health cancer risk assessment methodologies. Calculated emission rates for arsenic are most strongly related to control efficiency achieved for particulates. Measurements are currently being performed by the Electric Power Research Institute (EPRI) and the US Department of Energy (DoE) to establish the concentration of the trace element emissions to air from a variety of power plants. The CORE (Comprehensive Risk Evaluation) methodology is being applied by EPRI to evaluate the exposure and ensuing risk from power stations. A database assembled for over 600 power stations is used with preliminary emissions data to test the use of probabilistic techniques and guide more detailed CORE analysis. In the present work, a simplified first estimate of the probabilistic arsenic exposure is calculated using arsenic fuel content distributions, plant operations and control system capabilities, dispersion calculations, and receptor scenarios. Dispersion analysis of the arsenic emissions is performed using EPA's ISC2 Long Term Air Quality Model with site-dependent stack parameters and meteorology. These results will be used to guide more refined CORE arsenic exposure calculations, help to establish research priorities to improve the exposure assessment, and understand the key uncertainties for arsenic risk assessment. The initial results indicate only a small number of the power stations analysed result in computed arsenic concentrations at levels requiring further, more detailed investigation.

Introduction

Arsenic emissions from fossil fuel electric power generating stations are a key contributor of health risk assessments carried out for those facilities. The contribution of arsenic to the total computed cancer risk as a function of fuel (oil or coal by type) has been assessed (Levin *et al.*, 1991; Gratt and Levin, 1993). The purpose of this paper is to present the results of a simplified probabilistic analysis of the arsenic exposure from electric power generating stations. This analysis is intended to help prioritize research needs and serve as guidance for refined exposure analysis from arsenic emissions.

The Electric Power Research Institute's (EPRI) PISCES Program is currently establishing a database for emission factors for all significant air toxics based on fuel, process and stack measurements (Chu *et al.*, 1993). In the present analysis the magnitude of arsenic emissions from fossil-fueled power stations with units >25 MWe is estimated by fuel type and control system (Gratt *et al.*, 1993). The probabilistic analysis is based on assuming the concentration of arsenic in the fuel follows a lognormal distribution, and control system performance has a triangular distribution. The dispersion modelling calculations were performed using the Environmental Protection Agency (EPA) air quality model ISC2 using meteorology at the nearest National Weather Service station reporting at least 5 years of data reduced to STAR (STability ARray) format. The exposure calculations were based on the maximum concentration from the dispersion results. Results are summarised for 401 coal-fueled stations, and 54 oil-fueled stations. The use of the exposure results indicates only a small number of stations should be of concern for regulatory risk assessments.

Background

Risk assessments due to electric power generation facilities are performed to estimate the concentration of compounds dispersed into the environment and to evaluate the risk to human health resulting from exposure to those concentrations. Risk assessments using probabilistic models attempt to make the estimate more scientifically robust by conveying the uncertainty and variability in the variables analysed. Risk assessments involve investigation into all health impacts. Currently, both cancer and noncancer impacts are estimated. For electric power generation, cancer is the dominant concern for public exposure because of the nonthreshold modelling approach. A simple (such as might be used for a regulatory application) risk assessment algorithm for chronic cancer uses an annual emission rate, Q_c, of a chemical c, in units of mass per unit time to estimate the risk. Letting R_c be the excess lifetime cancer probability (LCP) for chemical c,

$$R_c\ (\mathrm{LCP}) = Q_c \cdot (\chi/Q) \cdot Uc \cdot MP_c \qquad (1)$$

here (χ/Q) is the dilution factor that converts the source rate Q into concentration c (mass per unit volume), Y_c is the cancer risk factor used to convert concentration to LCP, and MP_c is the multipathway factor for chemical c to account for non-inhalation pathways. For many chemicals, $MP \approx 1$ and is of little concern. For some chemicals, $MP >> 1$ and must be taken into account when evaluating the health risk (Constantinou and Seigneur, 1991). The dilution factor is usually obtained using an air dispersion, such as those recommended in EPA guidelines model (U.S. EPA, 1987). The product of Q_c and (χ/Q) is used as the simplest estimate of the exposure

Table 1 *Initial estimates for arsenic emission factors for coal- and oil-fueled power generating stations.*

Fuel	Type	Control System		Reference HHV (Btu/lb)	Reference Control Efficiency (%)	Emission Factor (lb/10^12Btu)	
		Particulate	Gas			Mean	Standard Deviation
Coal	Bituminous	ESP	None	13,016	99.5	4.8	1.03
		ESP	Scrubber	13,131	99.8	1.9	0.78
		Baghouse	None	13,016	99.3	6.3	0.95
		Baghouse	Scrubber	13,131	99.8	2.2	0.17
		Scrubber		13,131	97.6	31.9	0.95
		Other	None	13,016	95.0	48.5	0.74
		Other	Scrubber	13,131	95.0	69.1	0.41
	Sub-bituminous	ESP	None	11,905	98.6	3.3	0.91
		ESP	Scrubber	11,905	98.6	1.1	1.3
		Scrubber		11,905	98.6	4.7	1.3
	Lignite	ESP	None	7,086	99.5	6.1	1.2
		Scrubber		7,086	99.5	2.6	2.6
Oil	Residual	None	None	19,012	0	5.5	1.4
	Residual	ESP	None	19,012	Unknown	0.3	1.7

HHV = Higher Heating Value
ESP = Electrostatic Precipitator

and is the subject of this analysis of arsenic emissions from fossil fueled U.S. power stations.

Power station arsenic emissions

The first term in the right-side of equation (1) is the arsenic emission rate for the power station. The emission rates used for this study are summarised in Table 1. The coal stations studied have been divided into categories based on fuel type (*e.g.*, bituminous), particulate control system and flue gas control system. The oil stations are divided into two categories, uncontrolled and controlled. To simplify this analysis, stations using mixed fuels (*e.g.*, units burning coal and oil) were classified by the majority fuel and assumed to use that fuel 100% of the time.

The arsenic emission rates are used with the station power generation requirements, P, fuel higher heating value (HHV) and relative control efficiencies to estimate the arsenic emissions (lb yr^{-1}). This estimate is made using equation (2),

$$Q_C(\text{lb yr}^{-1}) = EF_C \cdot \frac{(1\text{-}CE)}{(1\text{-}CE_O)} \cdot \frac{(HHV_O)}{(HHV)} \cdot P \qquad (2)$$

where EF_C is the emission factor from Table 1, CE is the station control efficiency, CE_O is the reference control efficiency from Table 1, HHV is the higher heating value of the fuel, HHV_O is the reference higher heating value from Table 1, and P is the total thermal energy rate (Btu yr^{-1}) required to generate the annual electric power level in the year 1989. The emission factors are represented by lognormal distributions using the mean and standard deviations shown in Table 1. These values were calculated based on both literature and measured values for arsenic in the fuel, arsenic enrichment, and control system performance. Simple extrapolations were used to extend measured values to other similar and analogous station configurations. The measurement programs underway (Chu *et al.*, 1993; Sarkus and Hargis, 1993) are expected to significantly improve the quality of these initial estimates. The fuel content and measured emissions were assumed to be lognormally distributed with the sample mean equal to the distribution mean. Documented high and low extreme values were used to calculate the standard deviation of the distribution using an iterative technique to achieve an approximation for the ends of the distribution (*e.g.*, representations of the high and low as 90-95% and 10 to 5% percentiles, respectively) when feasible. The control system performance was modeled as a triangular distribution, with the most available point estimate chosen as the probable

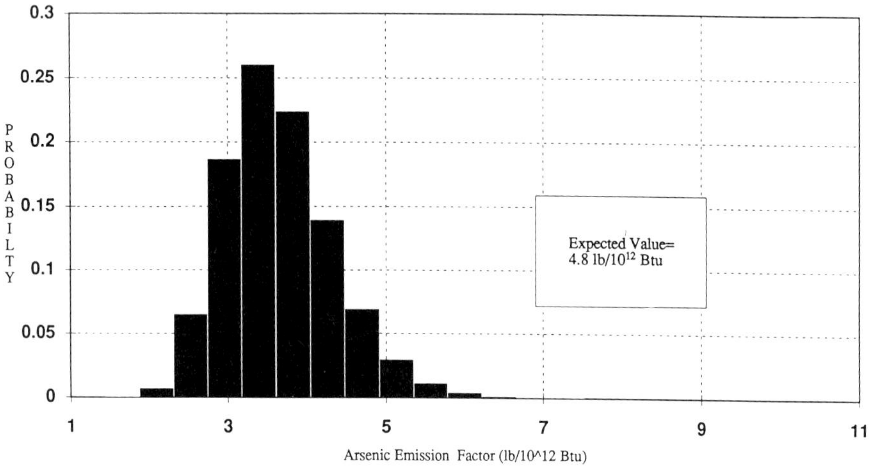

Figure 1 *Emission factor for bituminous coal-fueled power station with ESP particulate controlled (99.5%).*

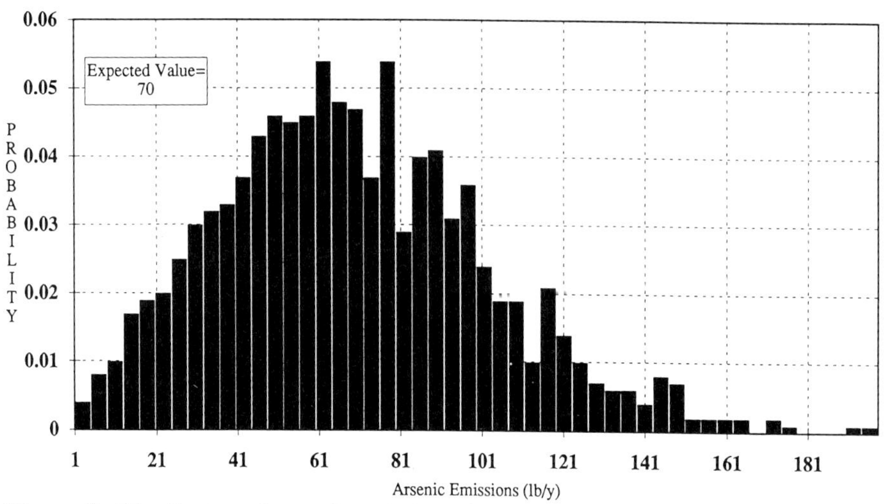

Figure 2 *Distribution of arsenic emissions from coal-fueled power station (410 MWe).*

Table 2 *Summary of dispersion modeling inputs used.*

Model	ISC2 (Long Term Mode)
Location	Station specific
Stack height	Station specific
Stack diameter	Station specific
Exit temperature	Station specific
Exit velocity	Station specific
Terrain	Assumed flat earth
Dispersion coefficients	Urban
Options	Regulatory default
Receptors	20 km x 20 km cartesian grid with 2 km spacing centered at the stack
Meteorological data	Nearest 5-year STAR summaries from 1984 NWS data tape

value. The high value was chosen as 99.99% and the low value adjusted to allow the mean to reach the value of the point estimate. This method may not be appropriate for low control efficiency (<90%).

Figure 1 shows the lognormally distributed emission factor for the entries corresponding to the first row of Table 1. The distribution is based on 1000 Monte Carlo simulations using 10 output ranges for the results with a fixed control efficiency of 99.5%. Figure 2 illustrates the use of this distribution in equation (2) to generate the emitted arsenic from a 410 MWe coal station. For this example, the control efficiency has been adjusted to the database value (UDI, 1989) with a triangular distribution as previously described (min = 98.5%, most probable = 99.5% and max = 99.99%). The result is a triangular distribution modified by a lognormal distribution. As an example for another fuel, Figure 3 is the arsenic emission from an uncontrolled oil-fueled plant, reflecting the input lognormal distribution modified by a scalar from the point-estimate values used in equation (2). The expected value of the arsenic emissions from the smaller capacity coal station are larger than the oil station because of the larger amount of coal required to generate the same thermal output.

The underlying distributions of the emission factor and control system are passed through to the resulting concentrations because the dilution factor (χ/Q) is treated as a scalar modification of emissions to calculate the concentration. The largest emission value is due to a single coal-fired power plant with an abnormally low tabulated control efficiency.

Key issues for the emissions calculation are the content of arsenic in fuel, the firing and operation of the combustion units, and the control system performance. The impact of enrichment and particle size in the risk assessment can be investigated using this type of analysis if appropriate distribution information is available. Enrichment refers to the concentration of the pollutant in emitted ash being greater than that in ash component of the bulk fuel. Arsenic is volatized during combustion and condenses on particulates, increasing the relative mass on the smaller-size particles. In coal combustion, the bulk of the arsenic emitted from electrostatically controlled stacks is bound to particles in the respirable range (about 1 to 10 microns) of concern in inhalation risk assessments (Kowalcyzk *et al.*, 1989). For uncontrolled oil combustion, the resulting particulate mass distribution vs. size has been observed

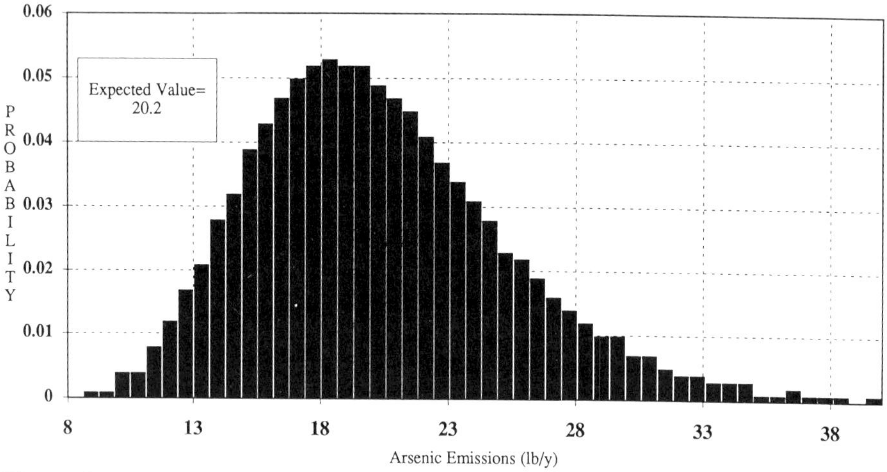

Figure 3 *Distribution of arsenic emissions from a large (>500 Mwe) uncontrolled oil-fueled station.*

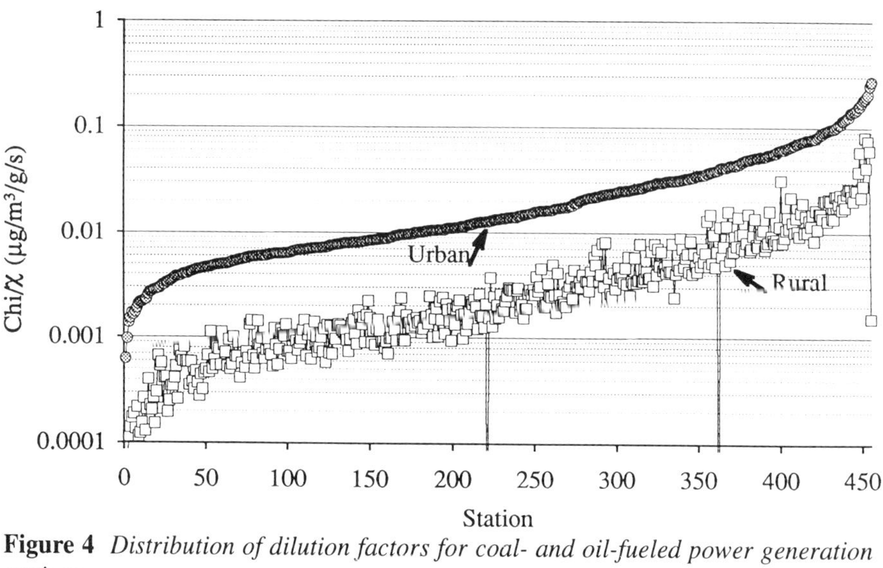

Figure 4 *Distribution of dilution factors for coal- and oil-fueled power generation stations.*

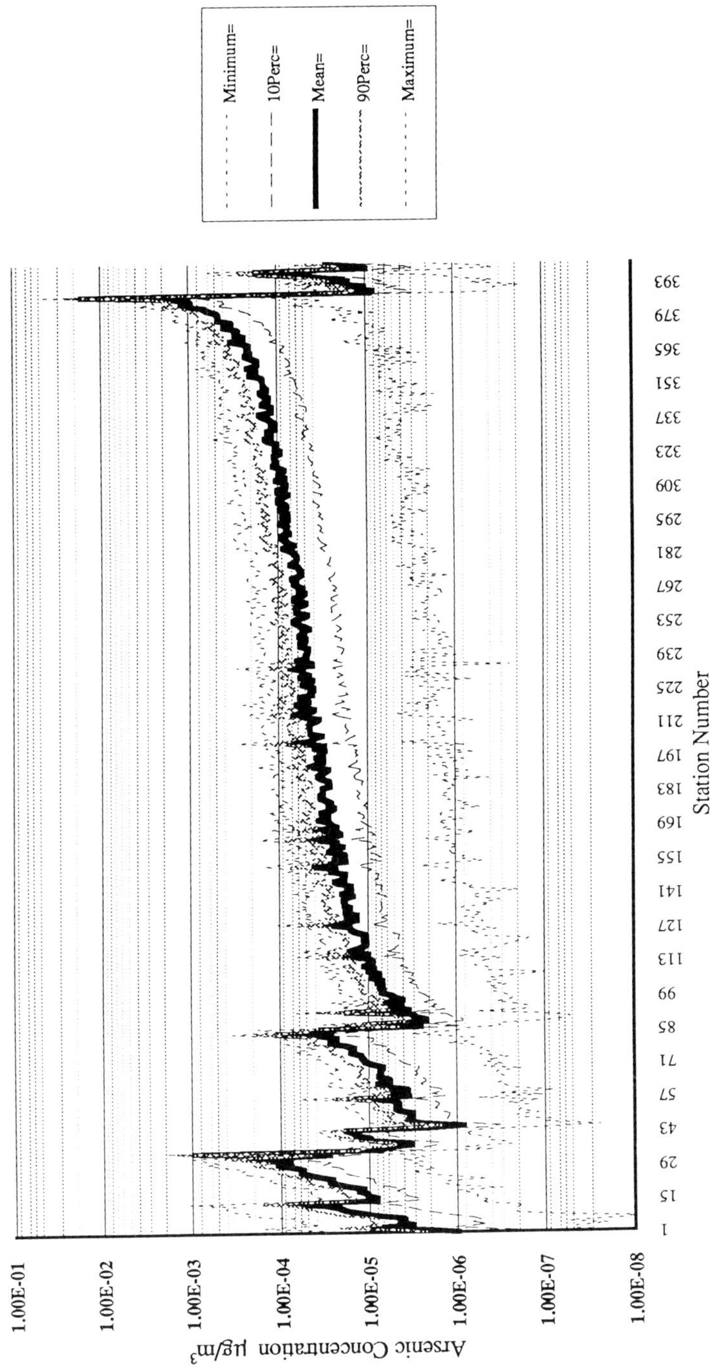

Figure 5 *Distribution of arsenic concentration from coal-fueled stations.*

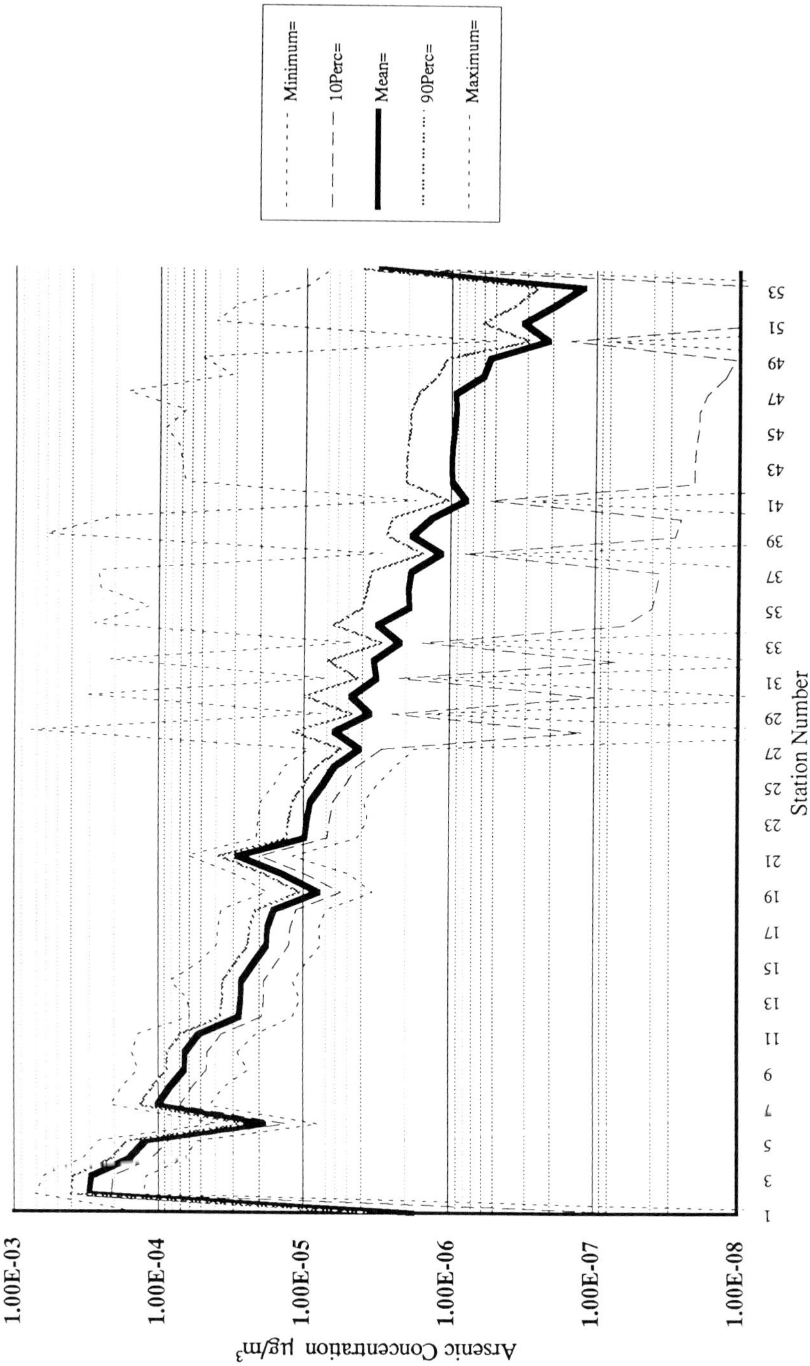

Figure 6 *Distribution of arsenic concentration from oil-fueled stations.*

to be bimodally distributed with the result that less than 50% of the mass is in the respirable range (Piper *et al.*, 1981). Additional measurements are needed for characterizing the distributions for enrichment and particle size analysis to be performed for a large variety of combustion system, control system and fuel types. A key parameter requiring better data to characterize the distribution of control efficiency by particle size.

Maximum arsenic concentrations

Atmospheric dispersion was calculated using ISC2 (US EPA, 1992) with site-dependent parameters. A 20 km by 20 km Cartesian grid centered on the stack at each generating station was used as the analysis domain. The receptor with the maximum ground level concentration is used in the risk assessment. This grid was divided into 100 4-km^2 zones using one receptor at the center of each grid cell. Flat terrain and urban dispersion coefficients were used for all sites. This may have a significant effect on the predicted ambient concentrations. Use of the Urban coefficients throughout was necessary because the site-specific modeling of downwash and terrain are beyond the scope of the current study, but will be considered in refined analysis (Room *et al.*, 1993). Table 2 summarizes the modeling inputs. Site-specific meteorological data were used at each station. These data are represented as multi-year measured data in the form of a joint frequency distribution of wind speed, direction and stability class used to calculate spatially dependent dilution factors. Only the maximum dilution factor (equivalent to maximum concentration) is used from this analysis, so the resulting probability distributions are transformed by only a scalar multiplication. As an example for a coal-fueled station with $\chi/Q = 0.128$, Figure 2 with the appropriate scalar multiplier represents the distribution of the arsenic concentrations(mean of $1\text{x}10^{-3}$ μg m^{-3}). An analogous result for an oil-fueled example uses Figure 3 ($\chi/Q = 0.076$) to obtain a similar distribution for an oil-fueled station (mean of $1.9\text{x}10^{-5}$ mg m^{-3}). Figure 4 shows the dilution factors for the 454 power stations.

Figure 5 shows the distribution of maximum ground-level arsenic concentrations levels from the 401 coal-fueled power stations. The mean values are shown with additional lines indicating the range of the values for the individual distribution of each station. This result can be used to establish which stations are candidates for further study and to guide further analysis. An example of the method to select the stations for further study is to use equation (1) by selecting a reference risk level and the currently tabulated unit risk factor (Anderson, 1983), U_c, for arsenic, $4.3\text{x}10^{-3}$ Lifetime Cancer Probability/(μg m^{-3}), and $MP_c=1$ to solve for a reference concentration, Q_c times (χ/Q), to count the number of stations above this reference level. For example, if the reference level is 10^{-4} μg m^{-3}, 75 stations (19%) will have mean concentrations above this level, whereas if the reference level is 10^{-3} μg m^{-3}, only one station's mean exceeds this value. Similar results can be generated for the 90th percentile values or even the maximum values. These results can also be used to evaluate the potential impact on background arsenic concentrations. The simplified analysis for arsenic indicates a small number of stations of concern. The combination of high emission (Q) and high dilution factor (χ/Q) determine these outcomes.

Figure 6 is the arsenic concentration resulting from the 54 oil-fueled stations. The low control efficiency sites show wide ranges in the resulting site distribution. The results are similar to the coal-stations, again with a small number of sites of potential concern.

Refined inhalation exposure

The probabilistic analysis for a simplified exposure assessment illustrated the feasibility and use of the risk assessment technique in the present work. As part of the CORE for electric generating stations, refined analysis will be performed (Chow and Levin, 1993). This analysis will go beyond the assumption of the simplified analysis and examine the validity of key assumptions using available data. Also, key uncertainties will be identified and used to help establish research needs, in air toxics for EPRI.

The refined analysis will include use of fuel, control system and emissions measurement, improved modelling of emissions, calculation of appropriate dispersion coefficient, and use of existing and new results in human exposure calculations.

A key assumption of the exposure portion of a simplified risk assessment is that there may be an individual at the location of the maximum concentration in the study area who is continuously exposed for a 70-year lifetime at this exposure level. This assumption will be investigated. Calculations will be performed based on human activity patterns to better understand the exposure to power-plant emissions (Hayes *et al.*, 1993). Residence duration will be evaluated using statistical techniques, while indoor *vs.* outdoor microenvironments will be considered using penetration ratios for the emissions of concern. Locations near and far from the plant will be evaluated, and compared. Variations in human parameters will also be addressed, including breathing rates and body weight.

Use of arsenic exposure in risk assessment

The valence state of arsenic in the power station emission is also of concern. In the future, it may be of importance to speciate the arsenic exposure. The bulk of the respiratory cancers attributed in humans appear to be related to arsenic trioxide (As III). The principal form of arsenic in drinking water is valence state V. There have been no epidemiological studies concerning the carcinogenicity of arsenic in flyash at the expected concentration levels. The inhalation evidence for carcinogenicity is based on exposures to arsenic that may be different from flyash arsenic in valence state, particle matrix and size, and solubility. Research is needed to characterize the relative distribution and concentrations of arsenic, and speciation of arsenic between arsenic III and arsenic V for flyash particles, for dose-response applications (Wyzga and Yager, 1991).

Implications and research needs

The simplified arsenic exposure assessment demonstrating probabilistic techniques indicates that a small set of fossil fueled electric power stations are of concern for further analysis. Research needed on arsenic emissions from power stations includes

validated measurement procedures at lower detection levels;
fuel arsenic concentration, including speciation;
enrichment of arsenic air emissions plume, by particle size class;
control system performance and efficiency by particle size and speciation;
particulate characterisation by size and speciation;
particulate behavior and fate in the environment as a function of particle size;
indoor/outdoor relationships;
bioavailability of particulate arsenic by size class and speciation; and
dose-response data relevant to community exposures.

Conclusions and recommendations

The distribution of the concentration arsenic from electric power stations has been estimated based on available data using a simplified analysis. Available data for emission factor distributions for specific fuels have been used with control system capabilities and site-specific dilution factors to calculate the distribution of maximum concentrations. The maximum arsenic concentration for coal-fueled stations ranges from 10^{-2} to 10^{-6} $\mu g\ m^{-3}$ with an overall mean of $1.5 x 10^{-4}$ $\mu g\ m^{-3}$. For oil stations, the range is from $3 x 10^{-3}$ to 10^{-7} $\mu g\ m^{-3}$ with an overall mean of $2.7 x 10^{-5}$ $\mu g\ m^{-3}$. Research needs to improve the exposure assessment for arsenic have been identified. These needs include improved measurements and speciation of the arsenic by particle size class in the fuel and flue gas emissions, and fly ash arsenic behavior and fate in the environment. Further efforts will refine this analysis with results from current and planned research and measurement programs. These efforts include fuel content measurements, station process measurements, emission measurements, site-specific dispersion coefficients, and exposure conditions. The measurements will allow probabilistic analysis and modelling to evaluate the impact of arsenic emissions from electric power stations without reliance on many overly-conservative assumptions.

Acknowledgments

This effort was performed under EPRI Research Project 3081-03.

References

Anderson, E.L. and the Carcinogen Assessment Group of the US Environmental Protection Agency. 1983. Quantitative approaches in use to assess cancer risk. *Risk Analysis*, **3**(4), 277-295.

Chow, W. and Levin, L. 1993. *EPRI research on hazardous air pollutants: An overview.*

Paper presented at Managing Hazardous Air Pollutants, 2nd International Conference, 13-15 July 1993, Washington D.C.

Chu, P., Nott, B. and Chow, W. 1993. *Results and issues from the PISCES field tests.* Paper presented at Managing Hazardous Air Pollutants, 2nd International Conference, 13-15 July 1993, Washington D.C.

Constantinou, E. and Seigneur, C. 1991. *Multimedia health risk assessment for power plant air toxic emissions.* Paper presented at the Managing Hazardous Air Pollutants: State of the Art Conference, Sponsored by Electric Power Research Institute, 4-6 November 1991, Washington, D.C.

Gratt, L.B. and Levin, L. 1993. *Health risk assessment of utility power stations under different assumptions: Emissions, dispersion and dose-response models.* Paper presented at Managing Hazardous Air Pollutants, 2nd International Conference, 13-15 July 1993, Washington D.C.

Gratt, L.B., Levin, L. and Silvers, A. 1993. *The role of arsenic emissions in power plant risk assessments: Uncertainties and research needs.* Paper 93-TA-36A.01 presented at the Air & Waste Management Association Annual Meeting, Denver, Colorado.

Hayes, S., Wilhelmi, J., Pye, S., Gates, L. and Levin, L. 1993. *Development of a modelling methodology for assessing exposure to acidic aerosols and gases.* Paper 93-RP-116B.03 presented at the Air & Waste Management Association Annual Meeting, Denver, Colorado.

Kowalcyzk, G.S., Gratt, L.B., Perry, B. and Yazdzik, E. 1989. Air emission risk assessment sensitivity analysis for a coal-fired power plant. In: Bonin, J. and Stevenson, D. (eds.), *Risk Assessment in Setting National Priorities, Advances in Risk Analysis,* Vol. 7,pp. 165-174. Plenum Press, New York.

Levin, L., Connor, K., Room, D. and Gratt, L. 1991. *Summary of case studies using EPRI's air toxics risk analysis framework: Results and research needs.* Paper presented at the Managing Hazardous Air Pollutants: State of the Art Conference, Sponsored by Electric Power Research Institute, 4-6 November 1991, Washington, D.C.

Piper, B.F., Hersh, S. and Mormile, D.J. 1981. *Particulate Emission Characteristics of Oil-Fired Utility Boilers,.* CS-1995, Research Project 1131-1. Electric Power Research Institute, Palo Alto, California .

Room, D., Kim, B. and Cohan, D. 1993. *Core project progress update: Methodology for estimating exposure to trace emissions.* Paper presented at Managing Hazardous Air Pollutants, 2nd International Conference, 13-15 July 1993, Washington D.C.

Sarkus, T.A. and Hargis, R.A. 1993. *Summary of air toxics monitoring activities being sponsored by DoE's Clean Coal Technology Demonstration Program.* Paper presented at Managing Hazardous Air Pollutants, 2nd International Conference, 13-15 July 1993, Washington D.C.

US EPA. 1987. *Guideline on Air Quality Models (Revised).* EPA Report 450/2-78-027R. Office of Air Quality Planning and Standards, Research Triangle Park, North Carolina.

US EPA. 1992. *Industrial Source Complex (ISC2) Dispersion Model* Vol. I *User Instructions.* EPA Report 450/4-92-008a. Office of Air Quality Planning and Standards, Research Triangle Park, North Carolina.

Utility Data Institute (UDI) database for 1989, *Environmental Directory of Power Plants.* Edison Electric Institute, 1991.

Wyzga, R.E. and Yager, J.E. 1991. *Carcinogenic risks of arsenic in flyash.* Paper presented at the Managing Hazardous Air Pollutants: State of the Art Conference, Sponsored by Electric Power Research Institute, 4-6 November 1991, Washington, D.C.

8 Ischemic heart disease induced by ingested inorganic arsenic

Chien-Jen Chen, Li-Ju Lin, Yu-Mei Hsueh, Hung-Yi Chiou, Kun-Fu Liaw, Shan-Fa Horng, Ming-Hsi Chiang, Chin-Hsiao Tseng and Tung-Yuan Tai

Institute of Public Health, National Taiwan University College of Public Health; Department of Public Health, Taipei Medical College; Department of Internal Medicine, National Taiwan University College of Medicine, Taipei, Taiwan

Abstract

In order to evaluate the association between ingested inorganic arsenic and ischemic heart disease, a series of epidemiologic studies have been carried out in the endemic area of blackfoot disease in Taiwan. Chronic arsenic exposure is significantly associated with the development of blackfoot disease. In an ecological correlational study based on 898,806 person-years and 172 ischemic heart disease deaths observed from 1973 to 1986, a dose-response relationship between arsenic concentration in drinking water and age-adjusted mortality from ischemic heart disease was observed. A total of 257 blackfoot disease patients and 753 matched healthy controls have been recruited and followed-up for more than seven years. A significantly increased ischemic heart disease mortality was observed for blackfoot disease patients and matched controls showing standardised mortality ratios (95% confidence intervals) of 937 (536–1,519) and 248 (139–409), respectively, compared with the general population in Taiwan (standardised mortality ratio = 100). Cox's proportional hazards regression analysis also showed a dose-response relationship between ischemic heart disease mortality and cumulative arsenic exposure after adjusted for age, sex, body mass index, and disease status of hypertension and diabetes mellitus. A case-control study including 78 patients of electrocardiogram-based ischemic heart disease and 384 normal residents was carried out in three villages where blackfoot disease was hyperendemic. Based on a multiple logistic regression analysis, cumulative arsenic exposure was found to be associated with ischemic heart disease in a dose-response relationship after adjusted for age, sex, body mass index, disease status of hypertension and diabetes mellitus, ratio between total cholesterol and high density lipoprotein-cholesterol, and cumulative alcohol consumption.

Introduction

Arsenic is widely distributed in nature and mainly transported in the environment by water. The general population is exposed to inorganic and organic arsenic through air, drinking water, food and beverages. Arsenic has been used for treatment of leukemia, psoriasis, chronic bronchial asthma, and as a tonic. Workers involved in the processing of copper, gold and lead ores, in the use and production of arsenic-containing pesticide, in the manufacturing and blowing glass, as well as in the chimney sweeping may have high exposure to air-borne arsenic (WHO, 1981; US Public Health Service, 1989).

Chronic exposure to inorganic arsenic has toxic effects on the cardiovascular system. An increased risk of both coronary heart disease and peripheral vascular disease has been related to the arsenic exposure. Environmental exposures to inorganic arsenic through drinking water have been found to be associated with an increased risk of peripheral vascular disease in Taiwan (Tseng, 1977), Chile (Borgono *et al.*, 1977), and Mexico (Cebrian, 1987). Peripheral vascular lesions have been described in Moselle vintners chronically exposed to arsenic through contaminated wine (Grobe, 1976). Altered blood vessel function and Raynaud's phenomenon has also been observed in copper smelter workers (Lagerkvist *et al.*, 1986).

Both environmental and occupational exposures to inorganic arsenic have been reported to be associated with an increased mortality from cardiovascular diseases. A significant excess mortality from cardiovascular disease has been found among copper smelter workers (Axelson *et al.*, 1978; Welch *et al.*, 1982), chimney sweeps (Hansen, 1983; Gustavsson *et al.*, 1987), and glass blowers (Wingren and Axelson, 1987) who were exposed to arsenic from working places. Environmental exposure to inorganic arsenic through drinking water has also been reported to increase the mortality from cardiovascular disease in Taiwan (Wu *et al.*, 1989), Chile (Zaldivar, 1980), and Japan (Tsuda *et al.*, 1990).

Blackfoot disease (BFD) is a unique peripheral vascular disease identified in the endemic area of chronic arseniasis in Taiwan (Wu *et al.*, 1961). Clinically the disease starts with numbness or coldness of one or more extremities and intermittent claudication. It progresses to ulceration due to minor traumas and ends with gangrene and spontaneous amputations (Tseng *et al.*, 1961). The underlying pathological change of BFD patients is the severe systematic atherosclerosis. BFD patients were reported to have an increased mortality from various cancers and cardiovacular disease (Chen *et al.*, 1988).

Most epidemiologic studies on arsenic-induced cardiovascular disease were occupational cohort studies and ecological correlation studies. The former may be subject to biases resulting from healthy worker effect and multiple exposure to various chemicals. The latter may have the problem of ecological fallacy. This report summarises the findings of a series of epidemiological studies aiming to evaluate the association between ingested inorganic arsenic and development of ischemic heart disease (ISHD) among residents in the endemic area of BFD in Taiwan.

Materials and methods

Ecological correlation study

Detailed description of study areas and materials was given in our previous study (Chen *et al.*, 1992). The study area was limited to 42 southwestern coastal villages in six townships including Peimen, Hsuechia, Putai, Ichu, Yensui and Hsiaying where the BFD was endemic. The mortality from ISHD was based on 172 ISHD (Codes 410–414 of the Eighth Revision of International Statistical Classification of Diseases, Injuries, and Causes of Deaths) deaths and 898,806 person-years from 1973 to 1986. The study population was stratified into four groups according to the median arsenic level of well water in each villlage, *i.e.*, <100 $\mu g\ L^{-1}$, 100–299 $\mu g\ L^{-1}$, 300–599 $\mu g\ L^{-1}$, and 600 or more $\mu g\ L^{-1.}$ In total, there were 13, 8, 15 and 6 villages, respectively, in the four groups with an arsenic level from the lowest to the highest. Age-adjusted ISHD mortality rates were calculated using the 1976 world population as the standard population.

Cohort follow-up study

A total of 257 BFD patients and 753 healthy controls matched on age, sex, and residential townships were recruited in 1985 from four BFD-endemic townships including Peimen, Hsuechia, Putai and Ichu. They were home-visit interviewed at the initial recruitment according to a structured questionnaire to obtain the information on sociodemographic characteristics, history of residential areas and water consumption, lifestyle habits including cigarette smoking and alcohol drinking, history of dietary consumption frequency, and personal and familial history of cancers and cardiovascular diseases. The follow-up of study subjects was carried out through the data linkage with household registration records and death certification datafile. Any events of birth, death, marriage and divorce, education, employment, and migration are mandatorily registered in the household registration offices in Taiwan. These registered records are double-checked through home-visit interviews by registration officials. The death certificates submitted to registration offices were sent to the Taiwan Provincial Department of Health for coding and compiling as a computer datafile. In other words, any resident in Taiwan can easily be traced through household registration and death certification systems. The vital status and causes of death of study subjects by June 30, 1992 were thus confirmed.

Age-sex-standardised mortality ratios (SMR) with their 95% confidence intervals of ISHD were calculated for BFD patients and matched controls, respectively. SMR was derived by dividing the observed number of ISHD deaths by the expected number of ISHD deaths which was obtained from the age-specific ISHD mortality of the general population in Taiwan and the number of cases or controls of the corresponding age groups. In other words, the general population in Taiwan was used as the standard population (SMR=100) for the indirect adjustment. In order to assess the association between the chronic arsenic exposure and the development of ISHD while the effects of other risk factors were controlled, Cox's proportional hazards regression analyses was used to estimate multivariate-adjusted relative risk and its 95% confidence interval.

Table 1 *Age-adjusted mortality from ischemic heart disease among residents in 42 villages of the BFD-endemic area in Taiwan, by median arsenic level in well water and sex, 1973-1986.*

Arsenic concentration in well water (μg L^{-1})	Age-standardized mortality (per 100,000 person-years)	
	Men	Women
<100	32.1	24.4
100-299	50.1	26.2
300-599	64.0	40.1
600+	110.7	55.1

Case-control study

In a community-based health survey carried out in Febuary 1993, there were 590 residents in three villages where BFD was hyperendemic participated. The condition of ISHD was diagnosed by the electrocardiogram (ECG) examinations and cardiovascular questionnaire interview. A total of 78 cases of ISHD as well as 384 controls who were ECG-normal and had no history of angina pectoris and myocardial infarction were identified. Blood pressures and antropometric characteristics of cases and controls were measured according to a standard protocol. The status of hypertension was defined as a systolic blood pressure > 160 mmHg, a diastolic blood pressure > 95 mmHg, and/or a history of currently treated hypertension. Oral glucose tolerance test was carried out to examine the serum level of glucose. The status of diabetes mellitus was defined as a fasting glucose level > 140 mg dL^{-1}, a 2-hours post-loading glucose level > 200 mg dL^{-1} or a history of diabetes currently treated with insulin or sulfonylurea. Serum lipid profiles including total cholesterol, high density lipoprotein (HDL)-cholesterol, low density lipoprotein-cholesterol, triglycerides, apo-A1, and apo-B were also tested. Standardized interviews based on a structured questionnaire were carried out to obtain the informtion on history of residential area and water consumption, life style characteristics, dietary habits, and personal and familial history of major cardiovascular diseases and cancers.

In the univariate analysis of the association between chronic arsenic exposure and ISHD, the age-sex-adjusted odds ratios and their 95% confidence intervals were calculated based on the regression coefficients and their standard errors obtained from the multiple logistic regression analysis. In order to estimate the multivariate-adjusted relative risk of developing ISHD for various chronic arsenic exposure indices, other risk factors associated with ISHD including age, sex, body mass index, disease status of hypertension and diabetes mellitus, ratio between total cholesterol/HDL-cholesterol level, and cumulative alcohol consumption were all included in the multiple logistic regression model.

Results

Table 1 shows the age-adjusted ISHD mortality among residents in 42 villages of the BFD-endemic area in Taiwan. There was a significant association between arsenic concentration in well water and ISHD mortality in a dose-response relationship. The

Table 2 *Standardised mortality ratio of ischemic heart disease for blackfoot disease patients and their age-sex-residential township-matched controls residing in the endenmic area of blackfoot disease.*

Study subjects	Standardised mortality ratio (95% confidence intertval)
Blackfoot disease patients	937 (536-1519)
Matched controls	248 (139- 409)

Table 3 *Multivariate-adjusted relative risk of developing ischemic heart disease for chronic arsenic exposure indices among blackfoot disease patients and matched controls residing in the endenmic area of blackfoot disease in Taiwan.*

Chronic arsenic exposure	Multivariate-adjusted relative risk (95% confidence interval)
Duration of residing in the blackfoot disease-endemic area (years)	
<25	1.0 (referent)
25-54	3.1 (0.9-11.1)+
55+	4.8 (1.6-14.2)**
Duration of consuming artesian well water (years)	
0	1.0 (referent)
1-19	2.6 (0.4-15.7)
20-29	5.3 (1.3-20.8)*
30+	3.6 (1.1-12.4)*
Cumulative arsenic exposure (ppm-years)	
0	1.0 (referent)
0.1-14.9	2.9 (0.8-11.2)
15.0+	4.9 (1.5-16.5)*

+ $0.05<p<0.10$, * $p<0.05$, ** $p<0.01$ while age, sex, body mass index, hypertension, and diabetes mellitus were also included in the Cox's proportional hazards regression model.

Table 4 *Multivariate-adjusted odds ratio of developing ischemic heart disease for various chronic arsenic exposure indices among residents in villages where the blackfoot disease was hyperendemic.*

Chronic arsenic exposure	Multivariate-adjusted odds ratio (95% confidence interval)
Duration of consuming artesian well water (years)	
0	1.0 (referent)
1-24	2.7 (0.8- 9.6)
25+	4.2 (1.1-15.7)*
Cumulative arsenic exposure (ppmyears)	
0	1.0 (referent)
0.1-14.9	1.6 (0.5- 5.3)
15.0+	3.6 (1.1-11.7)*

* $p<0.05$ while age, sex, body mass index, hypertension, diabetes mellitus, ratio of total cholesterol/HDLcholesterol level, and cumulative alcohol consumption were also included in the logistic regression model.

mortality age-adjusted to 1976 world population was 32.1 and 24.4 per 100,000, respectively, for men and women residing in villages where the arsenic concentration in well water was less than 100 $\mu g\ L^{-1}$. The age-adjusted mortality for men and women living in villages where the arsenic concentration in well water was 600 or more $\mu g\ L^{-1}$ were 110.7 and 55.1 per 100,000, respectively.

The age-sex-standardised mortality ratios (SMR) of ISHD for BFD patients and matched controls residing in the endemic area of blackfoot disease are shown in Table 2. After being followed-up for seven years, there were 16 BFD patients and 15 age-sex-residential township-matched controls died from ISHD. Compared with the general population in Taiwan as the referent (SMR=100), BFD patients had a SMR from ISHD of 937 (expected death number = 1.707) with a 95% confidence interval of 536–1,519 and matched controls had a SMR of 248 (expected death number = 6.051) with a 95% confidence interval of 139–409

The results of Cox's proportional hazards regression analysis illustrated a dose-response relationship between chronic arsenic exposure indices and ISHD mortality as shown in Table 3. All the three chronic arsenic exposure indices including duration of residing in the BFD-endemic area, duration of consuming artesian well water, and cumulative arsenic exposure in ppm-years were significantly associated with the ISHD mortality after adjustment for the effects of age, sex, body mass index, hypertension and diabetes. The multivariate-adjusted relative risks were 2.9 and 4.9, respectively, for those who had a cumulative arsenic exposure of 0.1–14.9 and 15.0 or more ppm-years as compared with those who had no exposure as the referent (relative risk = 1.0). After adjustment for the effects of chronic arsenic expsoure and other risk factors associated with ISHD, BFD patients still had a higher relative risk of developing ISHD than matched controls showing a multivariate-adjusted relative risk of 3.2 with a 95% confidence interval of 1.5–6.9.

In the case-control study carried out in the villages where BFD was hyperendemic, a dose-response relationship was also observed between ISHD prevalence and chronic arsenic exposure indices as shown in Table 4. After adjustment for the effects of age, sex, body mass index, hypertension, diabetes, ratio between total cholesterol and HDL-cholesterol, and cumulative alcohol consumption, the odds ratios of developing ISHD were 1.6 and 3.6, respectively, for those who had a cumulative arsenic exposure of 0.1–14.9 and 15.0 or more ppm-years as compared with those who had no exposure as the referent (odds ratio=1).

Discussion and conclusions

In this report a series of epidemiologic studies have shown a significantly increased ISHD mortality among residents in the BFD-endemic area, a significantly higher ISHD mortality for BFD patients than for age-sex-residence-matched controls, and a dose-response relationship between chronic arsenic exposure and development of ISHD.

Based on the strength, dose-response relation, consistency, and temporal correctness of the association between ISHD and cumulative arsenic exposure, it is reasonable to conclude the association is causal. The higher ISHD mortality observed in BFD patients than in matched controls after adjusted for the effect of chronic arsenic exposure suggests some residents in the endemic area may be more susceptible to the health hazards induced by ingested inorganic arsenic. It is worthwhile to evaluate the determinants of this susceptibility factor.

The mechanism for inorganic arsenic to induce atherosclerosis still remains to be elucidated. Inorganic arsenic may increase the ISHD risk through its effects on cardiovascular risk factors including hypertension and diabetes or on atherosclerosis directly. Further molecular and cellular epidemiologic studies are needed to explore possible mechanisms of the arsenic-induced atherogenicity.

Acknowledgements

These works were supported by grants from the National Science Council, Executive Yuan, Republic of China (NSC-80-0412-B002-17, NSC-81-0412-B002-122, NSC-82-0412-B002-262).

References

Axelson, O., Dahlgren, E., Janssen, C.D. and Rehnlund, S.O. 1978. Arsenic exposure and mortality: A case reference study from a Swedish copper smelter. *Br. J. Ind. Med.*, **35**, 8-15.

Borgono, J.M., Vincent, P., Venturino, H. and Infante, A. 1977. Arsenic in the drinking water of the City of Antofagasta: Epidemiological and clinical study before and after the intallation of the treatment plant. *Environ. Health Perspect.,* **19,** 103-105.

Cebrian, M.E. 1987. Some potential problems in assessing the effects of chronic arsenic poisoning in North Mexico. In: Preprints of papers presented at the American Chemical Society, Division of Environmental Chemistry, 194th Meeting, pp. 114-116.

Chen, C.J., Wu, M.M., Lee, S.S., Wang, J.D., Cheng, S.H. and Wu, H.Y. 1988. Atherogenicity and carcinogenicity of high-arsenic artesian well water: Multiple risk factors and related malignant neoplasms of blackfoot disease. *Arteriosclerosis,* **8,** 452-460.

Grobe, J.W. 1976. Peripheral circulatory disorders and acrocyanosis with arsenic (in German). *Berufs. Dermatosen.*, **24,** 78-84.

Gustavsson, P., Gustavsson, A. and Hogstedt, C. 1987. Excess mortality among Swedish chimney sweeps. *Br. J. Ind. Med.,* **44,** 738-743.

Hansen, E.S. 1983. Mortality from cancer and ischemic heart disease in Danish chimeny sweeps: A five-year follow-up. *Am. J. Epidemiol.,* **117,** 160-164.

Lagerkvist, B.E., Linderholm, H. and Nordberg, G.F. 1988. Arsenic and Raynaud's phenomenon: Vasospastic tendency and excretion of arsenic in smelter workers before and after the summer vacation. *Int. Arch. Occup. Environ. Health,* **60,** 361364.

Tseng, W.P., Chen, W.Y., Sung, J.L. and Chen, J.S. 1961. A clinical study of blackfoot disease in Taiwan: An endemic peripheral vascular disease. *Memmoir College Med. Natl. Taiwan Univ.*, **7,** 1-18.

Tseng, W.P. 1977. Effects and dose-response relationships of skin cancer and blackfoot disease with arsenic. *Environ. Health Perspect.,* **19,** 110-119.

Tsuda, T., Nagira, T. and Yamamoto, M. 1990. An epidemiological study on cancer in certified arsenic poisoning patients in Toroku. *Ind. Health,* **28,** 53-62.

US Public Health Service. 1989. *Toxicological profile for arsenic.* US Public Health Service, Washington DC.

Welch, K., Higgins, I., Oh, M. and Burchfiel, C. 1982. Arsenic exposure, smoking, and respiratory cancer in copper smelter workers. *Arch. Environ. Health,* **37,** 325-335.

Wingren, G. and Axelson, O. 1987. Mortality in the Swedish glassworks industry. *Scand. J. Work Environ. Health,* **13,** 412-416.

World Health Organization. 1981. *Environmental Health Criteria,* **18,** Arsenic. World Health Organization, Geneva.

Wu, H.Y., Chen, K.P., Tseng, W.P. and Hsu, C.L. 1961. Epidemiologic studies on blackfoot disease: I. Prevalence and incidence of the disease by age, sex, year, occupation and geographical distribution. *Memmoir College Med. Natl. Taiwan Univ.,* **7,** 33-50.

Wu, M.M., Kuo, T.L., Hwang, Y.H. and Chen, C.J. 1989. Dose-response relation between arsenic concentration in well water and mortality from cancer and vascular disease. *Am. J. Epidemiol.*, **130,** 1123-1131.

Zaldivar, R. 1980. A morbid condition involving cardiovascular, broncho-pulmonary, digestive and neural lesions in children and young adults after dietary arsenic exposure. *Zentralbl. Bakteriol.,* **170,** 4456.

9 Urinary excretion of arsenic species in a human population chronically exposed to arsenic via drinking water. A pilot study.

Luz María Del Razo J.[1], José Luis Hernández G.[1], Gonzalo G.García-Vargas[1], Patricia Ostrosky-Wegman [2], Cristina Cortinas de Nava[2] and Mariano E. Cebrián [1,3]*

[1]*Sección de Toxicología Ambiental, Departamento de Farmacología y Toxicología, CINVESTAV-IPN, AP 14-740, México, DF 07000, México*
[2]*Instituto de Investigaciones Biomédicas, UNAM, México, DF 04510, México*
[3]*Panamerican Center for Human Ecology and Health PAHO/WHO, Mexico*

Abstract
The objective of this pilot investigation was to study the pattern of arsenic (As) species in urine from individuals chronically exposed to high concentrations of As via drinking water in Region Lagunera, México, searching for evidence of an exposure-related saturation of the process of inorganic As methylation. The exposed group had 0.404 mg L^{-1} of total As in its drinking water, whereas the non-exposed group had 0.019 mgAs L^{-1}.
The main finding of this pilot study was a significant increase in the average proportion of monomethylarsonic acid (MMA) excreted in urine, accompanied by a significant decrease in the proportion of dimethylarsinic acid (DMA) excreted by exposed individuals. The combination of changes in opposite directions, could explain the similarities in the relative proportion of organic arsenicals (OAs) excreted by both groups. No data were found to suggest the presence of a saturation of the first methylation step. However, it appears that the second methylation step was impaired in the exposed individuals, since: (a) the significant increase in the proportion of MMA excreted in urine was accompanied by a significant decrease in the proportion of DMA, (b) the regression analysis of DMA(%) on MMA showed a significant inverse relationship. Further research is needed to fully characterize the findings here presented and their relationship to the presence of more overt signs of As toxicity.

*To whom correspondence should be addressed.

Introduction

The metabolism of inorganic arsenic (IAs) is considered a detoxication process. It has been studied in mammals and humans and involves reduction-oxidation reactions that interconvert As(V) and As(III), and methylation of As(III) to yield MMA and DMA (Crecelius, 1977; Tam *et al.*, 1979; Vahter, 1983). Studies on experimental animals and humans have suggested that after exposure to inorganic arsenic reaches certain level, the methylation process become less efficient as exposure increases (Vahter, 1981; Buchet *et al.*, 1981b). However, a recent review on the metabolism of As by human beings concluded that regardless of the exposure level to inorganic As (or more precisely, absorbed dose measured as IAs + MMA + DMA), there is always some IAs that remains unmethylated and that there is no evidence to suggest that this proportion raises with increasing dose (Hopenhayn-Rich *et al.*, 1993).

Most published studies on the metabolism of arsenic in humans, have been carried out in: (a) unexposed subjects (Foa *et al.*, 1984; Vahter and Lind, 1986); (b) volunteers who were administered single doses (Tam *et al.*, 1979; Buchet *et al.*, 1981a); (c) individuals who have ingested high quantities of arsenic, in suicide attempts or accidental exposure (Mahieu *et al.*, 1981; Lovell and Farmer, 1985); and (d) occupationally-exposed individuals (Vahter *et al.*, 1986; Yamauchi *et al.*, 1989; Farmer and Johnson, 1990). However, little information is available on the urinary pattern of As metabolites in humans exposed chronically to high concentrations of inorganic As *via* drinking water.

In some parts of the Region Lagunera, situated in the central part of northern México, chronic arsenic poisoning is endemic and various pathological effects attributed to As exposure have been reported in this area. Those include cutaneous manifestations (skin pigmentation changes, keratosis and skin cancer), peripheral vascular diseases (blackfoot disease), gastrointestinal disturbances (Cebrián *et al.*, 1983; García-Salcedo *et al.*, 1984; Cebrián, 1987), alterations in the coproporphyrin/uroporphyrin excretion ratio (García-Vargas *et al.*, 1991) and genotoxic effects (Ostrosky-Wegman *et al.*, 1991). The objective of this pilot investigation was to study the pattern of arsenic species in urine from individuals chronically exposed to high concentrations of As *via* drinking water, searching for evidence of an exposure-related saturation of the process of inorganic arsenic methylation.

Materials and methods

Group selection

Twenty-three individuals were selected from Santa Ana, Coahuila, a town which had 0.404 mg L^{-1} of total arsenic in its drinking water. Most As was in its inorganic form, 98% was As(V) (Del Razo *et al.*, 1990). The control group (22 individuals) was chosen from Lujan, Durango, whose As concentration was 0.019 mg L^{-1}, 93% was present as inorganic As(V). The individuals selected were between 20 and 55 years old and had lived in their respective towns for at least 15 years. No significant differences in age or sex distribution were observed between groups. Individuals who were exposed to pesticides and hormonal contraceptives in the 3 months before the study began, were excluded. Those exposed to alcohol in the last 5 days or suffering chronic alcoholism or generalized diseases were also excluded. First void urine samples were collected and stored in stoppered polyethylene bottles at -15°C during shipment and until analysis.

Chemicals

Stock solutions for arsenic compounds were prepared with sodium arsenate, sodium arsenite and dimethylarsinic acid purchased from Sigma Chemical Co. (St. Louis, MO). Monomethylarsonic acid was obtained from Vineland Chemical Co. (Vineland, NJ). All stock solutions contained 1 mg mL^{-1} of arsenic and were used to prepare working standards for analysis each day. Creatinine was purchased from Sigma Chemicals Co. Sodium borohydride was obtained from Alfa Products (Danvers, MA). The cation-exchange (AG 50WX-8, 100-200 mesh) resin was obtained from Bio-Rad (Richmond, CA). Freeze-dried urine standard reference material for toxic metals (SRM 2670) was obtained from the US National Institute of Standards and Technology (Gaithersburg, MD). All other chemicals were analytical grade. Distilled- deionized water was used for all analytical work and glassware was soaked in 10% nitric acid, rinsed with doubly distilled water and dried before use.

Arsenic analysis

An aliquot of urine (0.5-2.0 mL) was passed through an ion exchange column for the separation of IAs, MMA, and DMA according to the method of Tam *et al.* (1978). No attempt was made to separate trimethylarsenic compounds. The fractions collected were concentrated by evaporation and that containing DMA was wet digested according to Cox (1980). All fractions were pretreated with 1 mL of 20% (w/v) potassium iodide solution at room temperature 0.5 h before measurement. Fractions and urine samples were analysed with a Varian AA175 atomic absorption spectrophotometer equipped with a Model 65 vapor generation accessory. All measurements, were made using arsenic electrodeless discharge lamp (6 W), and background corrector. Since standard urine containing known amounts of As species is not commercially available, to assess the reliability of separation procedures, we analysed urine samples spiked with known amounts of IAs, MMA, and DMA (20:20:60), resulting in 3 different concentrations of total arsenic (TAs) 20, 456 and 1,043 μg L^{-1} (Vahter and Lind, 1986). Recoveries ranged from 90.7% to 113.3% with a precision between 0.5 and 12%. Freeze-dried urine standard reference material for toxic metals (SRM 2670) was analysed for TAs, an accuracy of 93-107% and 1-9% of precision were obtained. Creatinine concentrations, were measured according to Bousnes and Tawssky (1945).

Statistical procedures

Statistical comparisons between exposed and control groups were made using Mann-Whitney's 'U' test. Simple linear regression analysis and Pearson's correlation coefficient were used to study the relationships between arsenic species and total arsenic. The level of significance was set at $p < 0.05$.

Table 1 *Arsenic species excreted in urine from humans chronically exposed to arsenic. (geometric mean (range)).*

		CONTROL	EXPOSED
Total As	($\mu g\ g^{-1}$ creat.)	39.3 (12-104)	1,156[1] (456-1,981)
Inorganic As	($\mu g\ g^{-1}$ creat.)	5.6 (0.5-22)	187.9[1] (34-503)
	%	10.9 (1.4-29.5)	16.7 (9.3-25.5)
Organic As	($\mu g\ g^{-1}$ creat.)	35.1 (8.4-103.4)	834.8[1] (234.2-1,832)
	%	84.8 (70.4-98.6)	82 (74.5-96.8)
MMA	($\mu g\ g^{-1}$ creat.)	3.24 (0.7-16.6)	189.4[1] (44-645)
	%	7.12 (1.8-25)	18.42[1] (6.2-29.5)
DMA	($\mu g\ g^{-1}$ creat.)	30.56 (5.7-93.4)	680.21 (187-1,264)
	%	74.8 (46.8-94.6)	61.73[2] (47.9-81.8)
n		22	23

[1] $p = 0.00002$. Mann-Whitney's U test.
[2] $p = 0.002$. Mann-Whitney's U test.

Results

Age and sex distribution, length of residence in their respective communities and smoking habits were not significantly different between the groups compared. As expected, the urinary concentrations of arsenic species and consequently of total As, were significantly higher in the exposed group (Table 1). The sum of arsenic species accounted for at least 95% of total arsenic measured in urine. No significant differences in the proportion of inorganic and total organic arsenic compounds present in urine were observed (Table 1). However, a detailed study of the organic species showed that the exposed group had a significant increase in the proportion of MMA excreted in urine, accompanied by a significant decrease in the proportion of DMA excreted (Figure 1). Nonetheless, DMA was the major single arsenic species excreted by both groups. A linear regression analysis of the urinary concentrations of different As species showed significant direct relationships between MMA and IAs in both control and exposed groups, and between DMA and MMA in the exposed group (Table 2). The concentration of each species was also linearly related to TAs concentration in the exposed group.

As a first step in the search for indicators to assess the efficiency of As metabolism in humans, a linear regression analysis between the proportion of putative products and the concentration of putative substrates of the As metabolic pathway was performed. The regression of DMA(%) on MMA and that of DMA(%) on IAs showed a significant inverse relationship in the exposed group (Table 2), whereas no significant relationships were observed between MMA(%) and IAs (Figure 2) or between IAs(%) and TAs in both groups (Table 2). A similar logic was applied to study the ratios between urinary concentrations of putative products and substrates. In the exposed group, the ratio MMA/IAs was significantly higher, as compared to the control group. In contrast, the ratio DMA/MMA was significantly lower in the exposed individuals. However, no significant differences in the DMA/IAs or OAs/IAs ratios were observed (Table 3).

There were no significant differences in urinary creatinine concentrations between control (1.14 ± 0.80 mg L^{-1}) and exposed individuals (0.99 ± 0.57 mg L^{-1}).

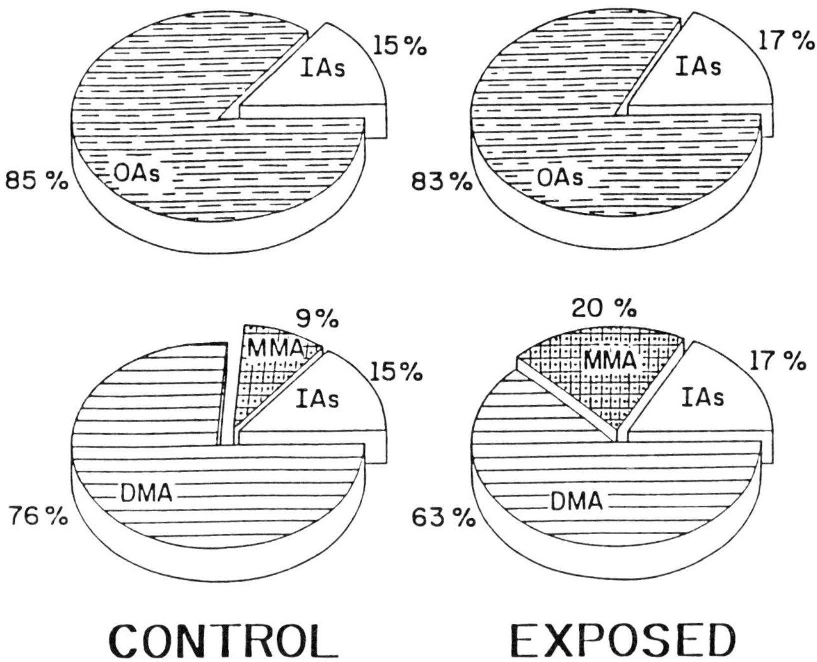

Figure 1 *Proportions of arsenic species in urine of individuals chronically exposed to arsenic via drinking water. Región Lagunera, México.*

Table 2 *Statistical analyses using simple linear regression.*

Dependent variable	Independent variable	Slope	Intercept	r	*p*
		Control (n = 22)			
MMA	IAs	0.342	2.1	0.467	0.028
DMA	MMA	0.768	34.6	0.140	0.535
IAs	TAs	0.004	7.13	0.017	0.939
MMA (%)	IAs	-0.001	9.4	-0.0009	0.991
DMA (%)	MMA	0.886	72.0	0.271	0.222
DMA (%)	IAs	0.545	72.11	0.227	0.308
IAs (%)	TAs	-0.049	16.85	0.153	0.498
		Exposed (n = 23)			
MMA	IAs	1.041	-24.1	0.741	0.00005
DMA	MMA	1.694	341.2	0.711	0.0001
IAs	TAs	0.166	19.49	0.856	0.000001
MMA (%)	IAs	0.019	15.3	0.357	0.103
DMA (%)	MMA	-0.025	67.9	-0.481	0.023
DMA (%)	IAs	-0.033	69.2	-0.453	0.034
IAs (%)	TAs	-0.001	19.45	-0.112	0.620

Discussion

The arsenic concentration in drinking water of the exposed group was about eight times higher than the maximum limit (0.05 mg L^{-1}) previously recommended by WHO (1984). Total As concentrations in urine samples from most control individuals were within the range considered normal for environmental exposure (2-100 μg L^{-1}), whereas the mean value in exposed individuals was about ten times the maximum value considered acceptable for non-occupational exposure (ATSDR, 1992).

The main finding of this pilot study was a significant increase in the average proportion of MMA excreted in urine, accompanied by a significant decrease in the proportion of DMA excreted by exposed individuals. The combination of changes in opposite directions, could explain the similarities in the relative proportion of OAs excreted by both groups. No data were found in the exposed or control individuals, to support the presence of a saturation of the first methylation step since: (a) about 15% of the total As excreted remained as IAs in both control and exposed populations, which is in agreement with the report by Hopenhayn-Rich *et al.* (1993); (b) the regression analysis of MMA(%) on IAs did not show a significant correlation; and (c) the increase in the MMA/IAs ratio was mainly caused by an increase in MMA concentrations, without significant changes in those of IAs, suggesting a back-log accumulation of MMA. Nonetheless, the possibility of an increased production of MMA could not be disregarded.

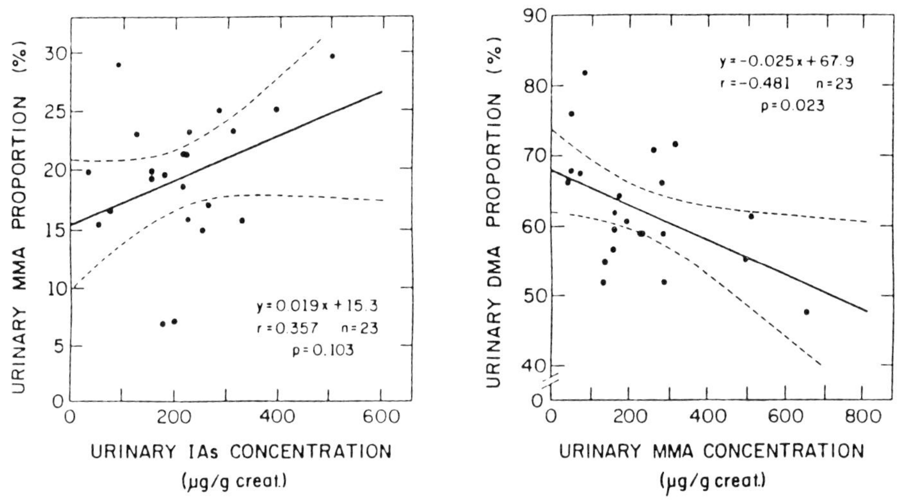

Figure 2 *Relationships between putative substrates and putative products of arsenic metabolism excreted in urine. Región Lagunera, México.*

Table 3 *Ratios of arsenic species in urine from individuals chronically exposed to arsenic (geometric mean (range)).*

	Control	Exposed
OAs/IAs	7.56 (2.3-71)	4.91 (2.9-30.5)
MMA/IAs	0.66 (0.1-6)	1.09[1] (0.3-6.9)
DMA/MMA	10 (1.8-45)	3.40[2] (1.6-12.9)
DMA/IAs	6.61 (1.6-65)	3.73[3] (2.1-23.6)
n	22	23

[1] $p = 0.010$. Mann-Whitney's U test.
[2] $p = 0.0001$. Mann-Whitney's U test.
[3] $p = 0.07$. Mann-Whitney's U test.

Our results suggest that the second methylation step was impaired in the exposed individuals, since: (a) the significant increase in the proportion of MMA excreted in urine was accompanied by a significant decrease in the proportion of DMA; (b) the regression analysis of DMA(%) on MMA showed a significant inverse relationship; and (c) The decrease in the DMA/MMA ratio was mainly caused by an increase in the relative excretion of MMA and, to a lesser extent, by reductions in that of DMA. Since concentrations in urine of exposed individuals were about 30 times higher than those found in the control group, a possible explanation for these changes is that high As(III) concentrations could inhibit the conversion of MMA to DMA in exposed individuals, as shown in rat liver slices by Georis *et al.* (1990), possibly by binding to the dimethyltransferase (Thompson, 1993), provided that hepatic IAs concentrations in the exposed individuals reach levels capable of producing this effect. In addition, a significant inverse relationship was found between DMA(%) and IAs in the exposed population (Table 2). This also suggest that the kinetic properties of the first methylation step, or its sensitivity to IAs, may be different from those of the second, supporting the idea that the methylation process is carried out by two different enzymes (Buchet and Lauwerys, 1985).

The pattern of As metabolites excreted in urine appears to depend on the characteristics of exposure, since exposure to single dosis resulted in increased proportions of IAs excreted, no significant changes in MMA and significant decreases in that of DMA (Yamauchi and Yamamura, 1979), whereas chronic exposure results in no significant changes in the proportion of IAs excreted, increases in MMA and decreases in that of DMA, as shown in this and other studies (Yamauchi *et al.*, 1989; Farmer and Johnson, 1990, as quoted by Hopenhayn-Rich *et al.*, 1993). The results obtained in chronically-exposed individuals are not directly comparable to those from studies on human volunteers (Buchet *et al.*, 1981b), where a saturation of the methylation process was suggested, since they were obtained 9 days after acute administration of 5 consecutive doses of As(III).

Until more is known on the toxicokinetics of arsenic in chronically-exposed populations, there is no evidence to suggest that measurement of total As concentration in urine is a good surrogate for doses in studies aimed to assess the efficiency of As metabolism. The ratios of urinary concentrations of putative products to substrates of the As metabolic pathway, appear to be a useful method to study alterations on the methylating ability of individuals chronically exposed to inorganic As, since TAs (which in itself is dependent from concentrations of IAs, MMA, DMA) is not used as an independent variable to assess changes in the proportion of As species present in urine. Physiologically based toxicokinetic models would help testing the usefulness of these ratios.

Further research on better parameters to assess the efficiency of As metabolism in chronically exposed populations is needed, as well as additional studies in a larger sample, to fully characterize the findings here presented and their relationship to the presence of more overt signs of As toxicity.

Acknowledgements

This study was supported by a research grant from Consejo Nacional de Ciencia y Tecnología, Mexico. (No. PCSACNA-051077). Additional support was provided by a Pan American Health Organization grant No. DRC/HPE/RG/MEX/88-97. The authors thank Dr. Jacobo Finkelman ECO/PAHO for his encouragement and support, and Dr. Hirochi Yamauchi for the gift of TMA and MMA standards.

References

ATSDR (Agency for Toxic Substances and Disease Registry). 1992 *Toxicological Profile for Arsenic*. US Public Health Service, Atlanta G.A.

Bousnes, R.W. and Tawssky, H. 1945. On the colorimetric determination of creatinine by the Jaffe reaction. *J. Biol. Chem.*, **158**, 581-584.

Buchet, J.P, Lauwerys, R. and Roels, H. 1981a. Comparison of the urinary excretion metabolites after a single oral dose of sodium arsenite, monomethylarsonate, or dimethylarsinate in man. *Int. Arch. Occup. Environ. Health,* **48**, 71-79.

Buchet, J.P, Lauwerys, R. and Roels, H. 1981b. Urinary excretion of inorganic arsenic and its metabolites after repeated ingestion of sodium metaarsenite by volunteers. *Int. Arch. Occup. Environ. Health,* **48**, 111-118.

Buchet, J.P. and Lauwerys, R. 1985. Study of inorganic arsenic methylation by rat liver *in vitro*: Relevance for the interpretation of observations in man. *Arch. Toxicol.*, **57**, 125-129

Cebrián, M.E. 1987. Some potential problems in assessing the effects of chronic arsenic exposure in North Mexico. In: Johnson J.D. (ed.), *Preprints of papers presented at the 194th Meeting of American Chemical Society. Division of Environmental Chemistry,* 1-4 September 1987. Vol. 27 pp.114116.

Cebrián, M.E, Albores, A., Aguilar, M. and Blakely, E. 1883. Chronic arsenic poisoning in the North of Mexico. *Human Toxicol.*, **2**, 121-133.

Cox, D.H. 1980. Arsine evolution electrothermal atomic absorption method for the determination of nanogram levels of total arsenic in urine and water. *J. Anal. Toxicol.*, **4**, 207-211.

Crecelius, E.A. 1977. Changes in the chemical speciation of arsenic following ingestion by man. *Environ. Health Perspect.*, **19**, 147-150.

Del Razo, L.M., Arellano, M.A. and Cebrián, M.E. 1990. The oxidation states of arsenic in well water from a chronic arsenicism area of northern Mexico. *Environ. Poll.*, **64**, 143-153.

Farmer, J.G. and Johnson, L.R. 1990. Assessment of occupational exposure to inorganic arsenic based on urinary concentrations and speciation of arsenic. *Br. J. Ind. Med.*, **47**, 342-348.

Foa, V., Colombi, A., Maroni, M., Buratti, M. and Calzaferri, G. 1984. The speciation of the chemical forms of arsenic in the biological monitoring of exposure to inorganic arsenic. *Sci. Total Environ.*, **34**, 241-259.

Georis, B., Cardenas, A., Buchet, J.P. and Lauwerys, R. 1990. Inorganic arsenic methylation by rat tissue slices. *Toxicol.* **63**, 73-84

García-Salcedo, J., Portales, A., Blakely, E. and Díaz, R. 1984. Estudio transversal de una cohorte de pacientes con vasculopatia por intoxicación crónica arsenical en poblados de los municipios de Francisco I. Madero y San Pedro Coahuila, México. *Revista de la Facultad de Medicina (Torreón)*, **1**, 12-16.

García-Vargas, G., García-Rangel, A., Aguilar-Romo, M., García-Salcedo, J., Del Razo, L.M. and Ostrosky-Wegman P., Cortinas de Nava C. and Cebrián M.E. 1991. A pilot study on the urinary excretion of porphyrin in human populations chronically exposed to arsenic in Mexico. *Hum. Exp. Toxicol.*, **10**, 189-193.

Hopenhayn-Rich, C., Smith, A.H. and Goeden, H.M. 1993. Human studies do not support the methylation threshold hypothesis for the toxicity of inorganic arsenic. *Environ. Res.*, **60**, 161-177.

Lovell, M.A and Farmer, J.G. 1985. Arsenic speciation in urine from humans intoxicated by inorganic arsenic compounds. *Human Toxicol.*, **4**, 203-214.

Mahieu, P., Buchet, J.P., Roels, A. and Lauwerys, R. 1981. The metabolism of arsenic in humans acutely intoxicated by As_2O_3. Its significance for the duration of BAL Therapy. *Clin. Toxicol.*, **18**, 1067-1075.

Ostrosky-Wegman, P., Gonsebatt, M.E., Montero, R., Vega, L. Barba, H., Espinosa, J., Palao, A., Cortinas, C., Garcia-Vargas, G., Del Razo, L.M. and Cebrián, M.E. 1991. Lymphocyte proliferation kinetics and genotoxic findings in a pilot study on individuals chronically exposed to arsenic in Mexico. *Mut. Res.*, **250**, 477-482.

Tam, G.K.H., Charbonneau, F.B., Pomroy, C. and Sandi, E. 1979. Metabolism of inorganic arsenic (^{74}As) in humans following oral ingestion. *Toxicol. Appl. Pharmacol.*, **50**, 319-322.

Tam, G.K.H., Charbonneau, S.M., Bryce, F. and Lacroix, G. 1978. Separation of arsenic metabolites in dog plasma and urine following intravenous injection of ^{74}As. *Anal. Bioch.*, **86**, 505-511.

Thompson, D.J. 1993. A chemical hypothesis for arsenic methylation in mammals. *Chem-Biol Interact.*, **88**, 89-114.

Vahter, M. 1981. Biotransformation of trivalent and pentavalent inorganic arsenic in mice and rats. *Environ. Res.*, **25**, 286-293

Vahter, M. 1983. Metabolism of arsenic. In: B.A. Fowler (ed.), *Biological and Environmental effects of arsenic*, pp.171-198. Elsevier, Amsterdam.

Vahter, M., Friberg, L., Rahnster, B., Nygren, A. and Nolinder, P. 1986. Airborne arsenic and urinary excretion of metabolites of inorganic arsenic among smelter workers. *Int. Arch. Occup. Environ. Health*, **57**, 79-91.

Vahter, M. and Lind, B. 1986. Concentrations of arsenic in urine of the general population in Sweden. *Sci. Tot. Env.*, **54**, 1-12.

WHO (World Health Organization). 1984. *Guidelines for Drinking-Water Quality*, Vols.1 and 2. Geneva, Switzerland.

Yamauchi, H. and Yamamura, Y. 1979. Dynamic change of inorganic arsenic and methylarsenic compounds in human urine after oral intake as arsenic trioxide. *Ind. Health Kaw.*, **17**, 79-83.

Yamauchi, H. Takahashi, K., Mashiko, M. and Yamamura, Y. 1989. Biological monitoring of arsenic-exposed, gallium arsenide and inorganic arsenic-exposed workers by determination of inorganic arsenic and its metabolites in urine and hair. *Am. Ind. Hyg. Assoc. J.*, **50**, 606-612.

10 Long term follow-up of patients treated with medicinal arsenic

Peter Sasieni[1], Susan Evans[2] and Jack Cuzick[1]

[1]*Department of Mathematics, Statistics and Epidemiology,Imperial Cancer Research Fund, P.O. Box 123, Lincoln's Inn Fields, London WC2A 3PX, UK*
[2]*Royal Liverpool Hospital, Liverpool, UK*

Abstract

A cohort of 478 patients treated with Fowler's solution in England during the period 1945-69 has been passively followed until the end of 1990. The cumulative dose of arsenic received by the majority of the cohort was similar in magnitude to the exposure that one would receive from drinking water containing high levels (25-50 g L^{-1}) inorganic arsenic for forty years. A significant excess of bladder cancer mortality occured (observed/expected ratio = 5/1.6; $p = 0.05$). No excess was found for other causes of death. In a subcohort of 142 patients examined for signs of arsenicism around 1970, all 11 subsequent cancer deaths occured in those with signs of arsenicism ($p = 0.0009$). At the time of examination 15 patients (10%) had skin cancer.

Introduction

A strong association between ingested inorganic arsenic and subsequent skin cancer has been well documented (Bergoglio, 1964; Yeh, 1973; Hutchinson, 1888; Neubauer, 1947; Evans, 1976). Evidence of a connection with internal malignancies is less convincing, but has been growing. Epidemiological studies and case reports from Taiwan, (Chen *et al.*, 1986; Wu *et al.*, 1989; Chiang *et al.*, 1988; Chiang *et al.*, 1993) Argentina (Astolfi *et al.*, 1981) and elsewhere (Tsuda *et al.*, 1990) suggest a possible link between bladder cancer and exposure to arsenic-contaminated drinking water.

Methodology

A mortality analysis has been carried out on a cohort of 478 patients given Fowler's solution (potassium arsenite) for between 2 weeks and 12 years between 1945 and 1969 in rural Lancashire (North West England). The patients, most of whom were prescribed Fowler's solution for a variety of skin conditions, were identified from pharmacy notes. Follow-up to January 1991 has been achieved by flagging patients in the National registry. Patients were censored at their 85th birthday because of the unreliability of death certificates beyond that age. Expected values are based on age-, sex- and calendar-year-adjusted rates for England and Wales. Dose levels of arsenic were computed by multiplying prescribed daily dose by the duration of treatment. They are shown in Figure 1.

Thirty-one percent of patients received less than 250 mg of arsenic, 29% received at least 1,000 mg of arsenic and only 3% had 10 grammes or more. Further details appear in previous reports (Cuzick *et al.*, 1982; Cuzick *et al.*, 1992).

In 1969, Dr Evans, a dermatologist, examined a subcohort of 142 patients for the presence of arsenical keratoses, hyperpigmentation and skin cancer.

Results

A total of 188 deaths have been recorded before the age of 85; 52 patients were alive at age 85 (of whom 30 have since died), 19 have emigrated and 31 have been lost to follow-up. Standardised mortality ratios (SMR's) for various causes of death are given in Table 1. The SMR's for all causes and for all neoplasms are slightly below one. The only significant excess (among causes considered) was for bladder cancer (5 cases observed, 1.6 expected, $p = 0.05$). The rate for all causes and for all neoplasms was particularly low in the first 5 years after exposure and if this is omitted the SMR for all causes is 0.92, all cancers 1.05 and bladder cancer 3.5.

Analysis by dose is not particularly illuminating (Table 2). A non-significant

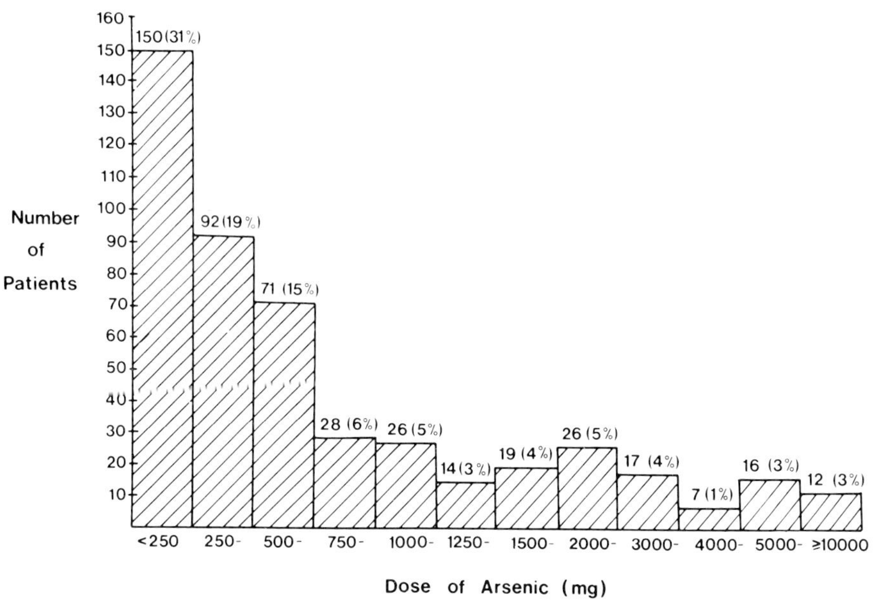

Figure 1 *Distribution of 478 patients by cumulative dose of arsenic.*

Table 1 *Cause-specific mortality in a cohort of 478 patients treated with Fowler's solution.*

Cause of death	Observed no.	Expected no.	SMR	95%CI
All causes	188	209.1	0.90	0.77-1.03
All neoplasms (140-239)	47	49.3	0.95	0.70-1.3
Digestive organs (150-157)	19	15.9	1.19	0.7-1.9
Stomach (151)	5	5.0	0.99	0.3-2.3
Respiratory system (160-163, 140-141, 143-149)	14	14.0	1.00	0.5-1.7
Skin (172-173)	1	0.4	2.44	0.08-14
Bladder (188)	5	1.6	3.07*	1.01-7.3
Liver and gallblader (155-156)	1	0.8	1.23	0.04-7
Hematopoietic tissue (200-208)	1	2.6	0.38	0.01-2
All circulatory disease (390-443, 445-458)	97	106.5	0.91	0.74-1.1
Ischemic heart disease (410-414)	49	57.4	0.85	0.6-1.1
Cerebrovascular disease (430-438)	20	27.6	0.72	0.4-1.1

*Two-sided $p = 0.05$
SMR, standardized mortality ratio; CI, confidence interval.
Numbers in parentheses, *International Classification of Diseases*, Eighth Revision, code

deficit is found in the lowest dose category and a weak trend with increasing dose is found ($p = 0.03$). However, the low rates in the lowest exposure category contribute substantially to this effect and when the linear trend is fitted without intercept it is insignificant ($p = 0.4$). The number of bladder cancer deaths was small, but the SMR in those with a cummulative dose of 500 mg or more was 5.0 (4 observed, 0.8 expected, $p = .02$).

A further bladder cancer death occurred in a man aged 85 and one patient had bladder cancer listed as a secondary cause of death. These cancers have not been included in the above analysis. Details of all seven cases of bladder cancer are given in Table 3.

Of the 142 individuals examined for signs of arsenicism (keratoses, hyperpigmentation or skin cancer), 49% had at least one sign; 45% had keratoses. There was a strong relationship between dose and signs as shown in Figure 2 (Wilcoxon, $p = .001$). A third of patients with under 250 mg of arsenic had signs, compared to 75% of those exposed to over 2,000 mg. Table 4 relates the mortality

Table 2 *Mortality by cumulative dose of arsenic in a cohort of 478 patients treated with Fowler's solution.*

Cause of death	Cumulative dose (mg)	Observed no. of deaths	SMR	95%CI
All causes	<500	84	0.79	0.6-0.98
	500-999	39	0.98	0.7-1.3
	1,000-1,999	30	1.15	0.8-1.6
	≥2,000	35	0.95	0.7-1.3
All neoplasms	<500	16	0.65	0.4-1.05
	500-999	10	1.04	0.5-1.9
	1,000-1,999	9	1.50	0.7-2.8
	≥2,000	12	1.34	0.7-2.3
Bladder cancer	<500	1	1.20	0.04-7
	500-999	2		
	1,000-1,999	1	5.00	2.0-15
	≥2,000	1		

*Results of trend tests: with intercept - all causes: $\chi^2 = 1.8$, $p = 0.2$; all neoplasms: $\chi^2 = 4.7$, $p = 0.03$; without intercept - all causes: $\chi^2 = 0.4$, $p = 0.5$; all neoplasms $\chi^2 = 0.7$, $p = 0.4$.
SMR, standardized mortality ratio; CI, confidence interval.

of the subcohort to the presence of signs of arsenicism. All 11 cancers in the subcohort have appeared in patients with signs of arsenicism whereas 7.1 were expected ($p = 0.2$). No cancers were found in patients without any signs although 6.3 were expected ($p = 0.004$). The probability that all cancers would appear in the group with signs by chance is extremely small ($p = 0.0009$ 1-sided).

In a separate study (Cuzick *et al.*, 1984) cases with bladder and lung cancer and hospital controls were examined for palmar keratoses. For males the average prevalence in controls aged 50 or more was 41% whereas for lung and bladder cancer patients the prevalence was 80% and 94% respectively. Keratoses were generally less common in women, but the difference between cases and controls was still marked. Overall the age- and sex-adjusted odds ratios for the two cancers were 4.2 and 11.7 respectively.

Discussion

Of the 5 bladder cancer deaths included in the analyses, 3 were within 10 years of exposure, whereas 2 were more than 20 years after using Fowler's solution (Table 3).

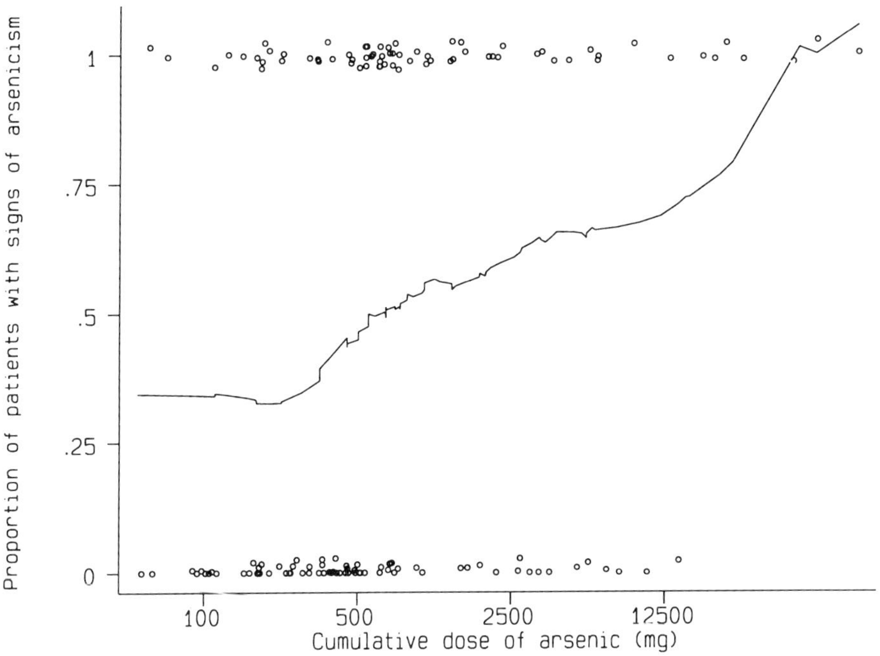

Figure 2 *Relationship between signs of arsenicism on examination, and cumulative dose of arsenic ingested in a subcohort of 142 patients.*

The additional bladder cancer mortality in a man aged 85 years, occurred 34 years after exposure. Our numbers are too small to say with certainty that no other cancers are associated with low exposure to arsenic, but the confidence intervals can be used to indicate the maximal odds ratios consistent with our data. The striking association between signs of arsenicism and subsequent cancer deaths may be due to "signs" being a marker for a susceptibility gene for bladder and possibly other cancers. We would suggest that individuals known to have been exposed to high levels of arsenic should be considered for bladder cancer screening particularly if they exhibit palmar keratoses.

Figures from the EPA (Science Applications, 1987) suggest that while the vast majority of drinking water in the US contains arsenic in concentrations of less than 5 g L^{-1}, 2.5 million Americans drink water with arsenic concentration of over 25 g L^{-1}; 350,000 of whom are exposed to arsenic at 50 g L^{-1}. Extrapolation based on 500 L yr^{-1} for 40 years gives cumulative doses of arsenic between 500 mg and 1,000 mg similar to the levels received in our cohort. These should be contrasted with the levels of arsenic in some Taiwanese and Argentinian well water (350 - 1,140 g L^{-1}). Extrapolating the excess bladder cancer deaths observed in our cohort to the US population suggests that of the order of 500 additional bladder cancer deaths each

Table 3 *Details of patients with bladder cancer mentioned on death certificate.*

Age at first exposure	Sex	Total dose of arsenic (mg)	Duration of dose (months)	Time from first exposure to death (years)	Age at death
69	F	504	2.5	7.0	76
45	F	1,901	17	34.0	79
48	M	224	2	25.8	73
27	F	963	17	7.7	34
67	M	3,324	28	6.2	73
50	M	11,940	32	34.9	85
65	M	1,800	2	14.7	79*

*The primary cause of death of this patient was occlusion of cerebral arteries, the bladder cancer was given as the secondary cause.

Table 4 *Mortality in a subcohort of patients examined for signs of arsenicism (n = 142).*

Cause of death	Signs	Observed no	Expected no.	SMR	95% CI*
All causes	Present	31	28.4	1.09	0.74-1.55
	Absent	12	26.6	0.45	0.23-0.79
	Total	43	55.0	0.78	0.57-1.05
All neoplasms	Present	11	7.1	1.55	0.8-2.8
	Absent	0	6.3	0	0-0.6
	Total	11	13.5	0.82	0.4-1.5
All circulatory disease	Present	15	14.4	1.04	0.6-1.7
	Absent	11	13.6	0.81	0.4-1.4
	Total	26	28.0	0.93	0.6-1.4

*SMR, standardized mortality ratio; CI, confidence interval.

year are caused by arsenic in drinking water. The small numbers in our study and the crude extrapolation however must result in wide confidence intervals and the actual number may be between 0 and 2,000 and still be consistent with our data.

References

Astolfi, E., Maccagno, A., Fernandez, J.C.G., *et al.*1981. Relation between arsenic in drinking water and skin cancer. *Biol. Trace. Elem. Res.*, **3**, 133-143.

Bergoglio, R.M. 1964. Cancer mortality in zones of arsenical waters of the Province of Cardoba, Argentine Republic. *Prensa Med. Argent.*, **51**, 994.

Chen, C.J., Chuang, Y.C., You, S.L. *et al.* 1986. A retrospective study on malignant neoplasms of bladder, lung and liver in the blackfoot disease endemic area in Taiwan. *Br. J. Cancer*, **53,** 399-405.

Chiang, H.S., Guo, H.R., Hong, C.L. *et al.* 1993. The incidence of bladder cancer in the black foot disease endemic area in Taiwan. *Br. J. Urol.* **71,** 274-278.

Chiang, H.S., Hong, C.L., Guo, H.R., *et al.* 1988. Comparative study on the high prevalence of bladder cancer in the blackfoot disease endemic area in Taiwan. *Taiwan I. Hsueh. Hui. Tsa. Chih.,* **87,** 1074-1080.

Cuzick, J., Evans, S., Gillman, M. *et al.* 1982. Medicinal arsenic and internal malignances. *Br. J. Cancer*, **45**, 904-911.

Cuzick, J., Harris, R. and Mortimer P.S 1984. Palmar keratoses and cancers of the bladder and lung. *Lancet*, **1**, 530-533.

Cuzick, J., Sasieni, P. and Evans, S. (1992). Ingested arsenic, keratoses, and bladder cancer. *Am. J. Epidemiol.*, **136,** 417-421.

Evans, S. 1976. Arsenic and Cancer. M.D.Thesis, University of Liverpool.

Hutchinson, J. 1888. On some examples of arsenic-keratoses of the skin and of arsenic-cancer. *Trans. Pathol. Soc. Lond.*, **39** 352.

Neubauer, O. 1947. Arsenical cancer: A review. *Br. J. Cancer*, **1,** 192-251

Science Applications International Corporation, 1987. Estimated National Occurrence and Exposure to Arsenic in Public Drinking Water Supplies. Revised draft prepared for US Environmental Protection Agency under Contract No.68-01-766 EPA, Washington.

Tsuda, T., Nagira, T., Yamamoto, M. *et al.* 1990. An epidemiological study on cancer in certified arsenic poisoning patients in Toroku. *Ind Health*, **28,** 53-62.

Wu, M.M., Kuo, T.L, Hwang, Y.H. *et al.* 1989. Dose-response relation between arsenic concentration in well water and mortality from cancers and vascular diseases. *Am. J. Epidemiol.*, **130,** 1,123-32

Yeh, S. 1973. Skin cancer in chronic arsenicism. *Hum. Pathol.*, **4,** 469-485.

11 Epidemiological study designs to address potential high bladder cancer risks from arsenic in drinking water

Allan H. Smith*, Claudia Hopenhayn-Rich, Mary Lou Biggs, Lee Moore, Jill Dale, Marcella Warner, Michael Bates and Robert Engel

Biomedical and Environmental Health Sciences, School of Public Health, University of California, Berkeley, CA 94720, USA

Abstract

Bladder cancer risks are markedly elevated in areas of Taiwan with high arsenic levels in drinking water. Limited data from studies elsewhere provide some support for a causal association, but confirmatory studies are needed. In this paper we outline various study designs which could be used to provide further information, including ecological studies, cohort studies, case-control studies, and biomarker studies. Advantages and disadvantages of each study design are given and illustrated with features of actual studies, including ones in progress or being planned by our group in California, Nevada, and northern Chile. Traditional epidemiological studies suffer from problems due to the rarity of bladder cancer, the long latency between exposure and disease manifestation, and relatively small numbers of people exposed at high levels. The main advantages of biomarker studies using exfoliated bladder cells are that small studies can be conducted rapidly, accurate exposure data can be obtained, and the effect of intervention by providing water low in arsenic can be assessed. However, the interpretation of biomarker studies is open to question. It is concluded that the potential risks of bladder cancer from ingesting arsenic in drinking water warrant a concerted epidemiological approach using a variety of different study designs.

Introduction

Arsenic ingestion and bladder cancer

Inorganic arsenic is known to be a cause of lung cancer by inhalation and a cause of skin cancer via ingestion. Epidemiological evidence indicates that ingestion of arsenic may also cause lung, bladder, kidney, and liver cancer (Bates *et al.*, 1992). Of these sites, the highest relative risks are for bladder cancer.

---

*To whom correspondence should be addressed

The primary sources of evidence suggesting a causal association between bladder cancer and ingested arsenic are the epidemiological studies conducted in southwestern Taiwan. The population in this area was exposed to high levels of arsenic in their drinking water and several different studies have documented clear dose-response increased risks for kidney, bladder, lung, and liver cancers (Chen *et al.*, 1986; Chen *et al.*, 1988; Wu *et al.*, 1989; Chen and Wang, 1990). The highest relative risks were found for bladder cancer mortality. At the highest level of exposure, averaging 800 μg As L^{-1}, relative risk estimates for bladder cancer mortality were 28.7 for men and 65.4 for women. Clearly increased risks were also reported for lower levels of exposure, including consumption of water containing an average of 170 μg As L^{-1}.

Additional evidence supporting the hypothesis that arsenic causes bladder cancer comes from two cohort studies of individuals exposed to high levels of arsenic. A British study followed a group of patients treated with Fowler's solution, an arsenic-containing medication used to treat various disorders, and observed an elevated rate of bladder cancer (Cuzick *et al.*, 1992). A Japanese study observed an excess of urinary tract cancers in a cohort of arsenic-poisoned individuals (Tsuda *et al.*, 1990). Both studies will be described in greater detail in a later section.

Finally, reports from Argentina also suggest an association between bladder cancer and arsenic ingestion. Several regions in Argentina have a history of high levels of arsenic in drinking water with many reports documenting cutaneous alterations known to be caused by arsenic. In particular, the province of Cordoba has had several areas considered endemic for arsenic induced skin lesions. Although no epidemiological studies of bladder cancer have been conducted, there are three reports which suggest increased bladder cancer risks in the affected regions (Bergoglio, 1964; Biagini, 1972; Besuchio *et al.*, 1980). Taken individually, none of the three studies provides strong evidence for a causal association between arsenic and bladder cancer, due to lack of an appropriate study design. However, taken together, the consistency of the findings lends support to the arsenic bladder cancer hypothesis.

Considered overall, the weight of the evidence from the Taiwan studies, and the supportive results from others, point to an association of bladder cancer with arsenic ingestion. Except for the studies in Taiwan, however, the interpretation of most of the relevant studies is limited by small numbers of cases. Clearly, further studies examining the relationship between arsenic and bladder cancer are needed to confirm causality and to quantify risks.

Estimation of cancer risks from arsenic in drinking water

Estimates indicate that about 347,000 people in the US are currently receiving water from public supplies with levels of arsenic that exceed the standard of 50 μg L^{-1} and about 2.5 million receive water containing over 25 μg L^{-1} (Science Applications International Corporation, 1987). Using data from the Taiwanese studies, we conducted a cancer risk assessment of ingestion of inorganic arsenic in drinking water in the US based on mortality from internal cancers (Smith *et al.*, 1992). Assuming consumption of one liter of water per day, and using US background rates

Table 1 *Comparison of epidemiological study designs to investigate bladder cancer risks from ingested arsenic.*

Study design	Advantages	Disadvantages
Ecological	Uniformity of exposure by geographical location	Migration
	Availability of information on bladder cancer rates	Confounding
Cohort	Study validity due to population being defined before the outcome	Low power for a rare disease Need very high exposure Difficulty in identifying population cohorts
Case control	Study power Individual exposure information Individual confounding factor data	Assessment of exposure in distant past Migration reduces power
Biomarker	Short time between exposure and measurable outcome Smaller number of people required for study Can study effects of intervention	Uncertainties in choice of biomarker Unknown relationship to cancer risk

of cancer, it was estimated that at the current EPA standard of 50 $\mu g\ L^{-1}$ the lifetime risk of dying from cancer of the liver, lung, kidney, or bladder could be in the neighbourhood of 13/1,000. This estimate is comparable to risks from environmental tobacco smoke and radon in homes, two of the highest known environmental risks for cancer. Thus, the study of arsenic ingestion from drinking water and risks of internal cancer, including bladder cancer, should receive high priority for investigation.

Epidemiological study designs

Various study designs could be employed to further investigate bladder cancer risks from ingestion of arsenic in drinking water, including ecological, cohort, case-control, and biomarker studies. In this section we describe each of the study designs, pointing out advantages and disadvantages of each (summarized in Table 1). Examples of each type of design are described, based on actual studies which have been conducted, or which are being planned.

Ecological studies

Ecological studies are frequently problematic in assessing evidence for causation. However, since arsenic in drinking water is geographically based, and since most people drink significant amounts of tap water, ecological studies have a potential role in examining the relationship between arsenic and bladder cancer.

A clear association was found in ecological studies in Taiwan. For example, Chen and Wang (1990) examined the ecological correlation between arsenic levels in water and mortality from various cancers in 314 precincts and townships in Taiwan. They found a significant association between arsenic levels in well water and age-adjusted mortality from cancers of the liver, nasal cavity, lung, skin, bladder, and kidney. For bladder cancer, they calculated that the age-adjusted mortality per 100,000 person-years for every 100 $\mu g\ L^{-1}$ increase in arsenic level of well water was 3.9 for males and 4.2 for females.

We have examined the ecological relationship between the population weighted mean arsenic concentration in public drinking water supplies in 30 US counties and mortality from bladder and kidney cancer. Levels of arsenic in water in the US are considerably lower than in Taiwan, but 15 counties with mean arsenic levels greater than or equal to 10 $\mu g\ L^{-1}$, and 15 counties with mean levels between 5 and 10 $\mu g\ L^{-1}$ have been identified (Engel, 1993). In addition, for 17 Nevada counties we also examined incidence data. An increasing trend of bladder cancer mortality with arsenic level was found among females but not among males. Futhermore, cancer incidence data among females did not support the mortality findings.

While the relative uniformity of exposure to arsenic in drinking water by geographical location is clearly in favor of ecological studies, there are also factors mitigating against them. The main one in countries such as the US is the effect of people changing residences. With respect to arsenic in drinking water, a move to a different geographic area, such as a different county, can have a profound effect on exposure. Consequently, current mortality in a county is influenced by migration in and out. Recent migrants into a high-exposure county will not have a long enough latency from first exposure to manifest effects. Similarly, persons moving from a high exposure county to a low one may manifest effects in the new county, therefore "diluting" the effect of the exposure and making it harder to detect. For this reason, ecological studies may not be very informative except in stable populations with high levels of exposure.

Cohort studies

The particular advantage of cohort studies is that they involve identifying subsequent disease rates in a defined population. Unfortunately, cohorts must be very large to study rare diseases, or the risks must be very high. Two cohort studies that took advantage of very high exposures are described below.

Tsuda *et al.* (1990) studied a cohort of arsenic-poisoned individuals from a Japanese mining village. Arsenic was intermittently mined and refined during the period 1920 to 1962, resulting in extensive arsenic contamination of the air and water in the nearby village. A cohort of 141 arsenic-poisoned patients were identified among former mine workers and residents in the local community, with causes of death identified up to 1989. Two deaths from urinary tract cancers were observed, versus 0.26 expected (SMR = 7.66, 95% CI 1.36 - 27.95).

Cuzick *et al.* (1992) followed a cohort of 478 patients treated with Fowler's solution (potassium arsenite) from 1948 to 1969. In the most recent follow-up of the cohort to 1991, 5 bladder cancer deaths were observed with 1.6 expected, based on rates for England and Wales (SMR = 3.07, 95% CI 1.01-7.3). For those patients who had received a cumulative dose of more than 500 mg of arsenic, the SMR based on four bladder cancer deaths was 5.00 (2.10 - 15.0).

Cohort studies have also been undertaken on smelter workers, but consistently elevated rates of bladder cancer have not been found. Welch *et al.* (1982) observed an increase in bladder cancer in their study of copper smelter workers in Montana (SMR = 2.0, 95% CI 1.0 - 3.4), however, others did not (Lee and Fraumeni, 1969; Pinto *et al.*, 1978; Wall, 1980; Lubin *et al.*, 1981; Enterline and Marsh, 1982). There are several possible explanations for this failure to detect an increase. One is that the smelter studies have low statistical power to detect an elevation in bladder cancer rates. A second possibility is that the doses from inhalation may have been insufficient to produce an epidemiologically detectable bladder cancer risk. Estimates of exposures derived from urinary arsenic measurements demonstrate that smelter workers have lower cumulative exposures than all but one of the other populations studied where increased bladder cancer risks were shown (Bates *et al.*, 1992). Finally, bladder cancer has not been systematically looked at in smelter studies. Data concerning bladder cancer rates in workers with the highest cumulative exposure to arsenic are not available in most of the published studies.

Because sufficiently high exposures to large populations are rare, cohort studies are likely to have a relatively limited role in studying arsenic and bladder cancer. In addition, in contrast to occupational cohort studies, it is difficult to identify and trace cohorts in the general population who have been exposed to arsenic in the past.

Case-control studies

Case-control studies are particularly valuable for rare disease with a high prevalence of exposure in a study population. The two major strengths of this study design are statistical power and the ability to obtain individual data on exposure and confounding factors. For these reasons, case-control studies have been very successful in the study of arsenic bladder cancer risks in Taiwan, where more than 100,000 people were exposed to arsenic in drinking water ranging from 10 to 1,820 $\mu g\ L^{-1}$ (Tseng, 1989). Chen *et al.* (1986) conducted a case-control study to evaluate the relationship between arsenic levels in artesian well water and several internal cancers, including bladder cancer. They found a strong positive dose-response relationship between bladder cancer and years of artesian well use.

A case-control study conducted by our group examined the association between arsenic levels in US drinking water supplies and bladder cancer (Bates, 1991). The study was based on data from the National Bladder Cancer Study, a very large case-control study involving nearly 3,000 bladder cancer cases and 6,000 controls recruited in 10 states. Exposure data were obtained from state authorities and, for a subset of two states, from the National Cancer Institute. Arsenic levels in drinking water in the states involved were quite low and interpretation of the findings is therefore problematic. Further case-control studies incorporating more highly exposed populations would be more informative.

We have recently investigated the feasibility of conducting a case-control study of bladder cancer and ingestion of arsenic in water in California and Nevada. The

Table 2 *Summary of study power calculations for a California/Nevada case control study.*

For z alpha 1.645
Expected odds ration of 2.0
Expected proportion of controls exposed for at least 20 years=0.12
Number of cases=187
Number of controls per case, 1, 2, 3 and 4

Controls per case	z beta	Power
1	0.805	79%
2	1.206	89%
3	1.381	92%
4	1.480	93%

proposed study area includes three counties with the highest average arsenic levels in the USA and adjoining counties with low arsenic levels (less than 10 μg L^{-1}) to provide a significant exposure contrast. We have estimated that about 27% of the study population, or about 58,700 people, drink water with levels of arsenic of 90-100 μg L^{-1}. If the study is carried out, an estimated total of 187 bladder cancer cases, diagnosed between 1988 and 1993, will be identified from the California and Nevada tumor Registries. Three controls will be selected for each case (total 561), frequency matched by sex and 5-year age group. Structured personal telephone interviews will be conducted by trained interviewers. The main interview questions will concern life-time residential history which will be linked to data on water arsenic concentrations. Questions will be asked concerning tap water consumption for periods of residence in locations supplied by arsenic containing water. Information will also be obtained concerning confounding factors, in particular cigarette smoking which may also be synergistic with arsenic in causing bladder cancer. Studies in Taiwan have shown significantly increased risks of bladder cancer associated with drinking water containing around 170 μg $L^{-1}$of arsenic. The proposed study has over 90% statistical power to detect predicted increased risks from intake of water containing around 100 μg L^{-1} (see Table 2).

Biomarker studies

The particular advantages of studies using biomarkers of effect are that they can be conducted quickly on relatively small numbers of subjects. In addition, it is possible to study the effects of interventions, such as reducing exposure (Table 1).

We have proposed the use of the micronucleus assay in exfoliated bladder cells as a biomarker of effect to study arsenic exposed individuals (Smith *et al.*, 1993). Arsenic has been shown to be a potent clastogen, and micronuclei can form as a result of clastogenic activity. Micronuclei can be measured in exfoliated bladder cells so the effect of arsenic can be studied directly in the target organ of interest. In studies of other risk factors for bladder cancer, those at risk have shown significantly

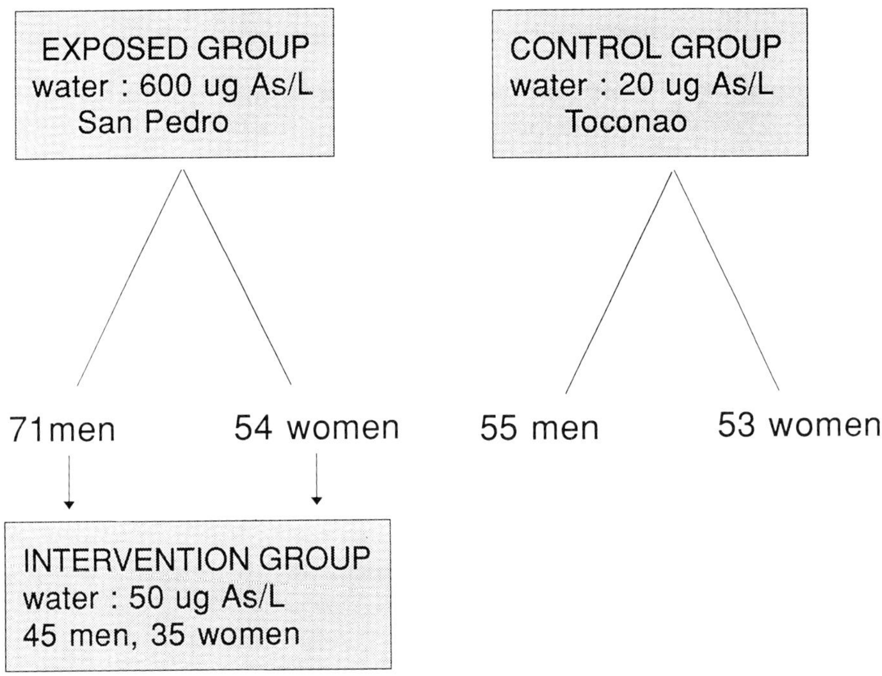

Figure 1 *Chile biomarker study group.*

elevated numbers of micronuclei in exfoliated bladder cells (Fontham *et al.*, 1986; Rosin and Anwar, 1992). Samples are obtained non-invasively (exfoliated bladder cells can be collected from urine), thus increasing the likelihood of study participation. Finally, due to the rapid turnover rate of bladder epithelial cells, the effects of intervention can be studied within a short time.

We have employed the micronucleus assay in exfoliated bladder cells as a biomarker in the study of two populations exposed to high levels of arsenic in drinking water. In the first study, subjects were residents of a small community in Nevada drinking water from private wells with arsenic levels ranging up to 5,000 μg L^{-1}(Warner *et al.*, submitted for publication). Exposed subjects (n=18) were defined as individuals with water arsenic levels ≥500 μg L^{-1} As. Unexposed subjects (n=18) were defined as individuals with water arsenic levels >10 μg $L^{-1.}$ To reduce the potential for confounding, subjects were individually matched on age, sex, and smoking status. A questionnaire was administered to each subject to obtain information concerning daily fluid consumption, occupation, smoking status, and diet. Urine samples were collected from each subject for arsenic speciation studies and to collect exfoliated bladder cells for the micronucleus assay.

We have conducted a larger biomarker study in northern Chile. We identified a town of about 1,500 people in the Atacama desert, many of whom drink water containing 600 μg As L^{-1}. We selected a neighboring town of 300 people, where the water contains 20 μg As L^{-1}, to provide a non-exposed comparison group. We selected men and women from these two towns for the investigation of fluid intake patterns, urinary arsenic analyses for exposure assessment and methylation patterns, and for measuring micronucleus frequency in bladder and buccal cells. As shown in Figure 1, the study group consisted of 125 people from the exposed town (71 men and 54 women) and 108 from the control town (55 men and 53 women). Urine samples and buccal cells were collected from all participants. Each participant was also administered a questionnaire by personal interview, covering such topics as fluid intake, smoking history, diet, medical history, and others.

The Chile biomarker study also included an intervention phase during which 40 homes (45 men, 35 women) in the high exposure town were provided with water containing low levels of arsenic. Following a period of two months, urine, buccal cells, and interview data were again collected from the intervention participants.

Conclusion

Each of the study designs presented has a role in assessing the relationship of arsenic to bladder cancer. The potentially high risks associated with arsenic ingestion warrant using each of the available study designs. However, it is likely that the most informative studies will be case-control studies and biomarker studies. Both of these study designs offer advantages over others by virtue of their enhanced study power and individual assessment of exposure and confounding factors. Additionally, biomarker studies can be done quickly, with small numbers of people, and the results of interventions can be determined rapidly. Case-control studies will be important in causal inference concerning arsenic and bladder cancer. Biomarker studies may supplement such causal inference, and have a particular advantage that the causal association between exposure and production of the biomarker can be confirmed in intervention studies.

Acknowledgements

Support for this work was provided by NIEHS grant P42-ES04705, NIEHS grant P30-ES01896, the California Occupational and Environmental Health Center, and the Health Effects Component of the University of California Toxic Substances Program.

References

Bates, M.N., Smith, A.H. and Hopenhayn-Rich, C. 1992. Arsenic ingestion and internal cancers: A review. *Am. J. Epidemiol.*, **135,** 462-476.

Bates, M.N. 1991. Case-control study of bladder cancer and arsenic in US Drinking Water Supplies. PhD thesis, University of California, Berkeley.

Bergoglio, R.M. 1964. Mortality from cancer in regions of arsenical waters of the province of Cordoba Argentine Republic. *Prev. Med. Argent.*, **51,** 994-998.

Besuschio, S.C., Perez Desanzo, A.C. and Croci, M. 1980. Epidemiological associations between arsenic and cancer in Argentina. *Biol. Trace Elem. Res.*, **2,** 41-55.

Biagini, R.E. 1972. Hidroarsenicismo cronico y muerte por canceres malignos. *La Semana Medica*, **141,** 812-816.

Chen, C.J., Chuang, Y.C., You S.L.and Lin, H.Y. 1986. A retrospective study on malignant neoplasms of bladder, lung and liver in Blackfoot Disease endemic area in Taiwan. *Brit. J. Cancer.*, **53,** 399-405.

Chen, C.J., Kuo, T.L.and Wu, M.M. 1988. Arsenic and cancers (letter). *Lancet*, **i,** 414-415.

Chen, C. and Wang, C. 1990. Ecological correlation between arsenic level in well water and age-adjusted mortality from malignant neoplasms. *Cancer Res.*, **50,** 5470-5475.

Cuzick, J., Sasieni, P.and Evans, S. 1992. Ingested arsenic, keratoses, and bladder cancer. *Am. J. Epidemiol.*, **136**, 417-421.

Engel, R.R. 1993. Ecological study of arsenic in drinking water and mortality in US counties. PhD thesis, University of California, Berkeley.

Enterline, P.E. and Marsh, G.M. 1982. Cancer among workers exposed to arsenic and other substances in a copper smelter. *Am. J. Epidemiol.*, **116,** 895-911.

Fontham, E., Correa, P., Rodriquez, E. and Lin, Y. 1986. Validation of smoking history with the micronuclei test. In: *Mechanisms in Tobacco Carcinogenesis, Banbury Report.*,**23**, 113-119.

Lee, A.M. and Fraumeni, J.F. 1969. Arsenic and respiratory cancer in man: an occupational study. *J Natl Cancer Inst.* **42,** 1045-1052.

Lubin, J.H., Pottern, L.M., Blot, W.J., Tokudome, S., Stone, B.J. and Fraumeni, J.F. 1981. Respiratory cancer among copper smelter workers: recent mortality statistics. *J. Occup. Med.*, **23,** 779-784.

Pinto, S.S. 1978. Mortality experience of arsenic-exposed workers. *Arch. Environ. Health.*, 325-331.

Rosin, M. and Anwar, W. 1992. Chromosomal damage in urothelial cells from Egyptians with chronic Schistomsoma haematobium infections. *Int. J. Cancer*, **50,** 539-43.

Science Applications International Corporation. 1987. Estimated National Occurrence and Exposure to Arsenic in Public Drinking Water Supplies (Revised Draft) (Prepared for EPA, contract No. 68-01-766). Washington, D.C.

Smith, A.H., Hopenhayn-Rich, C., Bates, M.N., Goeden, H.M., Hertz-Picciotto, I., Duggan, H., Wood, R., Kosnett, M. and Smith, M.T. 1992. Cancer risks from arsenic in drinking water. *Env. Health. Perspec.*, **97,** 259-267.

Smith, A.H., Hopenhayn-Rich C., Warner, M., Biggs, M.L., Moore, L. and Smith, M.T. 1993. Rationale for selecting exfoliated bladder cell micronuclei as potential biomarkers for arsenic genotoxicity. *J. Toxicol. Environ. Health.*, **40,** 227-238.

Tseng, W.P. 1989. Blackfoot disease in Taiwan: a 30-year follow-up study. *Angiology*, **40,** 547-58.

Tsuda, T., Nagira, T., Yamamoto, M. and Kume, Y. 1990. An epidemiological study on cancer in certified arsenic poisoning patients in Toroku. *Industrial Health.*, **28,** 53-62.

Wall, S. Survival and mortality pattern among swedish smelter workers. *Int. J. Epidemiol.*, **9,** 73-87.

Warner, M.L., Moore, L.A., Fanning, E., Smith, M.T. and Smith, A.H. Submitted for publication. Micronuclei in exfoliated bladder cells from chronic arsenic ingestion in drinking water in Nevada.

Welch, K., Higgins, I., Oh, M. and Burchfiel, C. 1982. Arsenic exposure, smoking, and respiratory cancer in copper smelter workers. *Arch. Environ. Health.*, **37,** 5-335.

Wu, M-M., Kuo, T-L., Hwang, Y-H. and Chen, C-J. 1989. Dose-response relation between arsenic concentration in well water and mortality from cancers and vascular diseases. *Am. J. Epidemiol.*, **130,** 1123-1132.

12 Arsenic in drinking water and urinary cancers: a preliminary report

How-Ran Guo[1*], Han-Sun Chiang[2], Howard Hu[1,3], Stuart R. Lipsitz[4] and Richard R. Monson[1]

[1]*Department of Environmental Health, Harvard School of Public Health, 677 Huntington Avenue, Boston, MA 02115, USA*
[2]*Department of Urology, Taipei Medical College, 252 Wu-Hsin Street, Taipei, Taiwan, Republic of China*
[3]*Channing Laboratory, Department of Medicine, Brigham and Women's Hospital, Harvard Medical School, 200 Longwood Avenue, Boston, MA 02115, USA*
[4]*Department of Biostatistics, Harvard School of Public Health, 655 Huntington Avenue, Boston, MA 02115, USA*

Abstract
A high incidence of urinary bladder and kidney cancers has been observed along the southwestern coast of Taiwan since the 1970s. Arsenic in well water has been proposed as the major cause of the excess cancer risk. We conducted an ecologic study to evaluate the association between the concentration of arsenic in well water and the incidence of urinary cancers. Previous measurements of arsenic in well water were collected from 243 townships in Taiwan. Based on data from the National Cancer Registration Program, we calculated standardized incidence rates for each township. The data were analyzed by regression models using multiple variables to describe exposure status; each variable denoted the proportion of wells in a specific arsenic-concentration category in each township. The results suggest an association between high arsenic concentration in well water and incidence of both urinary bladder cancer and transitional cell kidney cancer, but not of renal cell kidney cancer. The findings also suggest that a non-linear dose-response relationship between arsenic ingestion and cancers may exist.

Introduction

A high incidence of cancers has been observed since the 1960s among residents in an area along the southwestern coast of Taiwan (Yeh, 1963, Chen *et al.*, 1979). In

---

*To whom correspondence should be addressed at: National Institute for Occupational Safety and Health, 4676 Columbia Parkway, Mailstop R-21, Cincinnati, Ohio 45226, USA.

comparison with the general population of Taiwan, the residents were noted to have significantly higher risks for the following cancers of (in descending order of the standardized mortality ratio; SMR) urinary bladder, kidney, skin, lung, liver, and colon (Chen *et al.*, 1985). Among these cancers, skin cancers which are typically seen in arsenic-intoxicated patients were noted first, and dose-response correlations were observed between arsenic concentration in well water and cancer mortalities (Chen *et al.*, 1985; 1988; 1990). Accordingly, arsenic in well water was proposed as a major cause of cancers in that area.

Most of the investigations mentioned above were ecologic studies in which the degree of arsenic exposure for a population living in a defined area (township or village) was represented by a single figure, such as the mean or median arsenic concentration in well water in that area. Our study was also an ecologic study; however, a different analytical approach was developed in which the arsenic exposure in a given area was described by multiple variables to exploit use more fully the exposure data. Furthermore, previous studies in Taiwan were mostly based on information from death certificates, which have low specificity for cancer type and may not be able to estimate accurately the incidence of diseases that are less consistently fatal, such as bladder cancer. In our study, the use of data from the National Cancer Registration Program permitted the analysis of cancer incidence–as opposed to mortality–as well as the identification of specific cell types of cancers.

Methodology

Measurements of arsenic in well water were obtained from a nationwide survey conducted by the Taiwan Provincial Institute of Environmental Sanitation (Lo *et al.*, 1977) using the standard mercuric bromide stain method (American Public Health Association, 1985). In this procedure, a standard solution of 1.28 ppm arsenic was first prepared and then diluted to a series of standard solutions with arsenic levels ranging from 0.64 ppm to 0.01 ppm; the detection limit was estimated to be 0.001 ppm. For each sample, the arsenic concentration was estimated by comparison with the standard solutions. A designation of "trace" was assigned to samples with concentrations between the 0.01 ppm standard solution and the blank control. A value of 0 ppm was assigned to samples for which the result was compatible with the blank control. Data are available for 243 townships (precincts) with a total population of around 11.4 million. The distribution of exposure levels in the study population is shown in Table 1.

Cancer incidence was calculated based on data from the National Cancer Registration Program operated by the Department of Health of Taiwan. Sex, age, diagnoses, and township of residence were reported for each case. Both clinical and pathological diagnoses were coded according to the International Classification of Diseases for Oncology (ICD-O; World Health Organization, 1976). Cases diagnosed between January 1, 1980, and December 31, 1987, were included; those with ICD-O codes from 188.0 to 188.9 were defined as bladder cancers, and those with ICD-O codes 189.0 or 189.1 as kidney cancers. A total of 1,972 bladder cancer patients and 726 kidney cancer patients (including 239 transitional cell carcinomas and 210 renal

Table 1 *Estimated population (unit: 1,000) exposed to each arsenic level in well water.*

	Arsenic level (ppm)									
	0	trace	0.01	0.02	0.03 -0.04	0.05 -0.08	0.09 -0.16	0.17 -0.32	0.33 -0.64	>0.64
Males	4,680	343	140	131	196	167	158	84	47	25
Females	4,276	316	131	122	182	154	146	77	43	23
Total	8,956	659	271	253	378	321	304	161	90	48

Table 2 *Pathological diagnoses of urinary cancer patients.*

Pathology	Males		Females	
	Cases	%	Cases	%
Bladder cancers				
Transitional cell	1,185	88	363	83
Others	156	12	75	17
Subtotal	1,341	100	438	100
No pathology	120		73	
Total	1,461		511	
Kidney cancers				
Transitional cell	158	39	81	40
Renal cell	156	39	54	26
Others	87	22	70	34
Subtotal	401	100	205	100
No pathology	80		40	
Total	481		245	

cell carcinomas) were identified. (Table 2) Data on population were obtained from the Department of Internal Affairs. Sex-specific standardized incidence rates (SIRs; per 100,000 during the study period) were calculated for each township, and the world standard population in 1976 was used as the standard population to adjust the distribution of age (Waterhouse *et al.*, 1976).

The effects of smoking on cancer incidence were assessed by the analysis of data on cigarette sales, which is a monopoly business in Taiwan operated by the Tobacco and Alcohol Monopoly Bureau. The township-specific sales records collected from the bureau in a previous study (Tsai, 1987) were used to estimate the average number of cigarettes smoked per capita per day in each township. An urbanization index assigned by Wu (1986) to each township based on 19 socioeconomic factors was used to assess the association between degree of urbanization and cancer incidence.

The data were analysed using multiple linear regression models with the SIR of a

Table 3 *Rate difference (RD)[1] estimated for bladder cancer by Model 1 (conventional approach)[2].*

Variable	Male		Female	
	RD	p[3]	RD	p
Intercept	24.65		7.63	
Mean [As][4]	225.41	**	162.11	**
Urbanization	4.48	**	2.57	**
Cigarette	2.00	N/S	2.07	N/S

** Indicates $p < 0.01$; N/S indicates not significant at = 0.05.
[1] Per 100,000 during the eight-year study period.
[2] $p < 0.01$ for F test for the significance of model in both sexes.
[3] p value of t-test for the significance of rate difference.
[4] Arsenic concentration.

given cancer as the dependent variable and the arsenic exposure, the average number of cigarettes sold per capita per day, and the urbanization index as independent variables. The data were first analysed by the conventional approach (Model 1) with the mean arsenic concentration in well water for each township as the only arsenic-exposure variable:

$$SIR = \alpha + \beta X + \gamma U + \delta T \qquad \text{(Model 1)}$$

where X= mean arsenic level, U= urbanisation index, T= average number of cigarettes sold per capita per day, α= background rate, ß= the rate difference (RD) for a 1-ppm increase in the arsenic level,γ= the RD for a 1-unit increase in the urbanisation index, and δ= the RD for each additional cigarette sold per capita per day. For the calculation of the means, a value of 0.005 ppm was assigned to the "trace" category.

Data were then analysed again using a regression model with a series of variables to describe the arsenic exposure (Model 2). Each variable denoted the percentage of wells in each township with arsenic levels in a designated category. On the basis of the arsenic concentrations of the standard solutions used, arsenic levels in well water were classified into 10 categories: 0 ppm (undetectable), trace, 0.01 ppm, 0.02 ppm, 0.03-0.04 ppm, 0.05-0.08 ppm, 0.09-0.16 ppm, 0.17-0.32 ppm, 0.33-0.64 ppm and > 0.64 ppm. The "undetectable" category was used as the reference category:

$$SIR = \alpha + \beta_1 X_1 + \beta_2 X_2 + \ldots + \beta_9 X_9 + \gamma U + \delta T \qquad \text{(Model 2)}$$

where X_i= the percentage of wells in the township with arsenic levels falling into category i, (*i.e.*, X_1= percentage with trace level, X_2= percentage with 0.01 ppm, and so on), β_1= the RD for a 1% increase in wells with trace level of arsenic, β_2= the RD

Table 4 *Rate difference (RD)*[1] *estimated for bladder cancer by Model 2 (new approach)*[2].

Variable	Male		Female	
	RD	p^3	RD	p
Intercept	25.91		9.01	
Trace	-0.27	N/S	-0.25	0.01
0.01 ppm	0.19	N/S	-0.06	N/S
0.02 ppm	0.59	N/S	1.48	**
0.03-0.04 ppm	0.19	N/S	-0.30	N/S
0.05-0.08 ppm	-1.09	N/S	-1.07	**
0.09-0.16 ppm	1.11	N/S	1.50	**
0.17-0.32 ppm	-0.20	N/S	-0.37	N/S
0.33-0.64 ppm	-0.63	N/S	-0.05	N/S
> 0.64 ppm	6.09	**	3.31	**
Urbanization	3.75	**	1.67	0.04
Cigarette	1.40	N/S	1.73	N/S

** Indicates $p < 0.01$; N/S indicates not significant at = 0.05.
[1] Per 100,000 during the eight-year study period.
[2] $p < 0.01$ for F test for the significance of model in both sexes.
[3] p value of t-test for the significance of rate difference.

for a 1% increase in wells with 0.01 ppm arsenic, and so on. Both regression analyses were weighted according to the population in each township.

Results

Although both approaches showed statistically significant associations between the arsenic concentration in well water and bladder cancer, the associations estimated by Model 2 (new approach) were more complex. In males, no statistically significant association was observed until the highest exposure category (> 0.64 ppm). In females, no consistent trend was apparent in either positive or negative RDs (in comparison to background rates). The highest exposure category, however, was strongly associated with positive RDs in both males and females (Tables 3 and 4). Similar association patterns as in bladder cancer were observed in transitional cell kidney cancer using Model 1 (conventional approach) (Table 5). When Model 2 was applied, no consistent trend in either positive or negative RD was observed in males or females. In males, the exposure category 0.33-0.64 ppm was associated with the highest positive RD, while the highest exposure category was also strongly associated with a positive RD (Table 6). In females, the highest exposure category

Table 5 *Rate difference (RD)*[1] *estimated for transitional cell kidney cancer by Model 1 (conventional approach)*[2].

Variable	Male		Female	
	RD	p^3	RD	p
Intercept	2.28		1.16	
Mean [As][4]	46.97	**	43.36	**
Urbanization	0.49	0.02	0.30	N/S
Cigarette	-0.33	N/S	-0.82	N/S

** Indicates $p < 0.01$; N/S indicates not significant at = 0.05.
[1] Per 100,000 during the eight-year study period.
[2] $p < 0.01$ for F test for the significance of model in both sexes.
[3] p value of t-test for the significance of rate difference.
[4] Arsenic concentration.

was associated with the highest positive RD (Table 6). No association between the arsenic level in well water and renal cell kidney cancer was suggested by either Model 1 or Model 2 (Tables 7 and 8).

Urbanization appeared to be associated with all three types of urinary cancers (bladder cancer, transitional cell kidney cancer, and renal cell kidney cancer) among males in both models (Tables 3 to 8). In females, however, only the bladder cancer showed an association with urbanization index, and this association was observed in both models. The variable for cigarette sales showed no association with any of the urinary cancers studied. Except for renal cell kidney cancer among males, the background rates estimated by Model 1 were somewhat lower than those estimated by Model 2.

Discussion

As early as the 1950s, the occurrence of bladder cancer among arsenic intoxicated patients was noted (Sommers *et al.*, 1952). Besides Taiwan, a high incidence of urinary cancers associated with arsenic ingestion has been also reported in Japan (Tsuda *et al.*, 1990) and in Argentina (Besuschio *et al.*, 1980), and among Fowler's solution (potassium arsenite) users (Cuzick *et al.*, 1992). The results of studies of copper smelter workers exposed to arsenic were less consistent. Enterline *et al.* (1982) observed excess mortality from kidney cancers (SMR= 131.6, not statistically significant), but not from urinary cancers over-all (SMR= 63.0). Welch *et al.* (1982) found excess urinary cancers (SMR's between 195 and 197), but the SMRs were similar for workers heavily exposed to arsenic and the other workers. In most studies of other populations with occupational exposure to arsenic, no excess urinary cancers were observed (Tokudome *et al.*, 1976, Mabuchi, *et al.*, 1980, Sobel *et al.*, 1988). However, the populations observed in occupational studies were much

Table 6 *Rate difference (RD)[1] estimated for transitional cell kidney cancer by Model 2 (new approach)[2].*

Variable	Male RD	Male p^3	Female RD	Female p
Intercept	2.57		1.82	
Trace	-0.04	N/S	-0.05	0.04
0.01 ppm	-0.04	N/S	0.03	N/S
0.02 ppm	0.24	0.03	-0.03	N/S
0.03-0.04 ppm	-0.02	N/S	-0.01	N/S
0.05-0.08 ppm	-0.20	N/S	-0.08	N/S
0.09-0.16 ppm	0.31	0.02	-0.01	N/S
0.17-0.32 ppm	-0.37	0.03	0.05	N/S
0.33-0.64 ppm	0.80	**	-0.09	N/S
> 0.64 ppm	0.40	**	1.30	**
Urbanization	0.43	0.05	0.18	N/S
Cigarette	0.37	N/S	-1.08	N/S

** Indicates $p < 0.01$; N/S indicates not significant at = 0.05.
[1] Per 100,000 during the eight-year study period.
[2] $p < 0.01$ for F test for the significance of model in both sexes.
[3] p value of t-test for the significance of rate difference.

Table 7 *Rate difference (RD)[1] estimated for renal cell kidney cancer by Model 1 (conventional approach)[2].*

Variable	Male RD	Male p^3	Female RD	Female p
Intercept	2.80		1.29	
Mean [As][4]	-5.29	N/S	-3.95	N/S
Urbanization	0.95	**	0.26	N/S
Cigarette	-1.55	N/S	-1.49	N/S

** Indicates $p < 0.01$; N/S indicates not significant at = 0.05.
[1] per 100,000 during the eight-year study period.
[2] $p < 0.05$ for F test for the significance of model in males, $p > 0.05$ in females.
[3] p value of t-test for the significance of rate difference.
[4] Arsenic concentration.

Table 8 *Rate difference (RD)*[1] *estimated for renal cell kidney cancer by Model 2 (new approach)*[2].

Variable	Male		Female	
	RD	*p*[2]	RD	*p*
Intercept	2.87		1.39	
Trace	0.05	N/S	-0.01	N/S
0.01 ppm	-0.05	N/S	-0.07	N/S
0.02 ppm	0.10	N/S	0.02	N/S
0.03-0.04 ppm	-0.14	N/S	-0.04	N/S
0.05-0.08 ppm	0.01	N/S	0.09	N/S
0.09-0.16 ppm	-0.05	N/S	-0.01	N/S
0.17-0.32 ppm	-0.01	N/S	-0.05	N/S
0.33-0.64 ppm	-0.08	N/S	-0.03	N/S
> 0.64 ppm	0.09	N/S	0.01	N/S
Urbanization	1.03	**	0.29	N/S
Cigarette	-1.70	N/S	-1.50	N/S

** Indicates $p < 0.01$; N/S indicates not significant at = 0.05.
[1] Per 100,000 during the eight-year study period.
[2] $p > 0.05$ for F test for the significance of model in both sexes.
[3] *p* value of *t*-test for the significance of rate difference.

smaller than those observed in community studies, and the exposure routes were different.

As has been recognised by the US Environmental Protection Agency (US EPA), the investigations in Taiwan have provided the "best data available" for cancer risk assessment of ingested arsenic (US EPA, 1984). However, the only case-control study in Taiwan provided no data on arsenic level (Chen *et al.*, 1986), and all the other epidemiologic studies were ecologic studies which used the mean or median arsenic concentration in well water to represent the arsenic exposure level in a given area. Such an approach assumes a linear non-threshold dose-response effect. The results of our study, in which an alternative analytical approach was used, suggest that the linear model may under-estimate the background risk (estimated as the intercept in both models). Moreover, our new approach, while still ecologic in nature, exploits the data more fully by attempting to allow the data to manifest the shape of the dose-response relationship curve.

Although significant RDs for bladder and transitional cell kidney cancers were observed for some lower arsenic-exposure categories, the signs fluctuated between negative and positive, and the RDs did not reach their peak until the two highest exposure categories. One speculation is that there might be a threshold somewhere above 0.17 ppm for the occurrence of excess urinary cancers. In fact, some experimental and epidemiological evidence suggests a threshold for the carcinogenicity of arsenic (Stohrer, 1991).

On the other hand, other risk factors may also have contributed to the dose-response patterns observed. For example, some mutagenic humic substances found in well water along the southwestern coast of Taiwan were reported to be ecologically associated–in a dose-response manner–with the occurrence of bladder cancers (Lu *et al.*, 1986).

The positive association between urbanization index and bladder cancer probably indicates that unmeasured occupational and environmental exposures related to urbanization may contribute to the occurrence of bladder cancer. This association was more prominent among males, and bladder cancer is known to be correlated with many occupational risk factors.

On the other hand, the failure to demonstrate an association between cigarette sales and urinary cancers in our analyses does not exclude a correlation. Rather, it may imply that cigarette sales in a given geographic area are not a good indicator of actual cigarette consumption by its residents.

In most of the countries in the world, the proportion of all kidney cancers accounted for by renal pelvis tumors (mostly transitional cell carcinomas) is around 7% to 8%. However, the figure in Taiwan has been documented at about 47% (Hsieh *et al.*, 1979). Our findings confirm the relatively high incidence of transitional cell kidney cancer in Taiwan and suggest that arsenic exposure may contribute to this phenomenon.

Conclusions

Our analyses demonstrate that the use of multiple variables, instead of a single variable, to describe the exposure status of a unit population in ecologic studies may enhance the accuracy of risk assessment. On the other hand, our findings raise the question of whether a non-linear dose-response relationship exists between arsenic exposure and urinary cancers. Further research based on individual exposure data is necessary to better define the dose-response relationship between arsenic and cancer.

Acknowledgements

This study was partially supported by a grant from the National Science Council, Republic of China and National Institute of Environmental Health Sciences grant 2-P30-ES-00002. We would also like to thank Dr. Chien-Jen Chen and Dr. Hong-Hsin Lee for their help in carrying out this study.

References

American Public Health Association. 1985. *Standard Methods of the Evaluation of Water and Water Waste.* American Public Health Association. Washington, D.C.

Besuschio, S.C., Perez Desanzo, A.C. and Croci, M. 1980. Epidemiological association between arsenic and cancer in Argentina. *Biol. Trace Element Res.*, **2**, 41-55.

Chen, C-J., Chuang, Y-C., Lin, T-M. and Wu, H-Y. 1985. Malignant neoplasms among residents of a blackfoot disease endemic area in Taiwan: high-arsenic artisan well water and cancers. *Cancer Res.*, **45**, 5895-5899.

Chen, C-J., Chuang, Y-C., You, S-L., Lin, T-M. and Wu, H-Y. 1986. A retrospective study on malignant neoplasms of bladder, lung and liver in blackfoot disease endemic area in Taiwan. *Br. J. Cancer*, **53**, 399-405.
Chen, C-J., Kuo, T-L. and Wu, M-M. 1988. Arsenic and cancers. *Lancet*, **1**, 414-415.
Chen, C-J. and Wang, C-J. 1990. Ecological correlation between arsenic level in well water and age-adjusted mortality from malignant neoplasms. *Cancer Res.*, **50**, 5470-5474.
Chen, K-P., Wu, H-Y., Yeh, C-C. and Cheng, Y-J. 1979. *Color Atlas of Cancer Mortality by Administrative and Other Classified Districts in Taiwan Area: 1968-1976.* NSC Special Publication No. 2. National Science Council, Taipei, Taiwan.
Cuzick J, Sasieni P, and Evans S. 1992. Ingested arsenic, keratoses, and bladder cancer. *Am. J. Epidemiol.*, **136**, 417-421.
Enterline, P.E. and Marsh G.M. 1982. Cancer among workers exposed to arsenic and other substances in a copper smelter. *Am. J. Epidemiol.*, **116**, 895-911.
Hsieh, Y-F., Ling, G-C. and Lu, Y-B. 1979. Incidence of tumor of the renal pelvis in Taiwan. *J. Formosan Med Assoc.*, **78**, 749-753.
Lo, M-C., Hsen, Y-C. and Lin B-K. 1977. *Second Report on the Investigation of Arsenic Content in Underground Water in Taiwan.* Taiwan Provincial Institute of Environmental Sanitation. Taichung, Taiwan.
Lu, F-J., Guo, H-R., Chiang, H-S. and Hong, C-L. 1986. Relationships between the fluorescent intensity of well water and the incidence rate of bladder cancer. *J. Chinese Oncol. Soc.*, **2**, 134-143.
Mabuchi, K., Lilienfeld, A.M. and Snell, L.M. 1980. Cancer and occupational exposure to arsenic: A study of pesticide workers. *Prev. Med.*, **9**, 51-77.
Sobel, W., Bond, G.G., Baldwin, C.L. and Ducommun, D.J. 1988. An update of respiratory cancer and occupational exposure to arsenical. *Am. J. Ind. Med.*, **13**, 263-270.
Sommers, S.C. and McManus, R.G. 1952. Multiple arsenical cancers of skin and internal organs. *Cancer*, **6**, 347-359.
Stohrer, G. 1991. Arsenic: Opportunity for risk assessment. *Arch. Toxicol.*, **65**, 525-531.
Tokudome, S. and Kuratsune, M. 1976. A cohort study on mortality from cancer and other causes among workers at a metal refinery. *Int. J. Cancer*, **17**, 310-317.
Tsai, S-F. 1987. Ecological Correlation Study on major Risk Factors of Various Malignant Neoplasms in Taiwan; 1972-1983. Master's Thesis, National Taiwan University
Tsuda, T., Nagira, T. Yamamoto, M. and Kume, Y. 1990. An epidemiological study on cancer in certified arsenic poisoning patients in Toroku. *Ind. Health*, **28**, 53-62.
US EPA. 1984. Health Assessment Document for Inorganic Arsenic: Final Report. EPA Publication EPA-600/8-83-021F. United States Environmental Protection Agency, Washington, D.C.
Waterhouse, J., Shanmugaratnam, K., Muir, C. and Powell, J (eds.). 1976. *Cancer Incidence in Five Continents,* Vol. 3. IARC Scientific Publication 15. International Agency for Research on Cancer. Lyon, France.
Welch, K., Higgins, I., Oh, M. and Burchfiel, C. 1982. Arsenic exposure, smoking, and respiratory cancer in copper smelter workers. *Arch. Environ. Health*, **37**, 325-335.
World Health Organization. 1976. *International Classification of Diseases for Oncology.* 1st edn. World Health Organization. Geneva, Switzerland.
Wu, S-Y. 1986. *Report on the Degree of Urbanization of Precincts and Townships in Taiwan.* Bureau of Statistics, Executive Yuan, Taipei, Taiwan.
Yeh, S. 1963. Relative incidence of skin cancer in Chinese in Taiwan: with special reference to arsenical cancer. *N.C.I. Monogr.*, **10**, 81-107.

13 Arsenic in drinking water and cancers: a brief descriptive review of Taiwan studies

How-Ran Guo[1*], Chien-Jen Chen[2] and Howard L. Greene[3]

[1]*Department of Environmental Health, Harvard School of Public Health, 655 Huntington Avenue, Boston, MA 02115, USA*
[2]*Institute of Epidemiology, College of Public Health, National Taiwan University, 1 Jen-Ai Road, 1st Section, Taipei, Taiwan, Republic of China*
[3]*ARCO, 515 South Flower Street, Los Angeles, CA 90071, USA*

Abstract
The association between arsenic in drinking water and cancers has been intensively studied in Taiwan since the 1960s. The data generated from these studies are generally considered to be the best available for quantitative cancer risk assessment for ingestion of arsenic. Although the studies differ in arsenic exposure data used and populations studied, these differences have not been clearly described in the literature. The purpose of this paper is to clarify the differences among the most important epidemiologic studies on arsenic in Taiwan and to summarise their findings.

Introduction

A high prevalence of skin cancer has been reported among residents in an area along the southwest coast of Taiwan since the 1960s (Yeh *et al.*, 1963; Tseng *et al.*, 1968). Subsequently, high incidence of urinary bladder cancer (Chiang *et al.*, 1993; Guo *et al.*, 1993) and high mortality from several internal cancers (Chen *et al.*, 1985, 1988, 1990, 1992; Wu *et al.*, 1989) were reported.

The studies on skin cancer were motivated by studies of blackfoot disease (BFD), an obstructive peripheral vascular disease prevalent in the southwest coast area of Taiwan (Yeh and How, 1963). This disease was named for its most striking clinical feature the black discoloration of extremities, mostly feet, resulting from the development of gangrene (Tseng, 1988). The cause of blackfoot disease was not clear, and the first case report attributed this disease to potassium intoxication (Kao *et al.*, 1954). In 1958, the first skin cancer case among blackfoot disease patients

*To whom correspondence should be addressed at: National Institute for Occupational Safety and Health, 4676 Columbia Parkway, Mailstop R-21, Cincinnati, Ohio 45226, USA.

Table 1 *Characteristics of the major epidemiologic studies in Taiwan on arsenic in drinking water and cancers.*

Studies	Study area	Population		Exposure data		Outcome	Study period
		Age	Size	Surveyor	Categories[1]		
Tseng *et al.*[2] (1986)	37 villages	All	40,421	Kuo	3	Prevalence	1958-1965
	Ma-tsu Island	All	4,978	Unknown	1	Prevalence	1958-1965
	5 villages	All	2,522	Yeh	2	Prevalence	1958-1965
Chen *et al.* (1988)	6 townships	All	899,811py[3]	Kuo	3	Mortality	1973-1986
Wu *et al.* (1989)	42 villages	≥20	898,806py	Kuo	3	Mortality	1973-1986
Chen and Wang (1990)	314 townships	All	15,000,000[4]	Lo *et al.*	Mean	Mortality	1972-1983
Chen *et al.* (1992)	6 townships	All	898,806py	Kuo	4	Mortality	1968-1982
Brown and Chen (1993)	60 villages	All	1,355,915py	Kuo	11	Mortality	1973-1986
Guo *et al.* (1993)	243 townships	All	11,400,000[4]	Lo *et al.*	10	Incidence[5]	1980-1987

[1]Number of arsenic exposure categories.
[2]Only 33 villages had exposure data.
[3]Person-years.
[4]Estimated population.
[5]Cumulative incidence rate over the eight-year period.

was reported (Yeh *et al.*, 1958), which led to the proposal of arsenic as the cause of blackfoot disease and initiated studies on skin cancer as well as other cancers.

The skin cancers observed in this area, especially Bowen's disease, were those typically seen among arsenic intoxication patients (Yeh, 1963). Other typical skin changes caused by arsenic intoxication, such as hyperpigmentation and hyperkeratosis, were also prevalent in the endemic area (Yeh, 1963). High levels of arsenic were found in the water from many artesian wells, which were used by the inhabitants in the endemic area as sources of drinking water, and a positive dose-response relationship between arsenic concentration in well water and skin cancer was reported in an ecologic study by Tseng *et al.* (1968). These findings led to the hypothesis that arsenic in the drinking water was the cause of cancers, including skin cancer and internal cancers.

The data published by Tseng and his colleagues (Tseng *et al.*, 1968; Tseng, 1977) were generally considered to be the best available for a quantitative cancer risk assessment for ingestion of arsenic, and were used as the basis of the risk assessment by the US Environmental Protection Agency (US EPA, 1984). Other studies of arsenic exposed populations have also been conducted in Taiwan, but the differences among the studies, such as arsenic exposure data measurements and study populations, were not clearly described, either in the original reports or in review papers (Bates *et al.*, 1992; Smith *et al.*, 1992). The purpose of this review is to clarify the differences among different epidemiologic studies in Taiwan (Table 1), and to summarise the findings (Table 2). This review is limited to studies that were published in peer reviewed journals or presented in international conferences and provided exposure data with the potential for use in risk analysis. The study area, study population, exposure data used, cancer outcome measurement, and study

Table 2 *Findings of the major epidemiologic studies in Taiwan on arsenic in drinking water and cancers.*

Sites	Tseng *et al.*	Chen *et al.*[1] (1988)	Wu *et al.*[2]	Chen *et al.* (1990)	Chen & Wang	Brown & Chen	Guo *et al.*
Skin	+	+	+	+	NA	+	NA
Bladder	NA	+	+	+	+	+	+
Kidney	NA	+	+	+	+	NA	+
Lung	NA	+	+	+	+	+	NA
Liver	NA	+	+(m) 0(f)	+	+	+	NA
Prostate	NA	+	+	+	NA	NA	NA
Nasal Cavity	NA	NA	NA	+	NA	NA	NA
All cancer	NA	+	+	NA	NA	NA	NA

+: (Statistically) significant positive association.
0: No significant association.
NA: Not assessed; (m): males; (f): females.
[1]Study underlying cause of death only.
[2]Assess trends only.

period in each study are specified in Table 1. Table 2 summarises the major findings. In addition, surveys of well water arsenic concentrations that provide exposure data for the epidemiologic studies reviewed are discussed.

Surveys of well water arsenic concentrations

The survey most frequently employed for characterising arsenic levels in well water by epidemiologic research – including Tseng *et al.*'s study (1968, 1977) used by US EPA for cancer risk assessment – was conducted by Kuo using a modification of Natelson's method (Natelson, 1957). Although more than 250 wells were studied from 1964 to 1966, only the measurements of 126 wells in 29 blackfoot disease-endemic villages were reported (Kuo, 1968). Arsenic concentrations in individual wells ranged from 0.01 to 1.80 ppm (mg L^{-1}), with a mean of 0.50 ppm, and about half of the samples were between 0.40 to 0.70 ppm. The mean arsenic concentration in each village ranged from 0.05 to 0.83 ppm. Tseng and his colleagues used the data of 114 wells in 37 villages in their landmark study (Tseng *et al.*, 1968; Tseng, 1977), and Chen and his colleagues used the data from 155 wells in 42 villages of 6 blackfoot disease-endemic townships to conducted a series of studies (Chen *et al.*, 1988, 1992, 1993; Wu *et al.*, 1989) (Table 1). On average, a township in Taiwan has a total population of about 30,000 to 50,000 divided into around 20 villages. However, not all the villages in a blackfoot disease-endemic township are blackfoot disease-endemic villages.

The largest survey on well water arsenic concentration in Taiwan to date was carried out by the Taiwan Provincial Institute of Environmental Sanitation using the standard mercuric bromide stain method (American Public Health Association, 1982). This nationwide survey was mainly conducted from 1974 to 1976, and the measurement results of 83,565 wells in 313 townships (or precincts) were published in 1977 (Lo *et al.*, 1977). The report grouped wells by township and provided summary statistics, including the number of wells with arsenic concentrations above 0.05 ppm (Taiwan's regulatory standard), and the number of wells with arsenic concentration above 0.35 ppm (the minimum concentration to cause blackfoot disease proposed by Chen and Wu, 1962) in each township. The results indicated that 119 townships had wells with arsenic concentrations above 0.05 ppm, and 58 townships had wells with arsenic concentrations above 0.35 ppm. As in Kuo's

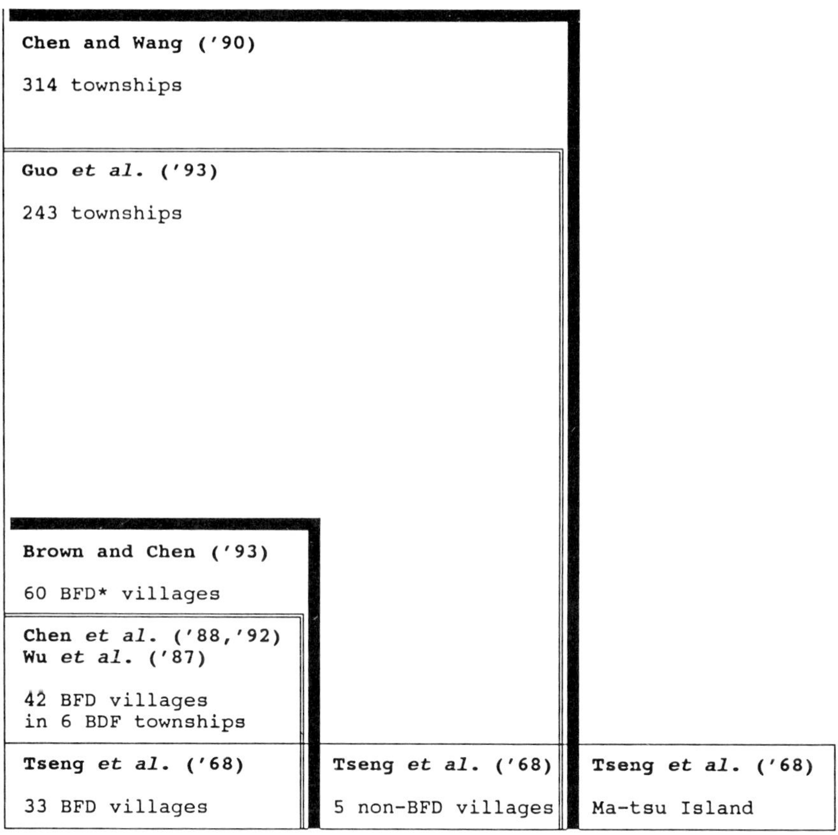

*BFD: blackfoot disease

Figure 1 *Geographic areas studied by each major epidemiologic study reviewed.*

survey, some of the results were not published. Based on the published and unpublished data of 83,656 wells in 314 townships from Lo *et al.*'s survey, Chen's group conducted two ecologic analysess (Tsai, 1987; Chen and Wang, 1990). Wang *et al.* (1989) also used the published data to perform an ecologic analysis. In all three analyses, researchers used township as the unit of analysis because more detailed information on exposure was not available in the survey reports.

More than a decade after Lo *et al.*'s survey, Lu *et al.* (1988) published the results of a survey on 1,189 wells in 226 villages in southwest Taiwan, although most of the wells were not in the blackfoot disease-endemic area. Recently, Brown and Chen (1993) conducted an ecologic analysis of four cancers based on the data from 18 villages in this survey as well as data from 42 villages in Kuo's survey (Brown and Chen, 1993) (Table 1). All 60 villages studied were in the six townships that had the highest prevalence of blackfoot disease.

Studies focused on blackfoot disease-endemic area

A total of 40,421 inhabitants in 37 villages were examined, and 428 skin cancer patients were observed by Tseng *et al.* (1968, 1977) used by the US EPA for skin cancer risk assessment (US EPA, 1984). Kuo's measurements of 114 wells in these villages (including 110 artesian wells and four shallow wells) were employed; however, only 33 of the 37 villages had exposure data. Based on the median arsenic concentration in each village observed by Kuo, the villages were grouped into four exposure categories, namely "low" (below 0.3 ppm), "mid" (0.3-0.6 ppm), "high" (above 0.6 ppm), and "undetermined." Age- and sex-specific prevalence was calculated for three age groups: 20-39 years, 40-59 years, and 60 years and over. For comparison, 4,978 residents of Ma-tsu Island with no detectable arsenic in the drinking water (from shallow wells), and 2,522 resident in 5 villages outside the blackfoot disease-endemic area with 0.001 to 0.017 ppm of arsenic in the drinking water (from shallow wells) were also examined around the same time (Table 1). A positive dose-response relationship between the arsenic level in well water and prevalence of skin cancer was observed in the 33 blackfoot disease endemic villages, and no cases of skin cancer were observed in the two comparison populations (Table 2).

Chen *et al.* (1988) also conducted an ecologic analysis in six townships of the blackfoot disease-endemic area. A total of 899,811 person-years were observed from 1973 to 1986, and 1,031 cancer deaths were identified. Sex-specific age-standardized mortality rates (ASMRs) were calculated for three different exposure groups (< 0.30 ppm, 0.30-0.59 ppm, and ≥ 0.60 ppm), based on the median well water arsenic concentration in each village observed in Kuo's survey (Table 1). In comparison with the general population in Taiwan, higher ASMRs were observed for liver cancer, lung cancer, skin cancer, prostatic cancer, bladder cancer, kidney cancer, and all cancers together at all levels of exposure, except for prostatic cancer in the lowest exposure group (< 0.30 ppm). Positive dose-response relationships were observed for each of the cancers mentioned (Table 2).

Soon after, Chen's group published a more detailed ecologic analysis (Wu *et al.*,

1989). The analysis was conducted on the residents of the same 42 villages as in the previous study (Chen *et al.*, 1988), but the observations were limited to residents 20 years old or older. A total of 889,806 person-years were observed between 1973 and 1986, and 1,152 cancer deaths were ascertained (Table 1). Sex-specific ASMRs were calculated for the same three levels of well water arsenic contents, and additional analyses were conducted for leukemia, nasopharyngeal cancer, esophageal cancer, stomach cancer, colon cancer, and cervical cancer. Statistically significant positive dose-response relationships were observed for bladder cancer, kidney cancer, skin cancer, lung cancer, and all cancers combined among both men and women, as well as liver and prostatic cancers among men. No statistically significant dose-response relationship was observed for liver or cervical cancer in women, for esophageal cancer, nasopharyngeal cancer, stomach cancer, colon cancer, or leukemia in either men or women (Table 2). The data were further analysed using the Armitage-Doll multistage model to estimate the risk for cancers of the liver, lung, bladder, and kidney due to ingested arsenic in drinking water (*Chen et al.*, 1992). On the basis of new data, one village previously classified into the group with a median arsenic concentration ≥ 0.60 ppm or more was reclassified into the group < 0.30 ppm. Although only those with cancer as the underlying cause of death were identified as cancer deaths in the previous analyses (Chen *et al.*, 1985, 1988; Wu *et al.*, 1989), all cases with primary cancers listed on the death certificate were identified as cancer deaths in the re-analysis (Table 1). Statistically significant dose-response relationships were observed for cancers of the liver, lung, bladder, and kidney (Table 2).

Recently, a larger ecologic analysis was conducted in 60 villages, including the previously studied 42 villages in the blackfoot disease-endemic area (Brown and Chen, 1993). Each village was classified into one of 11 exposure levels (< 0.01 ppm, 0.01-0.019 ppm, 0.02-0.029 ppm, 0.03-0.039 ppm, 0.04-0.049 ppm, 0.05-0.09 ppm, 0.1-0.29 ppm, 0.3-0.49 ppm, 0.5-0.69 ppm, 0.7-0.89 ppm, and 0.9+ ppm) of median well water arsenic concentration in each village based on Kuo's (1968) and Lu *et al.*'s surveys (1988), and SMRs from 1973 to 1986 were calculated for each level of expopsure (Table 1). Only cancers of bladder, liver, lung, and skin were studied. The observed dose-response relationships appeared non-linear. The mortality of bladder cancer, lung cancer, and lung cancer were strongly associated with median arsenic concentration only at exposure levels of 0.05-0.09 ppm and above (Table 2).

Nation-wide studies

The largest documented study on arsenic and cancers was an ecologic analysis carried out by Chen and Wang (1990). Cancer mortality was observed in 314 townships where Lo *et al.* (1968) had tested the arsenic levels in well water. ASMRs between 1972 and 1983 were calculated for 21 different cancers, including leukemia and cancers of the nasopharynx, esophagus, stomach, small intestine, colon, rectum, liver, pancreas, nasal cavity, larynx, lung, bone and cartilage, skin, breast, uterine cervix, ovary, prostate, bladder, kidney, and brain. The total population surveyed was around 15 million (Table 1). The mean well water concentration in each township observed was used as the exposure measurement. In addition to analysing data from all 314 townships, separate analyses were conducted using data from the 170 southwestern townships only. A significantly higher mortality associated with higher well water arsenic levels was observed for liver cancer, nasal cavity cancer, lung cancer, skin cancer, bladder cancer, kidney cancer and prostatic cancer (Table 2). When the analyses were limited to 170 southwestern townships, in general, a greater incremental risk for each 0.1 ppm increase in mean well water arsenic concentration was observed in comparison with the estimates based on data from all 314 townships.

Another nation-wide ecologic analysis was conducted on bladder cancer and kidney cancer in 243 townships where individual well measurement reports from Lo *et al.*'s survey were available (Guo *et al.*, 1993). Based on the standard solutions used in the mercuric bromide stain method, arsenic concentrations were grouped into 10 categories: 0 ppm (undetectable), trace (between the 0.001 ppm detection limit and 0.01 ppm), 0.01 ppm, 0.02 ppm, 0.03-0.04 ppm, 0.05-0.08 ppm, 0.09-0.16 ppm, 0.17-0.32 ppm, 0.33-0.64 ppm, and > 0.64 ppm. Cancer incidence between 1980 and 1987 was assessed based on data from the National Cancer Registration Program, and standardised incidence rates (SIRs) were calculated for each township (Table 1). The population studied was about 11.4 million, and 1,972 bladder cancer and 726 kidney cancer cases (including 239 transitional cell carcinomas and 210 renal cell carcinomas) were observed. The data were analysed by regression models using a series of variables to describe exposure status. Each variable denoted the proportion of wells in a specific arsenic concentration category in each township. For bladder cancers, going from the lowest to the highest arsenic categories, significant positive associations with some categories were observed alternating with insignificant or significant negative associations. The highest exposure category (> 0.64 ppm) was associated the highest increased risks. For kidney cancers, an association was suggested for transitional cell carcinoma and high arsenic levels, but not renal cell carcinoma, suggesting a cell type specificity for the carcinogenicity of arsenic.

Summary of the epidemiologic studies

Characteristics and findings of the epidemiologic studies reviewed are summarised in Tables 1 and 2 . Studies conducted in the blackfoot disease endemic area are not totally independent; some of the residents were included in all the studies (Figure 1), especially those who lived in the four hyper-endemic townships (the townships with highest incidence of blackfoot disease). Therefore, it is not surprising that the results show some consistency. The two nationwide studies are not totally independent either; all the townships studied by Guo *et al.* were included in Chen and Wang's study, and the observation periods had four years overlapping (1980-1983). However, Guo *et al.* used incidence data, but Chen and Wang used mortality data.

A statistically significant association with arsenic ingestion was relatively consistently observed for cancers of skin, bladder, and kidney by different research groups using different study designs on different study populations. Associations of arsenic ingestion with lung cancer and liver cancer were also observed quite consistently by the same research group in different studies with different study designs and study populations using mortality data. However, no confirmatory studies by other researchers in Taiwan were available. The association with prostatic cancer was observed in two studies with similar designs and study populations by the same research team but was not confirmed by the same researchers using a different study design in a different study population. No statistically significant associations were observed for other cancers, including colon cancer, esophageal cancer, nasopharyngeal cancer, stomach cancer, uterine cervix cancer, cancer of small Intestine, rectal cancer, laryngeal cancer, cancers of bone and cartilage, breast cancer, ovarian cancer, brain cancer, and leukemia.

Most of the studies in other countries were inconclusive primarily due to the limited size of study populations (Bates *et al.*, 1992; US EPA, 1984). Nonetheless, positive associations were observed between arsenic ingestion and skin cancer, liver angiosarcoma, and sometimes urinary cancers, regardless of the sources of exposure (Bates *et al.*, 1992). Positive associations between arsenic exposure through inhalation, mostly occupational exposures, and lung cancer were also frequently observed (US EPA, 1984). However, the association between inhaled arsenic and urinary cancer was less frequently observed (Guo *et al.*, 1993), and associations between inhaled arsenic and other cancers were seldom observed (US EPA, 1984; Bates *et al.*, 1992).

In comparison with studies in other areas, studies in Taiwan provide the best available data on ingested arsenic and cancer. However, all the key studies conducted were ecologic studies, and the very few case-control or cohort studies did not provide actual arsenic exposure data (Chen *et al.*, 1986, 1988). Because of the limitations inherent in ecologic studies, such studies are generally used for hypotheses generating, rather than hypotheses testing. Studies from other countries have frequently observed associations between arsenic ingestion and skin cancer and liver angiosarcoma. These associations may therefore be considered well established. Associations of arsenic ingestion and other cancers are less consistently observed, so further studies with exposure data at the individual level, such as case-control or cohort studies, are necessary to confirm the hypotheses generated from studies in Taiwan. Indeed, such studies are needed for more accurate characterization of dose-response assessment of arsenic exposure and cancer of each site. Associations observed in epidemiologic studies, however, are not necessarily causal relationships. Confirmatory studies, such as animal experiments, are desirable.

Acknowledgements

The authors would also like to thank Dr. Richard R. Monson and Dr. Robert L. Bornschein for their help in conducting this review.

References

American Public Health Association. 1982. *Standard Methods of the Examination of Water and Waterwaste*, 15th edn. American Public Health Association: Washington, D.C.

Bates, M.N., Smith, A.H. and Hopenhayn-Rich, C. 1992. Arsenic ingestion and internal cancers: A review. *Am. J. Epidemiol.,* **135**, 462-475.

Brown, K.G. and Chen, C.-J. 1993. On the observed dose-response for internal cancers and arsenic in drinking water in the blackfoot disease endemic region of Taiwan. Presented at the International Conference on Arsenic Exposure and Health Effects, New Orleans, LA, 1993.

Chen, C.-J., Chung, Y.-C., Lin, T.-M. and Wu, H.-Y. 1985. Malignant neoplasms among residents of a blackfoot disease-endemic area in Taiwan: High-arsenic artesian well water and cancers. *J. Cancer Res.,* **45**, 5895-5899.

Chen, C.-J., Chung, Y.-C., You, S.-L., Lin, T.-M. and Wu H.-Y. 1986. A retrospective study on malignant neoplasms of bladder, lung, and liver in blackfoot disease endemic area in Taiwan. *Br. J. Cancer,* **53**, 399-405.

Chen, C.-J., Kuo, T.-L. and Wu, M.-M. 1988. Arsenic and cancers (letter). *Lancet,* **2**, 414-415.

Chen, C.-J., Wu, M.-M., Lee, S.-S., Wang, J.-D., Cheng S.-H. and Wu, H-Y. 1988. Atherogenicity and carcinogenicity of high-arsenic artesian well water: Multiple risk factors and related malignant neoplasms of blackfoot disease. *Arteriosclerosis,* **8**, 452-460.

Chen, C.-J. and Wang, C.-J. 1990. Ecological correlation between arsenic level in well water and age-adjusted mortality from malignant neoplasms. *Cancer Res.,* **50**, 5470-5474.

Chen, C.-J., Chen, C.-W., Wu, M.-M. and Kuo, T.-L. 1992. Cancer potential in liver, lung, bladder and kidney due to ingested inorganic arsenic in drinking water. *Br. J. Cancer,* **66**, 888-892.

Chen, K.-P. and Wu, H.-Y. 1962. Epidemiologic studies on blackfoot disease 2: A study of source of drinking water in relation to the disease. *J. Formosan Med. Assoc.,* **61**, 611-618.

Chiang, H.-S., Guo, R.-H., Hong, C.-L., Lin, S.-M. and Lee, E.F. 1993. The incidence of bladder cancer in the black foot disease endemic area in Taiwan. *Br. J. Urol.,* **71**, 274-278.

Guo, H.-R., Chiang, H.-S., Hu, H., Lipsitz, S.R. and Monson R.R. 1993. Arsenic in drinking water and urinary cancers. Presented at the International Conference on Arsenic Exposure and Health Effects. New Orleans, LA, July, 1993.

Kao, T.-M., Kao, S.-J. and Hou, S.-T. 1954. Spontaneous gangrene and chronic potassium poisoning. Presented at the 46th Annual Meeting of the Formosan Medical Association, Taipei, Taiwan, 1954.

Kuo, T.-L. 1968. Arsenic content of artesian well water in endemic area of chronic arsenic poisoning. *Repts. Inst. Pathol. Natl. Taiwan Univ.,* **20**, 7-13.

Lo, M.-C., Hsen, Y.-C. and Lin K-K. 1977. *Second Report on the Investigation of Arsenic Content in Underground Water in Taiwan.* Taichung, Taiwan: Taiwan Provincial Institute of Environmental Sanitation.

Natelson S. 1957. *Microtechniques for the Routine Laboratory.* Charles C. Thomas, Springfield Il., USA.

Smith, A.H., Hopenhayn-Rich, C., Bates, M.N., Goeden, H.M., Hertz-Picciotto, I., Duggan, H.M., Wood, R., Kosnett, M.J. and Smith, M.T. 1992. Cancer risks from arsenic in drinking water. *Environ. Res.,* **97**, 259-267.

Tsai, S.-F. 1987. Ecological Correlation Study on Major Risk Factors of Various Malignant Neoplasms in Taiwan; 1972-1983. Master's Thesis, National Taiwan University.

Tseng, W.-P., Chu, H.-M., How, S.-H., Fong, J.-M., Lin, C.-S. and Yeh'S. 1968. Prevalence of skin cancer in an endemic area of chronic arsenicism in Taiwan. *J. Nat. Cancer Inst.*, **40**, 453-463.

Tseng, W.-P. 1977. Effects and dose-response relationships of skin cancer and blackfoot disease with arsenic. *Environ. Health Perspect.*, **19**, 109-19.

US EPA. 1984. *Health Assessment Document for Inorganic Arsenic: Final Report.* Washington, D.C.: United States Environmental Protection Agency. (EPA publication EPA-600/8-83-021F).

Wang, J.-D., Sheng, C.-Y., Yang, H.-L., Wu, T.-Y. and Lin, K.-K. 1989. Relationship between arsenic in drinking water and mortality rates of bladder cancer and renal cancer. *Reports on Blackfoot Disease Research,* **31**, 21-35.

Wu, M.-M., Kuo, T.-L., Hwang, Y.-H. and Chen C-J. 1989. Dose-response relation between arsenic concentration in well water and mortality from cancers and vascular diseases. *Am. J. Epidemiol.,* **130**, 1123-1132.

Yeh, S. 1963. Relative incidence of skin cancer in Chinese in Taiwan: with special reference to arsenical cancer. *Natl. Cancer Inst. Monogr.,* **10**, 81-107.

Yeh, S. and How, S.-W. 1963. A pathological study on the blackfoot disease in Taiwan. *Repts. Inst. Pathol. Natl. Taiwan Univ.*, **14**, 25-73.

14 Review of health assessments for US/Canada populations exposed to arsenic in drinking water

Jane L. Valentine

Department of Environmental Health Sciences, School of Public Health, University of California, Los Angeles, Los Angeles, California 90024-1772, USA

Abstract

A review was conducted of 11 population studies for health assessments to arsenic exposure. Arsenic concentrations in the water of exposed populations were all greater than 1.6 times the current US EPA drinking water standard of 50 μg L^{-1}. The maximum arsenic exposure for one population study was 2,450 μg L^{-1}. Variations in numbers of persons for an exposure gradient existed. Most notable were the highest elevations where few people were exposed making statistical assessments of health effects impossible. Greater numbers of individuals were studied for levels of exposure ranging from approximately 180 to 390 μg L^{-1}. Even in these instances population groups were all less than 300 persons.

Arsenic in the water supplies was found to contain a mixture of pentavalent and trivalent arsenic, the latter existing at 14 to 50%.

Health effects noted in the studies reviewed were increases in levels of arsenic in body samples, blood, hair, and urine. Increases of as much as 5 times in blood and 20 times in urine occured above documented amounts for unexposed population groups. Health effects associated with these increases were assessed through questionnaire response, physician-performed dermatologic and neurologic examinations. Seven out of the eight studies evaluating symptom response by questionnaire found no health effects. One study suggested that clinical signs and symptoms showed a pronounced increase when hair arsenic concentrations increase above 1 ppm (7 times the amount of arsenic in hair of unexposed populations). Dermatologic evaluations were made in four of the 11 community reports and no overt adverse conditions of the skin were noted. Peripheral neurologic examinations were reported for 4 of the community-based studies. Adverse effects were reported in two of the studies. Arsenic effects were most pronounced in the ulnar and sural sensory nerves.

Introduction

Health effects of arsenic exposure via drinking water have been documented in North America. Most recently Franzblau and Lilis (1989) presented findings of a variety of health problems including gastrointestinal disturbances, central and peripheral neurological effects, and skin rash in a male and female (husband and wife) living in upstate New York. Their drinking water contained 9,000 to 10,900 $\mu g\ L^{-1}$ arsenic. Urinary arsenic concentrations were reported to be 2,260 $\mu g\ L^{-1}$ after a three month exposure in the female and 488 $\mu g\ L^{-1}$ (random specimen) in the male. These concentrations were approximately 50 to 230 times the normal excretion of less than 10 $\mu g\ L^{-1}$ for non-arsenic-exposed persons. The male reportedly consumed less drinking water. An earlier report by Feinglass (1973) documented adverse health outcomes in 11 out of 13 exposed persons in western Minnesota. Well water used for drinking contained 11,800 to 21,000 $\mu g\ L^{-1}$. Similar findings of gastrointestinal disturbances and neuropathy were made, with the number and severity related to the amount of water consumed.

The above two reports were for family units, where one of the family units included a small group of employees living on their property. Larger population studies have been conducted using subsets of populations from major communities exposed to arsenic in their drinking water either from central reservoirs or individual wells. In these studies increased levels of arsenic in the drinking water resulted in increased amounts in biomedical tissue. Documentation of increased concentrations in blood, hair, and urine has been accompanied by searches for changes in health outcomes, including such diseases as neurological disorders (Southwick *et al.*, 1983; Valentine *et al.*, 1981; Hindmarsh *et al.*, 1977; Kreiss *et al.*, 1983), development of skin diseases (Harrington *et al.*, 1978; Southwick *et al.*, 1983; Valentine *et al.*, 1985, 1991, 1992), and development of internal cancers and adverse non-cancerous effects to internal organs and major systems of the human body (Vig *et al.*, 1984; Hindmarsh *et al.*, 1977; Southwick *et al.*, 1983 and Valentine *et al.*, 1985, 1992).

Consistent findings of adverse effects have not always been recorded. However, exposure levels in drinking water have varied, the duration of exposure was possibly not similar, and lifestyles may have varied with differences in occupation, nutrition etc. We present here a review of major studies conducted in the USA and Canada. The significance of such studies is related to those reported from other countries where lifestyle and experience have major differences and some similarities to the USA/Canadian experience.

Methodologies in reviewed studies

Eleven reports of population studies were reviewed (see Table 1). These studies were compared for water concentration exposures, body burden measurements, and health outcomes.

Laboratory methodologies

Various analytical methods were employed in the health studies reviewed. Analytical methodology employed in water and body burden measurements were atomic absorption spectrophotometry hydride generation (Valentine *et al.*, 1985, 1992, Southwick *et al.*, 1983; Harrington *et al.*, 1978), colorimetric method using silver diethyldithiocarbamate (Vig *et al.*, 1984; Goldsmith *et al.*, 1972; Hindmarsh *et al.*, 1977). Some reports did not state the analytical methodology used (Feinglass, 1973; Franzblau and Lilis, 1989). The two major analytical methods, atomic absorption spectrometry with hydride generation and silver diethyldithiocarbamate colorimetric procedure, had reproducibility of ± 5% of an EPA reference trace-metal standard (Kang and Valentine, 1977, Valentine *et al.*, 1979) and 12% for coefficient of variation (Hindmarsh *et al.*, 1977) respectively.

Questionnaire response

In the major health studies differing modes of questionnaire administration were employed to assess health status. Valentine *et al.* (1992) commented on the validity of mailed vs. interviewer-administered questionnaires. They state that improved data can be obtained by trained interviewers. However, self-administered mailed questionnaires do retain validity while being less costly. These methods could have a bearing on the quality of the data obtained.

Questionnaires were administered by trained interviewers in several studies (Valentine *et al.*, 1985, 1991; Hindmarsh *et al.*, 1977). Also, health responses to arsenic exposure were assessed by mailed and by self-administered questionnaires (Valentine *et al.*, 1992; Harrington *et al.*, 1978). In each case, questions were asked to elicit responses concerning dermatologic complaints, weight loss, fatigue, muscle weakness or wasting, emotional problems, tumors, major organ diseases, sensory and motor changes, socioeconomic status, and length of residence. In some studies the frequency of consumption of certain foods was elicited (Valentine and Faraji, 1986). In two, the assessment of 24-hour dietary recall was taken (Valentine *et al.*, 1993, 1994). A few of the studies employing questionnaire response to assess health status did not state how questionnaires were given (Vig *et al.*, 1984; Goldsmith *et al.*, 1972).

Clinical examinations

Physician-performed physical examinations were conducted to assess dermatologic conditions (Harrington *et al.*, 1978; Southwick *et al.*, 1983; Valentine *et al.*, 1989; Hindmarsh *et al.*, 1977) and neurologic examination for peripheral neuropathy or nerve conduction measurements (Harrington *et al.*, 1978; Valentine *et al.*, 1981; Southwick *et al.*, 1983; Hindmarsh *et al.*, 1977).

Nerve conduction measurements were made using surface electrodes for stimulating and recording. The instruments used for the measurements included a portable "Neurodiagnostic" machine (Valentine *et al.*, 1981), a Grass S44 stimulator using a Tektronix No. 2A61 preamplifier for muscle and nerve potentials and measuring using Tektronic No. 564B storage oscilloscope (Hindmarsh *et al.*, 1977), or a TECA4 (White Plains) electromyograph (Kreiss *et al.*, 1984; Southwick *et al.*, 1983). One study stated that the temperature of the room air was kept between 23° and 25°C (Hindmarsh *et al.*, 1977). One other report stated that the extremities were maintained at or above a temperature of 32°C (Valentine *et al.*, 1981). Southwick *et al.* (1983) reported that nerve temperatures were not above 30°C in all cases and that

Table 1 *Studies considered in this review.*

Study	Location	Duration	Concentration	Population Characteristics
Case Studies				
Feinglass (1973)	Western Minnesota	About 10 to 20 weeks	Well water with 11,800 μg L^{-1} to 21,000 μg L^{-1} due to grasshopper bait	13 exposed and 12 non-exposed
Franzblau and Lilis (1989)	Upstate New York	Once or twice each week for 2 months, one to two glasses of the well water, then 8 to 10 glasses for one additional month	9,000 μg L^{-1} to 10,900 μg L^{-1}	One female age 43 and one male age 45 who consumed fewer glasses
Community Studies				
Goldsmith *et al.* (1972)	Lassen County, California	(not given)	Individual wells, ≤10 μg L^{-1} to 1,400 μg L^{-1}	72 households of 171 adults from which 144 questionnaires, 108 hair samples, and 98 combined questionnaire, hair sample and water sample
Harrington *et al.* (1978)	Ester Dome, Alaska	85% of persons lived in area less than 10 years	Home wells, 1 μg L^{-1} to 2,450 μg L^{-1}, mean of 224 μg L^{-1}	119 exposed, 113 less exposed (< 50 μg L^{-1})
Hindmarsh *et al.* (1977)	Waverly, Nova Scotia	(not given)	< 50 μg L^{-1} to 1,400 μg L^{-1}	92 exposed, 21 controls
Kreiss *et al.* (1983)	Ester Dome, Alaska	At least two years, 74 month average	Home wells, 1 μg L^{-1} to 478 μg L^{-1}, mean of 347 μg L^{-1}	147 adults aged 17 to 57 (mean 36), 43.5% male

Table 1 *continued*

Study	Location	Duration	Concentration	Population Characteristics
Southwick *et al.* (1983)	West Millard County, Utah	At least five years	Two exposed towns: 180 μg L^{-1} and 270 μg L^{-1}, and 20 μg L^{-1} low-exposed control	145 exposed, 105 controls, 53% to 55% female, predominantly Mormon lifestyle
Valentine *et al.* (1979 and 1985)	Exposed: Edison, California; Fallon, Hidden Valley, Virginia Foothills, Nevada; controls: Sun Valley, Nevada, and Casper, Wyoming	At least one year, 4.3 year to 8.7 year average in exposed communities	Exposed: 51 μg L^{-1} to 393 μg L^{-1} (where 51 μg L^{-1} was only lowered from 100 μg L^{-1} four months prior); controls: 0.6 μg L^{-1} to 2.3 μg L^{-1}	131 exposed, 138 controls
Valentine *et al.* (1981)	(Same as 1979)			130 exposed, 95 controls, aged 8 to 81 years
Valentine *et al.* (1993)	Edison, California; northern Mexico	From 1962 or earlier to 1985, 16 years average residence to 1986-87	393 μg L^{-1} before 1985	6 males, 9 females in US, 21 males, 26 females in Mexico
Vig *et al.* (1984)	Exposed: Fallon, Nevada; control: Reno, Nevada	At least five years	Exposed: 50 μg L^{-1} to 2,275 μg L^{-1} mean 100 μg L^{-1} controls: 12 μg L^{-1}	105 exposed, 46.7% female; 86 controls, 41.9% female

Table 2 Hair *vs.* tapwater mean arsenic concentrations for 14 studied community samples

Tapwater As (mg L^{-1})	Hair As (μg g^{-1})	Community	Study
0.002	<0.10	Sun Valley, NV	Valentine *et al.*, 1985
<0.006	0.15	Fairfax, CA	Valentine *et al.*, 1979
0.008	0.06	Edison, CA	Valentine *et al.*, 1991
0.015	0.20	Fairfax, CA	Valentine *et al.*, 1991
0.08	0.51	Ester Dome, AK	Harrington *et al.*, 1978
0.10	0.48	Virginia Foothills, NV	Valentine *et al.*, 1979
0.10	0.57	Fallon, NV	Valentine *et al.*, 1979
0.10	0.6	Nova Scotia, Canada	Hindmarsh *et al.*, 1977
0.12	0.50	Hidden Valley, NV	Valentine *et al.*, 1979
0.18	1.21	Hinkley, UT	Southwick *et al.*, 1983
0.27	1.09	Deseret, UT	Southwick *et al.*, 1983
0.39	1.16	Edison, CA	Valentine *et al.*, 1983
0.39	3.0	Nova Scotia, Canada	Hindmarsh *et al.*, 1977
0.40	3.29	Ester Dome, AK	Harrington *et al.*, 1978

corrections were applied to measurements on participants with low temperatures. The physician who performed neurological examination in the Southwick *et al.* (1983) study also screened for threshold of touch, hypalgesia, temperature sense, and vibratory sense. Harrington *et al.* (1978) did not state how neurologic examinations were performed. One of the nerve conduction reports gave no indication of temperatures of extremities or the ambient temperature in which measurements were made (Kreiss *et al.*, 1983).

Results

Arsenic concentrations in water supplies

The reported tapwater concentrations of arsenic in the exposed populations were all greater than 1.6 times the current allowable drinking water standard for arsenic of 50 μg L^{-1}. The maximum arsenic exposure in a large population study was 2,450 μg L^{-1} (49 times the current EPA standard) (Harrington *et al.*, 1978). As individual wells were evaluated in the Harrington *et al.* studies (1978) there is no indication of the number of persons exposed at each concentration gradient and consequently to the highest level, 2,450 μg L^{-1}. Hindmarsh *et al.* (1977) in his study of water arsenic concentrations in the Canadian population gave a range of exposure, from < 50 to 1,400 μg L^{-1}, for individual well measurements. Similarly for the studies of Goldsmith *et al.* (1972), individual well water arsenic concentrations ranged from ≤ 10 to 1,400 μg L^{-1}. Thus although large population numbers were employed exposures to individual concentrations were for small population groups (possibly family units). The studies of Southwick *et al.* (1983) and of Valentine *et al.* (1979, 1985) report values of drinking water concentrations on far greater numbers of

Table 3 Urine *vs*. tapwater mean arsenic concentrations for 11 studied community samples.

Tapwater As (mg L^{-1})	Urine As (µg L^{-1})	Community	Study
<0.006	8.4	Fairfax, CA	Valentine *et al.*, 1979
0.008	9.6	Edison, CA	Valentine *et al.*, 1991
0.016	18.7	Fairfax, CA	Valentine *et al.*, 1991
0.10	45.0	Ester Dome, AK	Harrington *et al.*, 1978
0.10	48.5	Virginia Foothills, NV	Valentine *et al.*, 1979
0.10	54.2	Hidden Valley, NV	Valentine *et al.*, 1979
0.10	61.5	Fallon, NV	Valentine *et al.*, 1979
0.18	175	Hinkley, UT	Southwick *et al.*, 1983
0.27	211	Deseret, UT	Southwick *et al.*, 1983
0.39	176	Edison, CA	Valentine *et al.*, 1983
0.40	170	Ester Dome, AK	Harrington *et al.*, 1978

exposed persons. The concentrations of arsenic in these studies were much lower. Drinking water arsenic concentrations were reported at 180 and 270 µg L^{-1} in the two exposed communities in Utah (Southwick *et al.*, 1983). The studies conducted in the western United States by Valentine *et al.* (1979, 1985) gave approximate water arsenic concentrations ranging from 100 to 390 µg L^{-1} with three of the four exposed communities receiving the lower concentration. Populations studied where exposure to these concentrations occurred in Utah and in Nevada and California were all less than 300 persons.

The form of arsenic in these water supplies was assessed and found to contain pentavalent arsenic. However, trivalent arsenic was found to exist in most of these supplies at 14 to 50%.

Body burden measurements

In all the studies increases in arsenic concentrations in the body were found with increases in arsenic concentration in the drinking water. Since not all studies utilized the same biomedical tissues, some comparisons are difficult to make. Hair and urine showed to be the universal biomedical tissue and fluid in the studies employed. A comparison of hair concentrations to water concentrations in the studies is given in Table 2 and shown in Figure 1. These comparisons all show the body burden arsenic concentration increase with exposure increase. A regression of hair arsenic concentration with water arsenic concentration produced a correlation coefficient of 0.87 for n=12. Two of the comparisons were omitted due to one of the concentrations of a variable in each set being at less than detection limits. The correlation coefficient was statistically significant ($p < 0.001$). However, the degree of increase in the tissue between studies is variable which may reflect differences in sample preparation, hair segments used, analytical measurements, amount of water consumption, and possibly nutritional status.

Urinary excretion values were also compared (Table 3 and Figure 2). The values agreed much closer and seemingly the μg day^{-1} excretion values were more in agreement if the values of Southwick *et al.* (1983) were ignored. Correlation coefficients of 0.87 and 0.99 were obtained with and without the data of Southwick respectively. These coefficients were statistically significant ($p < 0.001$). The manuscript of Southwick did not state if 24-hour urine samples were collected. Harrington *et al.* (1978) reported collection of early morning urines. The studies of Valentine *et al.* (1979 collected 24-hour urine samples for arsenic determination seemingly providing a more accurate reflection of excretion. Of particular note is the water arsenic concentrations and associated urinary output for persons with very low arsenic exposure via drinking water. The water arsenic concentrations for Sun Valley, Casper, and Fairfax were 2.3, 0.6, and < 6.0 μg L^{-1} respectively. Urinary arsenic values for these participants were 4.6, 4.0, and 10.9 μg day^{-1} (Valentine *et al.*, 1985).

Harrington *et al.* (1978) and Valentine *et al.* (1979) give values typifying an excretion in urine of 28 to 38% of ingested arsenic. These latter values approximate the excretion percentages given by the monograph of the toxicological profile of arsenic developed by the ATSDR (1989). They state excretion percentages of 25% of the dose when citing the data of Mappes (1977) for a single dose of sodium arsenite. No fecal measurements of arsenic in any of the drinking water studies were performed.

Valentine *et al.* (1979, 1985) measured arsenic concentrations in blood. These values were found to average 0.39 μg/100 mL for three community groups of participants receiving approximately 100 μg L^{-1} arsenic in drinking water. Their ranges were 0.29 to 0.51 μg/100 mL in blood. Unexposed persons had levels of <0.1 μg/100 mL for two specifically matched communities of Sun Valley and Casper. One other community of participants from Fairfax near Edison, CA, had measured arsenic levels in blood of 0.49 μg/100 mL.

Although there was a deviant exposure response for the Fairfax population, all body tissues and fluids seem to give increasing concentrations reflecting the increase in exposures via drinking water. Due to the ease of collection and large amounts obtained, urinary arsenic seems to be the most useful for biological monitoring.

Health status resulting from arsenic exposure

The health status of arsenic-exposed individuals in the US/Canadian studies has been assessed through questionnaire responses and through the use of physician-performed examination. Questionnaire responses showed no effect from arsenic exposure with regard to adverse changes in health (Valentine *et al.*, 1985; Valentine *et al.*, 1992; Valentine *et al.*, 1991; Vig *et al.*, 1984; Harrington *et al.*, 1978; Goldsmith *et al.*, 1972; Southwick *et al.*, 1983). A significant positive correlation between arsenic exposure and the frequency of clinical findings was found by Hindmarsh *et al.* (1977) for his study of 92 persons. Hindmarsh suggests 1 ppm in hair as a warning level above which health effects due to arsenic exposures may become prevalent. Fourteen percent of those persons in his study with levels of arsenic in hair less than 1 ppm had symptoms and/or clinical signs *vs.* 38% with symptoms and/or clinical signs in those with greater than 1 ppm. Other studies did not combine symptoms and clinical examination results as one single indicator. If these comparisons had been made, the relationships in these studies may have been similar to those reported by Hindmarsh. A separate evaluation with only symptoms of arsenic exposure was not done in the Hindmarsh study. The overall findings of no health symptoms and signs in seven out of the eight reports of community studies where such studies were conducted by questionnaire differ from the two case studies of Feinglass (1973) and Franzblau and Lilis (1989), where symptoms of gastrointestinal disturbance and nail changes (Mees lines) were noted. This was most likely due to the high arsenic concentration in the water as they had drinking water exposures to arsenic at least 22 times higher than the community-based studies.

Dermatologic evaluations for skin disorders have been evaluated in four out of the 11 community reports. No overt adverse conditions of the skin were noted. No overt conditions of hypo- or hyperpigmentation, keratosis, or skin cancer were reported. These studies conducted by Hindmarsh *et al.* (1977), Harrington *et al.* (1978), and Valentine *et al.* (1991) revealed no increased prevalence of abnormalities in the skin after dermatologic physicians performed examinations. Southwick *et al.* (1983), however, reported twice the percent of specific dermatologic signs found in exposed (6.21% of 145) versus controls (2.86% of 105) from clinical examinations. These signs of arsenic ingestion included palmar and plantar keratosis, diffuse palmar and plantar hyperkeratosis, diffuse pigmentation, and arterial insufficiency. The numbers were small however. No significant associations of the arsenic skin toxicity signs were found with annual arsenic dose.

Peripheral neurologic examinations were conducted by Valentine *et al.* (1981), Kreiss *et al.* (1983), Harrington *et al.* (1978), Hindmarsh *et al.* (1977), and Southwick *et al.* (1983). In each of the studies, nerve conduction measurements were made to select nerves in the arm or leg. Nerve conduction slowing was not found in the studies of Southwick *et al.* (1983), Kreiss *et al.* (1983), or the studies of Harrington *et al.* (1978). However, decreased nerve conduction velocities were noted in the studies of Valentine *et al.* (1981) and Hindmarsh (1977). Valentine *et al.* (1981) reports that in their study most values were within the range of normal. However, within that range arsenic-exposed participants had nerve conduction measurements that were slower than those determined on control populations. They state that the differences were most prevalent in the measurement of ulnar motor and ulnar sensory latency. A comparison of percentages of measurements outside the UCLA normal ranges done for arsenic-exposed vs. nonexposed groups showed significant difference for the ulnar sensory nerves (Valentine *et al.*, 1981). A larger proportion of exposed females showed a non-significant sural sensory slowing (29.2% of exposed *vs.* 14.6% of control). Goldstein *et al.* (1975) has stated that it is thought that lead is more apt to produce a motor neuropathy, whereas other metals produce a symmetrical sensory or mixed sensorimotor neuropathy that affects the lower more than the upper limbs. Hindmarsh *et al.* (1977) determined that 10 of 33 people with high arsenic content wells had abnormal electromyographic findings whereas none of the 12 persons with arsenic exposure below 50 $\mu g\ L^{-1}$ had abnormal findings. For this manuscript, the data of Valentine *et al.* (1981) was reevaluated and grouped into all exposed and nonexposed for each of the nerve conduction

Table 4 *A comparison of nerve conduction measurement for arsenic-exposed and control populations.*

	Valentine *et al.*, 1981			Southwick *et al.*, 1983		
	Exposed (*n*)	Control (*n*)	*p*-value	Exposed (*n*)	Control (*n*)	*p*-value
Ulnar M. L.	2.93±0.41(130)	2.60±0.36(95)	<0.0001	2.53±0.59(38)	2.29±0.31(24)	0.08
Ulnar C. V.	58.3±5.35(130)	60.0±5.43(95)	0.10	64.1±8.5(38)	63.8±8.2(23)	>0.5
Ulnar S. L.	3.26±0.46(130)	2.94±0.36(95)	<0.0001	2.73±0.45(67)	2.63±0.33(58)	0.17
Peroneal M. L.	4.81±0.74(130)	4.62±0.80(95)	0.07	3.97±0.92(81)	3.99±0.90(67)	>0.5
Peroneal C. V.	50.8±4.74(130)	50.8±5.74(95)	1.0	52.7±8.4(81)	52.3±4.6(67)	>0.5
Sural S. L.	4.08±0.37(130)	3.90±0.55(95)	0.004	–	–	–

measurements made (Table 4). In each case nerve conduction measurements were slower in the exposed when compared to control measurements with the exception of the measurements made on the peroneal nerve where no difference was noted. These data of Valentine *et al.* (1981) when compared to those of Southwick *et al.* (1983) were similar in magnitude of the response but no difference was noted between exposed and control in the Southwick measurements.

Southwick *et al.* (1983) found no statistically significant differences in the nerve conduction measurements made for comparisons between exposed and non-exposed participants. Sural sensory nerve was most often affected in the Southwick studies. In the regrouping of the data from the paper of Valentine *et al.* (1981), the statistically significant differences were determined for some nerves when comparing all the exposed individuals to the combined controls. Ulnar motor latency, ulnar sensory latency, and sural sensory latency showed effects in response to arsenic exposure significantly different from those same nerves of control individuals. Valentine *et al.* (1981) in the same paper for differently grouped data found significant differences for ulnar sensory latency. The trend of adverse effect in the sural sensory nerve was there in females but was not statistically significant. This latter comparison was made on proportions of individuals in the exposed and control groups showing nerve conduction responses outside the normal range for separately evaluated males and females. The researchers concluded that low arsenic exposure may have an adverse affect on nerve conduction.

Discussion

Arsenic exposure via drinking water has been related to health effects in various studies in the US and Canada (Harrington *et al.*, 1978; Hindmarsh *et al.*, 1977; Valentine *et al.*, 1981, 1979, 1985, 1992, 1991; Goldsmith *et al.*, 1972; Southwick *et al.*, 1983; Kreiss *et al.*, 1984; Vig *et al.*, 1983). However in most of the reports, significant association to effects on skin, internal organs, cancer, and various clinical signs and symptoms have not been noted.

The above studies were conducted on community population groups experiencing arsenic exposure through drinking water at concentrations approximately 2 to 8 times the current limit of 50 $\mu g\ L^{-1}$. Case studies reported by Feinglass (1973) and by Franzblau and Lilis (1989) report consistent symptoms of gastrointestinal disturbances, nail changes, skin rash, and neurological involvement when drinking waters consumed contained 9,000 to 21,000 $\mu g\ L^{-1}$. These case study reports are for arsenic concentrations in tap water at 200 to 400 times the current EPA standard of 50 $\mu gAs\ L^{-1}$.

The maximum arsenic exposure for one population study was 2,450 $\mu g\ L^{-1}$ (Harrington *et al.*, 1978). Variations in numbers of persons for an exposure gradient existed. Most notable were the highest elevations where few people (possibly only family units) were exposed in the population studies making statistical assessments of health effects impossible. When greater numbers of individuals were employed levels of arsenic exposure ranged from approximately 180 to 390 $\mu g\ L^{-1}$ in drinking water. Even in these instances population groups were all less than 300 persons.

Various researchers have noted increases of arsenic in body samples, blood, hair, and urine. Increases of as much as 5 times in blood and 20 times in urine above documented amounts for unexposed population groups (Valentine *et al.*, 1979, Harrington *et al.*, 1978, Hindmarsh *et al.*, 1977, Southwick *et al.*, 1983).

Health effects associated with these increases were assessed through questionnaire response, physician-performed dermatologic and neurologic examinations. The generally low level of symptom and disease reporting in the US studies for questionnaire response and physician-performed examinations have differed from other reports outside of North America with the exception of the report by Cebrian *et al.* (1983) for studies conducted in Northern Mexico. The studies from northern Mexico (Cebrian *et al.*, 1983), the southwest coast of Taiwan (Tseng *et al.*, 1968; Chen *et al.*, 1985, 1986) have indicated skin disorders of hypo- and hyperpigmentation, skin cancer, and ischemia leading to gangrene development due to arsenic exposure from drinking water. Chen *et al.* (1985, 1986) have also documented cases of cancer of internal organs in Taiwanese populations exposed to arsenic in the endemic area studied earlier by Tseng *et al.* (1968). Such findings are not noted in the North American studies evaluated for similar arsenic exposure in drinking water supplies. However, differences in lifestyle for these populations in Mexico and Taiwan may exist. This difference could relate to differences in nutrient intakes, water consumption patterns, and certainly the overall exposure due to the more significant mobility of most US/Canadian populations.

Valentine *et al.* (1994) has evaluated the nutritional status of populations in two towns of northern Mexico in the arsenic endemic area. All measurements were made using the 24-hour dietary recall methodology. This methodology employed the use of Food Processor II Nutrition Diet Analysis System to calculate the nutrient intake for foods entered on the 24-hour recall questionnaire. The diets of the Mexican residents were found to consist of beans, tortillas, eggs, milk, cheese, potatoes, and coffee. Their intakes are similar for selenium, protein, zinc, and calories. These nutrients were also similar to intakes experienced by residents of Edison California where similar arsenic intakes were experienced. However, a difference in Vitamin A intake was noted for Mexican populations (exposed and not exposed) when compared to a population of residents from Edison, California. Vitamin A intake of Mexican residents was determined to be at 1/3 the US RDA requirement. Edison participants also consumed low amounts of Vitamin A but this consumption was determined to be at 2/3 the US RDA requirements.

The significance of Vitamin A and other nutrient intakes in protecting against the carcinogenic potential of arsenic is yet to be fully investigated. Differences in total grams of arsenic ingested and amount and duration of exposure may lead to a better understanding of arsenic and its effect on health. Although consistent findings in neurologic involvement have not been reported the two studies which reported possible effects indicate that caution in assessing exposure limits is justified.

References

Agency for Toxic Substances and Disease Registry (ATSDR). 1989. US Public Health Service. *Toxicological Profile for Arsenic.* Oak Ridge National Laboratory.

Cebrian, M.E., Albores, A., Aguilar, M. and Blakely, E. 1983. Chronic arsenic poisoning in the north of Mexico. *Hum. Toxicol.*, **2,** 121-133.

Chen, C.J., Chuang, Y.C., Lin, T.M. and Wu, H.Y. 1985. Malignant neoplasms among residents of a Blackfoot disease -- endemic area in Taiwan: High-arsenic artesian well water and cancers. *Cancer Res.,* **45,** 5895-5899.

Chen, C.J., Chuang, Y.C., You, S.L., Lin T.M. and Wu, H.Y. 1986. A retrospective study on malignant neoplasms of bladder, lung, and liver in a Blackfoot disease -- endemic area in Taiwan. *Br. J. Cancer,* **53,** 399-403.

Chen, C.J., Wu, M.M., Lee, S.S., Wang, J.O., Cheng, S.H. and Wu, H.Y. 1988. Atherogenicity and carcinogenicity of high-arsenic well water: Multiple risk factors and related malignant neoplasms of blackfoot disease. *Arteriosclerosis,* **8**, 452-460.

Feinglass, E.J. Arsenic intoxication from well water in the United States. 1973. *N. Engl. J. Med.,*_**288**, 828-830.

Franzblau, A. and Lilis, R. 1989. Acute arsenic intoxication from environmental arsenic exposure. *Arch. Environ. Health,* **44,** 385-390.

Goldsmith, J.R., Deane, M., Thom, J. and Gentry, G. 1972. Evaluation of health implications of elevated arsenic in well water. *Water Res.*, **6,** 1133-1136.

Goldstein, N.P., McCall, J.T. *et al. 1975.* In: Dyck, P.J., Thomas, P.K. and Lambert, E.H. (eds.), *Peripheral Neuropathy,* Vol. II. W.B. Saunders Co., Philadelphia.

Harrington, J.M., Middaugh, J.P., Morse, D.L.and Housworth, J. 1978. A survey of a population exposed to high concentrations of arsenic in well water in Fairbanks, Alaska. *Amer. J. Epidemiol.,*_**108**, 377-385.

Hindmarsh, J.T., McLetchie, O.R., Heffernan, L.P.M., Hayne, O.A., Ellenberger, H.A., McCurdy, R.F.and Thiebaux, H.G. 1977. Electromyographic abnormalities in chronic environmental arsenicalism. *J. Anal. Toxicol.,* **1**, 270-276.

Kang, H.K. and Valentine, J.L. 1977. Acid interference in the determination of arsenic by atomic absorption spectrophotometry. *Anal. Chem.*, **49,** 1829-1832.

Kreiss, K., Landrigan, P.J., Feldman, R.G., Niles, C.A., Chirico-Post, J., Sax, D.S., Boyd, M.H.and Cox, D.H. 1983. Neurological evaluation of a population exposed to arsenic in well water. *Arch. Environ. Health*, **38**,116-121.

Mappes, R. 1977. (Experiments on excretion of arsenic in urine) Versuche zur Ausscheidung von Arsen in Urin. *Int. Arch. Occup. Environ. Health*, **40**, 267-272.

Southwick, J.W., Western, A.E., Beck, M.M., Whitley, T., Isaacs, R., Petajan, J., and Hansen, C.D. 1983. An epidemiological study of arsenic in drinking water in Millard County, Utah. In: Lederer, William H. and Fensterheim, Robert J. (eds.) Arsenic: Industrial, Biomedical, Environmental Perspectives. Van Nostrand Rheinhold Co, New York.

Tseng, WP, Chu, H.M., How, S.W., Fong, J.M., Lin, C.S. and Yeh, S. 1968. Prevalence of skin cancer in an endemic area of chronic arsenicism in Taiwan. *J. Natl. Cancer Inst.*, **40,** 453-463.

Valentine J.L., Kang, H.K. and Spivey, G. 1979. Arsenic levels in human blood, urine, and hair in response to exposure via drinking water. *Environ. Res.*, **20**, 24-32.

Valentine, J.L., Campion, D.S., Schluchter, M.D. andMassey, F.J. 1981. Arsenic effects on human nerve conduction. In: McC. Howell, J., Hawthorne, J.M. and White, C.L. (eds), *Trace Element Metabolism in Man and Animals.* Australian Academy of Science,Canberra.

Valentine, J.L., Reisbord, L.S., Kang, H.K. and Schluchter, M.D. 1985. Arsenic effects on population health histories. In: Mills, C.F., Bremner, J. and Chesters, J.K. (eds.). *Trace Elements in Man and Animals - TEMA 5.* Commonwealth Agricultural Bureaux, Aberdeen.

Valentine, J.L. and Faraji, B. 1986. Nutritional status and toxic response: the development of a drinking water standard for arsenic. In: *International Symposium on Water-Related Health Issues.* American Water Resources Association. Bethesda, Maryland.

Valentine, J.L., Bennett, R.G., Borok, M.E. and Faraji, B. 1991. Environmental arsenic and skin toxicity. In: Momcilovic, B. (ed.), *Trace Elements in Man and Animals - Tema 7.* IMI. Institute for Medical Research and Occupational Health University of Zagreb. Zagreb, Yugoslavia.

Valentine, J.L., He, S-Y, Reisbord, L.S., and Lachenbruch, P.A. 1992. Health response by questionnaire in arsenic-exposed populations. *J. Clin. Epidemiol.*, **45**, 487-494.

Valentine, J.L., Cebrian, M.E., Garcia-Vargas, G.G., Faraji, B., Kuo, J., Gibb, H.G., Lachenbruch, P.A. 1994. Daily selenium intake estimates for residents of arsenic endemic areas. *Environ. Res.*, **64**:1-9.

Valentine, J., Cebrian, M., Faraji, B. and Garcia-Vargas, G. 1993. Dietary zinc intake in arsenic-endemic areas. In: Anke, M., Meissner, D.and Mills, C.F. (eds.). *Trace Elements in Man and Animals - TEMA-8.* Verlag Media Touristik. Gersdorf, Germany.

Vig, B.K., Figueroa, M.L., Cornforth, M.N. and Jenkins, S.H. 1984. Chromosome studies in human subjects chronically exposed to arsenic in drinking water. *Amer. J. Ind. Med.*, **65**, 325-338.

15 Observed dose-response for internal cancers and arsenic in drinking water in the blackfoot disease endemic region of Taiwan

Kenneth G. Brown* and Chien-Jen Chen

Kenneth G. Brown, Inc., P. O. Box 16608, Chapel Hill, NC 27516;Institute of Public Health, National Taiwan University, College of Public Health, Taipei, Taiwan, Republic of China

Abstract

Data from the largest ecologic study in Taiwan on the association of cancer mortality at several internal sites and arsenic concentration in drinking water from deep wells are evaluated using more detailed exposure classifications than previously published. Mortality data for cancer of the bladder, liver, lung, and skin, recorded between 1973 and 1986 in 60 villages in the southwest region of Taiwan endemic to blackfoot disease, are cross-classified by 5 year age intervals and 11 intervals for arsenic concentrations. With the possible exclusion of skin cancer (for which mortality rates are not a particularly good indicator), the observed dose-response curves exhibit considerable nonlinearity. The cancer mortality rates for bladder, liver, and lung are strongly associated with arsenic concentrations at exposure interval 0.05-0.09 mg L^{-1} and above. The dose-response from the 10 villages in the four exposure intervals below 0.05 is erratic. Closer examination of data for bladder cancer revealed that virtually all of the observed cancer mortality occurred in just two of the villages, which also accounted for the only significant prevalence of blackfoot disease among the ten villages. Comparison of bladder cancer mortality rates in the endemic area with Taiwan as a whole indicates a significant increase when all ten villages below 0.05 mg L^{-1} are included; when the two discrepant villages are excluded, the observed mortality rates are well below those of Taiwan. Although the sample size may be too small to conclude that there is no arsenic effect below 0.05 mg L^{-1} in the latter case, it may be noted that an absence of observable evidence is consistent with the outcome of studies on metabolism of arsenic in humans that suggest arsenic may be adequately methylated and detoxified below 0.05 mg L^{-1}. Further study, using the village as a sampling unit and comparing outcomes across all four cancer sites, may be instructive.

---

*To whom correspondence should be addressed.

Introduction

The US Environmental Protection Agency (EPA), in compliance with the Safe Drinking Water Act, is currently assessing the evidence on detrimental health effects of inorganic arsenic in drinking water as part of its requirements to set maximum contaminant level goals and to promulgate primary drinking water regulations for contaminants judged to have an adverse health effect. The current regulation of 0.05 ppm (=0.05 mg L^{-1}), set as a National Interim Primary Drinking Water Regulation in 1976 (EPA, 1976), became the National Drinking Water Regulation following an amendment to the Safe Drinking Water Act in 1986.

The EPA health assessment document (HAD) for inorganic arsenic of 1984 (EPA, 1984) addressed the risk of lung and skin cancer as endpoints from inhalation and ingestion, respectively. The primary source for the evidence of (nonmelanoma) skin cancer was from an epidemiologic study conducted in an area of Southwest Taiwan endemic to blackfoot disease (Tseng *et al.*, 1968; Tseng *et al.*, 1977). Disagreement within EPA over aspects of the 1984 HAD, particularly the magnitude of the estimates of risk, led the Risk Assessment Forum to convene a Technical Panel to address further the issues of concern (EPA, 1988). The panel's results, published as a special report of the Risk Assessment forum (EPA, 1988), include a re-analysis of the Tseng data with estimates of arsenic-induced skin cancer substantially lower than the prior EPA values (also see Brown *et al*, 1989). Based on the Risk Assessment Council's review of the Forum's report, the EPA administrator (Lee Thomas at that time) endorsed a cancer potency slope factor of 0.05 (mg L^{-1}) for calculation of lifetime risk of skin cancer from exposure to arsenic in drinking water.

At the time of the Forum's report, evidence was mounting of an association between ingested arsenic and the occurrence of cancer at internal organs (EPA, 1988; Appendix C). Many of the references were case reports, but two epidemiologic studies (Chen *et al.*, 1985; Chen *et al.*, 1986) were singled out for further attention. Tseng and coworkers (Tseng *et al.*, 1977) had noted that the most common cause of death in patients who had either skin cancer or blackfoot disease was reported to be carcinomas at various internal sites. Chen *et al.*, (1985) reported the results of continued investigation of the population of the endemic area first studied by Tseng. It was found that the mortality rates were elevated in the endemic area for cancer of the kidney, bladder, skin, lung, liver and colon. Chen *et al.*, (1986) reinforced the findings of arsenic-induced cancer in the lung, bladder, and liver, based on a case-control study. Both studies demonstrated a qualitative relationship between arsenic exposure and increased risk of cancer at several internal sites. The Forum's report (Appendix C) notes, however, that further details of the data were needed to assess the dose-response, specifically information on age and arsenic concentrations in the drinking water, and that EPA would attempt to obtain those additional data.

Those additional data of interest to EPA were first published in a letter to the *Lancet* (Chen *et al.,* 1988). Exposure concentrations were grouped into three categories, (<0.30, 0.30-0.59, >0.60 ppm of well-water concentration), consistent with the groupings used in the previous study (Tseng *et al.*, 1968). The authors note that various cancer mortality rates were significantly higher than for the general population of Taiwan, and that a significant dose-response relationship was observed between arsenic concentrations in drinking water from artesian wells and the mortality rates for cancers of the bladder, kidney, skin, prostate, lung, and liver. Smith *et al.* (1992) extrapolated the summary data in the Lancet letter to the USA and concluded that "the population cancer risks due to arsenic in US water supplies may be comparable to those from environmental tobacco smoke and radon in homes". The increased cost to reduce the arsenic concentrations in drinking water to a value below the current standard of 0.05 ppm could be quite high for the US as a whole, depending largely on the target concentration to be achieved. On the other hand, drinking water must be safe for consumption. Consequently, the problem is not whether long-term consumption of arsenic at high concentrations in drinking water increases the risk of cancer at some organ sites, but the magnitude of risk at relatively low concentrations found in drinking water in the USA and elsewhere.

It is common to obtain risk estimates by fitting a dose-response model from some family of distributions and then using the estimates or confidence bounds as guidance. That approach is not particularly well suited for our objectives because the outcome is dependent on the choice of model (the correct choice of model is unknown) and the approach itself is more descriptive of how well a selected model fits the data than it is informative. It is preferable, instead, to examine the observed dose-response curve by methods that are relatively assumption-free to see what information can be gained, possibly for fitting a suitable dose-response model at a later stage. The most fundamental question that needs to be addressed now, however, is whether the (true) dose-response curve is essentially flat up to some positive exposure concentration where it starts upward (as studies of arsenic metabolism might suggest) or whether the curve essentially starts upward at the origin, as often assumed for cancer endpoints based on certain paradigms of cancer mechanisms. Other sources of information, *e.g.*, epidemiologic data that describe observational relationships between cancer and exposure but not the mechanism of occurrence, also need to be examined since coherence across sources of evidence is an important determinant of the strength of a claim based on observational studies. For this purpose, the major epidemiologic studies on ingested arsenic and their outcomes are briefly reviewed in the next section. Attention is then focused on an analysis of the Chen ecologic study (to be referred to as the Taiwan study) to see what may be learned about the dose-response relationship at low arsenic concentrations in Taiwan.

An overview of epidemiologic study results

The epidemiologic studies of consumption of arsenic in drinking water have been reviewed in several sources, *e.g.*, Fielder *et al.* (1986), Hendricks (1992), EPA (1993), Petito and Beck (1990). There are areas in Chile, Argentina, Mexico, and Taiwan where the substrata contain high levels of arsenic that contaminate artesian wells used as a source of drinking water. A high incidence of dermatological lesions characteristic of arsenic toxicity (*e.g.*, hyperpigmentation and hyperkeratosis) has been found in populations where the arsenic concentration in drinking water was 0.3 to 0.4 ppm or above, but no evidence of arsenic toxicity have been observed at concentrations of 0.2 ppm or below (Fielder et al., 1986). A similar conclusion was reached by Hendricks (1992) who observes that in the studies from Taiwan, Chile and Mexico, where evidence of arsenic toxicity included cutaneous lesion, keratosis, and hyperpigmentation, and arsenic was associated with various cancers, the mean arsenic concentrations ranged from 0.4 to 0.6 ppm.

Table 1 *Standardised Mortality Ratio (SMR), bladder cancer 11 exposure classifications.*

	Females			Males		
Arsenic exposure (ppm)	Est.	95% Confidence interval	Signif. (*p*-value)	Est.	95% Confidence interval	Signif. (*p*-value)
Level 1 (<0.01)	0.00	0.0, 31.0	>0.2	0.00	0.0,16.4	>0.2
Level 2	22.26	8.9, 45.9	<<0.001	2.03	0.03, 11.3	>0.02
(0.01-0.019)	**0.00**	**0.0, 15.7**	**>0.2**	**0.00**	**0.0, 9.5**	**>0.2**
Level 3 (0.02-0.029)	0.00	0.0, 21.1	>0.2	0.00	0.0, 14.1	>0.2
Level 4	26.16	5.3, 76.4	<<0.001	9.15	1.03, 33.0	0.05
(0.03-0.039)	*	*	*	*	*	*
Level 5 (0.04-0.049)	0.00	0.0, 6.5	>0.2	1.01	0.01, 5.6	>0.2
Level 6 (0.05-0.09)	6.83	3.5, 11.9	<<0.001	3.12	1.5, 5.7	0.004
Level 7 (0.1-0.29)	6.45	3.2, 11.5	<<0.001	3.50	1.7, 6.4	<0.001
Level 8 (0.3-0.49)	21.06	12.7, 32.9	<<0.001	14.44	9.3, 21.3	<<0.001
Level 9 (0.5-0.69)	23.71	16.2, 33.5	<<0.001	12.65	8.5, 18.1	<<0.001
Level 10 (0.7-0.89)	80.57	46.9, 129.0	<<0.001	55.25	32.7, 87.3	<<0.001
Level 11 (0.9+)	88.92	44.3, 159.1	<<0.001	51.10	23.3, 97.0	<<0.001

Boldface shows changes if Villages 3-H and 3-5 are omitted
*No data.

By contrast, studies in the US have shown no evidence of adverse effects but they have been conducted in regions where the average exposure concentrations is below 0.2 ppm (aside from an epidemiological study in western Minnesota where 11 individuals experienced acute gastrointestinal disorders shortly after drinking water from a well with extremely high arsenic levels resulting from contamination by arsenical insecticides (Feinglass, 1973)). The limited sample sizes in the US however, make lack of detection of an effect inconclusive, *i.e.*, the sample sizes appear too small for adequate power to support the conclusion that there is no increased risk of cancer or other health hazards. As Hendricks (1992) observes, there are also other differences between factors in the US studies and those in countries where adverse effects were found, which may confound direct comparison. In the USA, exposure durations are comparatively short term (duration, and possibly accumulation, may be factors in exposure as well as the exposure concentration), nutrition is relatively good (which may facilitate the capacity to methylate/detoxify arsenic), and the socioeconomic status is high (better access to health care and other factors). A summary table comparing exposure levels and findings in epidemiologic studies may be found in Petito and Beck (1990; Table 1).

Skin lesions characteristic of arsenic toxicity are often used as an indicator of high exposure to arsenic, which raises the question of whether these conditions may be useful precursors to more severe effects, particularly cancer of the skin or elsewhere. The Technical Panel of the EPA Risk Assessment Forum considered the relationship between the occurrence of arsenic-related skin lesions such as hyperpigmentation and hyperkeratosis and the occurrence of skin cancer. Although such lesions may be a useful biomarker for exposure to arsenic, it was concluded that since some malignant skin lesions arise *de novo* the appearance of such lesions could not be interpreted as a precursor to skin cancer, and the occurrence of these lesions at exposure levels below those of arsenic-related skin malignancies should not be interpreted as evidence of a threshold phenomenon (EPA,1988; p 6). Nevertheless, some uncertainty remains. Cuzick and co-workers have been studying the occurrence of palmar keratoses in relation to bladder and lung cancer and other factors (Cuzick *et a*l., 1984; Cuzick *et al.*, 1992). In the more recent of those two publications, the authors conclude "Our original observation that cancer is associated with previous signs of arsenicism has been strengthened with further followup...We suspect that the appearance of signs of arsenicism after arsenic exposure is a marker for a susceptibility gene for bladder cancer and possibly other cancers".

Petito and Beck (1990) provide insight into the expected behavior of a dose-response curve for arsenic at low exposure levels by considering the implications from toxicological, biochemical, and genotoxic data, as well as the epidemiologic evidence described above. They list factors that may contribute to nonlinearity of the dose-response curves for arsenic-induced cancers (their Table 5), for both the population studied and for extrapolation to other populations. Their detailed assessment will not be reviewed here, except to note their emphasis on the capability of human detoxification of arsenic by methylation up to 250-1,000 micrograms per day and the significance of genotoxicity studies suggesting that arsenic does not interact directly with DNA. Both metabolic and genotoxic processes may operate as nonlinear or threshold processes.

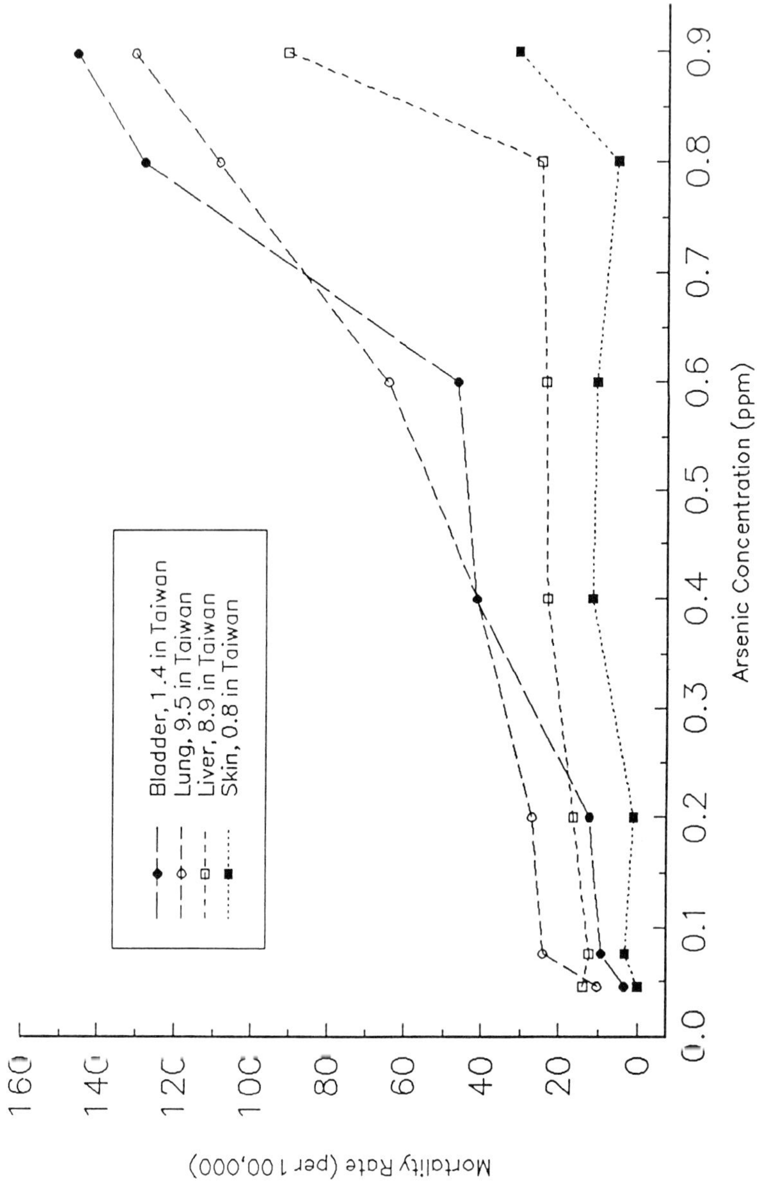

Figure 1a *Female mortality rates standardized to the world population of 1976, age 20+, for exposure levels 5-11 (0.04 - 0.9+, mg* L^{-1}*). Numerical values in the legend box are the corresponding mortality rates for all of Taiwan.*

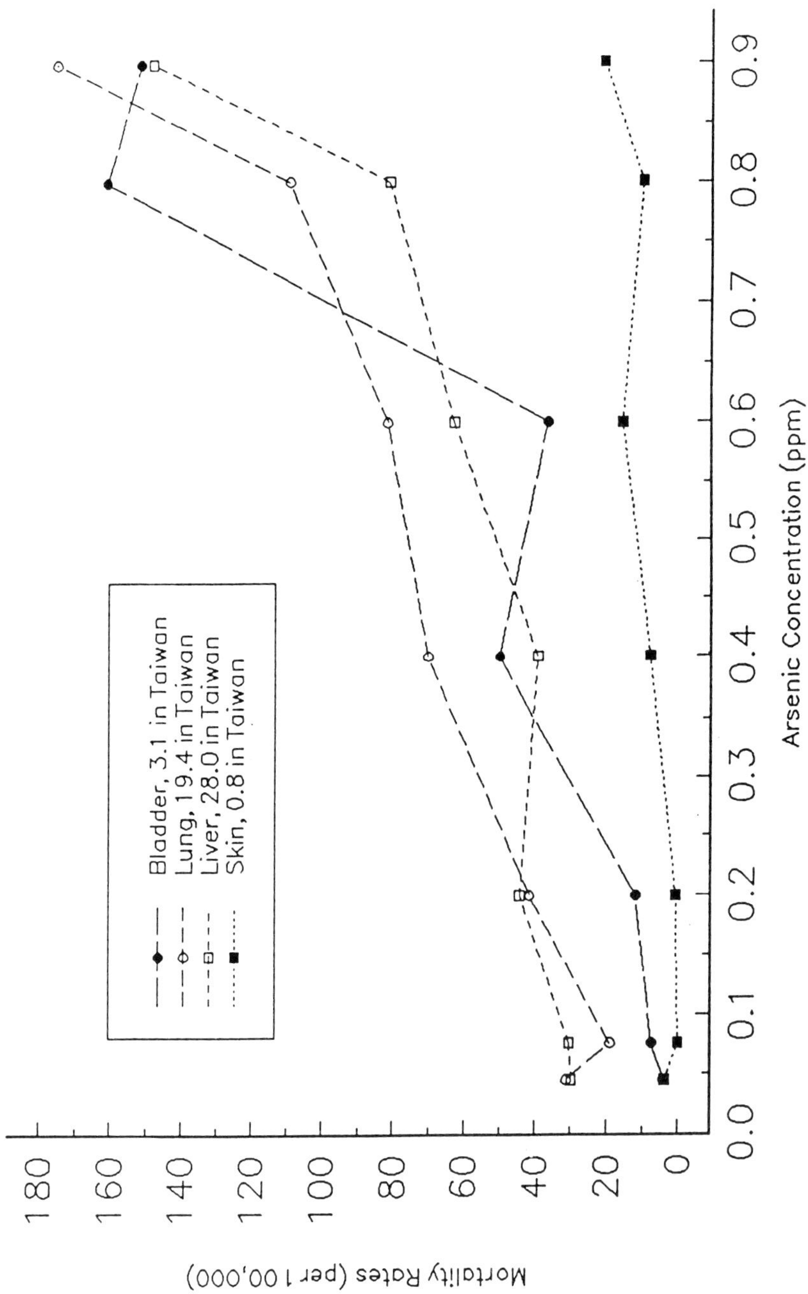

Figure 1b *Male mortality rates standardized to the world population of 1976, Age 20+, for exposure levels 5-11 (0.04 - 0.9+, mg L^{-1}) Numerical values in the legend box are the corresponding mortality rates for all of Taiwan*

Dose-response in the Taiwan study

In this section we investigate the shape of the observed dose-response curve for various cancers in relation to arsenic concentrations in drinking water for the Taiwan Study. The current dataset is somewhat larger than the previous summary reports in Chen *et al.* (1988, 1992) and Wu *et al.* (1989), and is grouped into eleven exposure classifications instead of three, which should provide improved definition of the shape of the observed dose-response curve. Data are provided for four cancer sites (lung, liver, bladder, and skin), both sexes, and three observation periods (Period 1: 1973-1977; Period 2: 1978-1982; Period 3: 1983-1986). For each of these 24 combinations, the data consist of the number of deaths and the number of person-years in the study population, cross-classified by age (five-year age intervals) and exposure level (11 intervals of the median arsenic concentration in well water for each study village). The data for each exposure interval consists of the composite of villages with arsenic concentrations in that interval, where the intervals and their numbered levels are : 1 (< 0.01 ppm), 2(0.01-0.019), 3(0.02-0.029), 4(0.03-0.039), 5(0.04-0.049), 6(0.05-0.09), 7(0.1-0.29), 8(0.3-0.49), 9(0.5-0.69), 10(0.7-0.89), 11(0.9+).

The mortality rates were age-standardized to the 1976 world population and plotted against the 7 exposure intervals from 0.04 to 0.09+ ppm, as illustrated in Figure 1. It is assumed that the population of each village was exposed from birth until 1970, after which the deep-water artesian wells were no longer used as a source of drinking water. Results could be combined across the three observation periods by age or by exposure duration. For the latter, a specific age group in Period 1 has the same exposure duration as the next older age group in Period 2 and the next older group from that one in Period 3 (*e.g.*, being of age 40-44 in Period 1 is equivalent to being of age 45-49 in Period 2 and age 50-54 in Period 3). Calculations were made both ways and the difference was negligible.

It is readily apparent from Figure 1 that the dose-response curves for bladder, lung, and liver are nonlinear, so we turn our attention to the low exposure range. We would like to be able to tell where the dose-response curve starts upward, in particular whether the slope of the curve is positive at the origin or whether the slope is zero over some range of low exposures before it becomes positive. Petito and Beck (1990, p.150) report that studies conducted on white adult US and European males indicate that changes in arsenic methylation and detoxification begin at oral intake levels of 0.2-0.25 mg/day. To rescale this to an equivalent for an adult Taiwan male, it will be assumed that an adult US male weighs 70 kg and an adult Taiwan male weighs 55 kg and consumes 3.5 liters of water per day (EPA, 1988, Appendix B). These values give a multiple of 0.2245 (= 55/(70x3.5)), leading to 0.045-0.056 ppm on our scale for drinking water in Taiwan. If the same results for arsenic methylation and detoxification apply to women after rescaling for weight then the corresponding range for Taiwanese women is somewhat higher, namely 0.071-0.090 (assuming an adult Taiwanese female weighs 50 kg and drinks 2 liters of water per day (EPA, 1988, Appendix B)). These calculations do not take into account differences in background rates of arsenic exposure in Taiwan and the US or other factors that might affect methylation and detoxification, *e.g.*, diet and other lifestyle

Table 2 *Standardised Mortality Ratio (SMR), bladder cancer 4 exposure classifications.*

	Females			Males		
Arsenic exposure (ppm)	Est.	95% Confidence interval	Signif. (p-value)	Est.	95% Confidence interval	Signif. (p-value)
Levels 1-5 (0.00-0.049)	7.79 **0.00**	3.7, 14.3 **0.0, 3.4**	<<0.001 **>0.2**	1.83 **0.54**	0.5, 4.7 **0.01, 3.0**	>0.2 **>0.2**
Levels 6-7 (0.05-0.29)	6.64	4.2, 10.0	<<0.001	3.30	2.0, 5.1	<<0.001
Levels 8-9 (0.3-0.69)	22.64	16.9, 29.8	<<0.001	13.40	10.1, 17.5	<<0.001
Levels 10-11 (0.7-0.9+)	83.65	55.6,120.9	<<0.001	53.79	35.4, 78.3	<<0.001

Boldface shows changes if Villages 3-H and 3-5 are omitted

characteristics. These uncertainties notwithstanding, we might anticipate a dose-curve for males to turn upwards in the neighbourhood of 0.05 ppm (the current drinking water standard for the US) and perhaps at a little higher concentration for females.

The dose-response data were omitted from the four lowest exposure levels in Figure 1 because the outcomes are erratic, making the graph difficult to read. The behavior of the dose-response data at these lowest exposure levels is best depicted in individual graphs of the cancer sites of interest, as illustrated for bladder cancer in Figure 2. The error bars in the graph show deviations of one standard error, except where the observed mortality rate is zero. The dotted line in the graphs is the maximum likelihood estimate of the curve using only the assumption that the true response curve is nondecreasing as exposure increases. This is a much weaker assumption than specifying the family of distributions to fit to the data, such as the multistage, Weibull, Gamma multi-hit, *etc*., virtually all of which model the response as nondecreasing. It is clear from Figure 2 that the mortality rates at levels 2 and 4 (centered at 0.015 ppm and 0.035 ppm, respectively) for females and at level 4 for males are markedly out of alignment with the pattern described by the remaining observations. The dotted-line is indicative of the elevating influence these points would have on estimation of the dose-response curve at low exposure.

Graphs for the other three cancer sites are not included here, but their outcomes are similar to Figure 2. There is some inconsistency, however, in the exact location

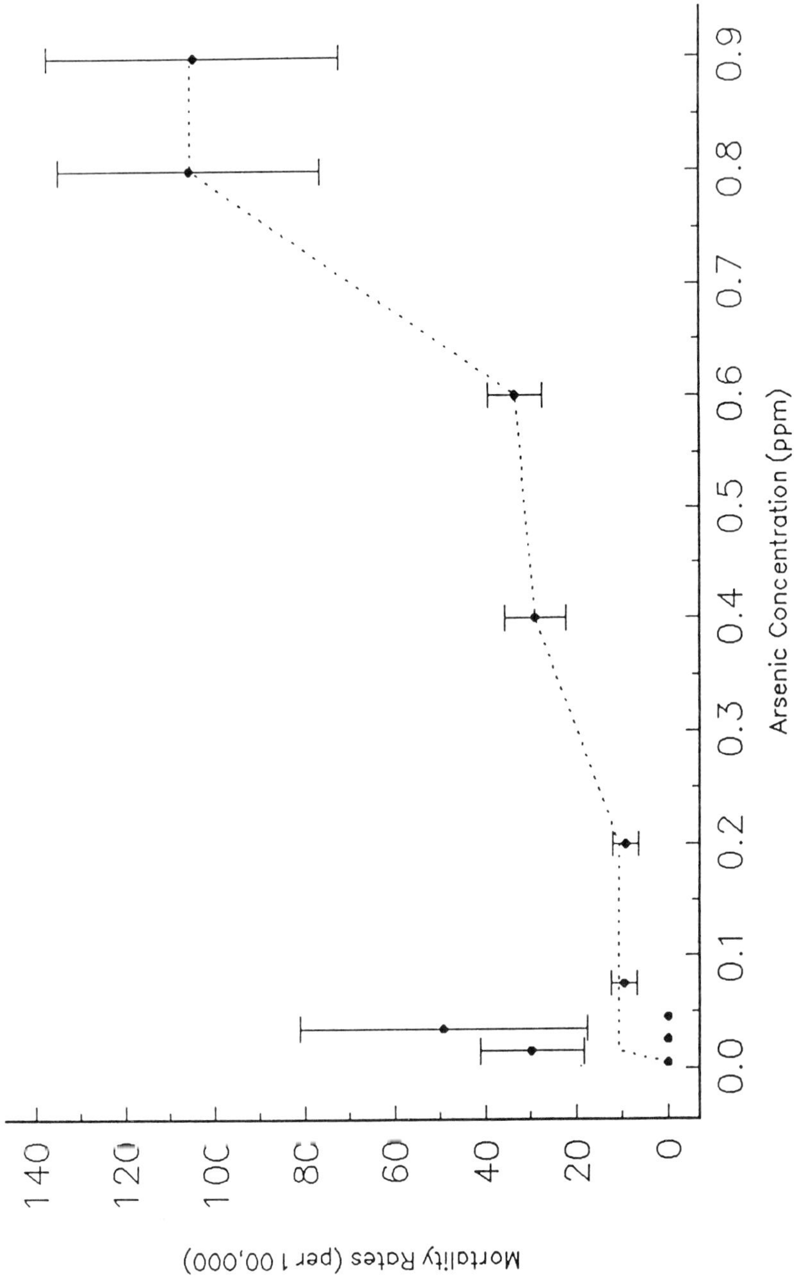

Figure 2a *Female bladder cancer mortality rates standardized to the world population of 1976, age 20+, for all exposure levels. Error bars are for one standard error. Dotted line is a plot of the maximum likelihood estimates if it is assumed that the true dose-response curve is nondecreasing.*

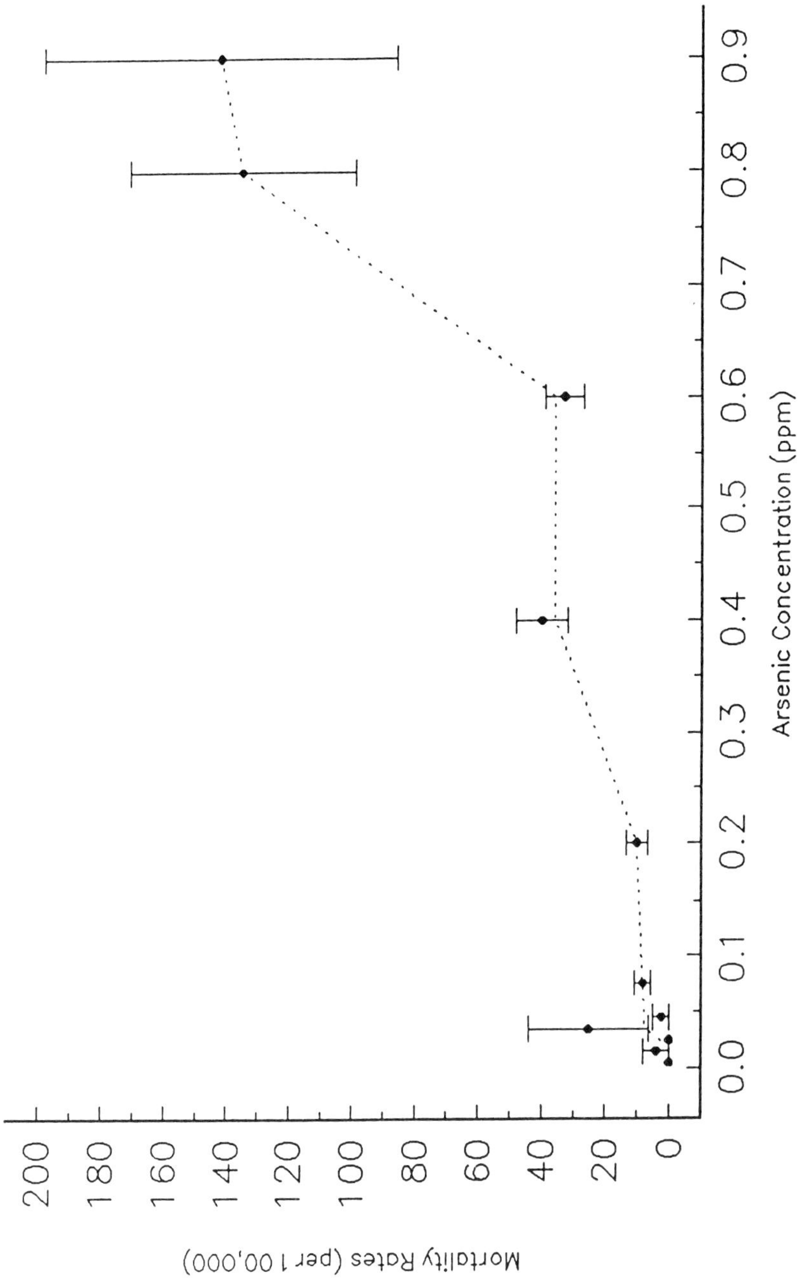

Figure 2b *Male bladder cancer mortality rates standardized to the world population of 1976, age 20+, for all exposure levels. Error bars are for one standard error. Dotted line is a plot of the maximum likelihood estimates if it is assumed that the true dose-response curve is nondecreasing.*

and magnitude of the "discrepant" points. The issue of interest is whether the discrepant points are occurring by chance or by data error (intended broadly to include limitations in representativeness of data used as surrogate measures of something else, *e.g.*, arsenic concentration as a surrogate for exposure to arsenic). This issue will be considered further for bladder cancer, which makes a good example because it has received considerable attention in recent literature on arsenic, there is a clear dose-response effect with arsenic exposure, and the background rate for bladder cancer in Taiwan as a whole, which will be used for comparison, is much lower than for the other three organ sites (skin cancer aside).

We examined whether any of the villages in level 2 or level 4 (3 villages in level 2 and 1 in level 4) might have any distinguishing characteristics, demographic or otherwise, that might explain the elevated response rate, but none were found aside from the possibility that the surrogate measure for arsenic exposure in a village, namely the median arsenic concentration for wells tested in the village, may not always be a reliable representative value for the village. Some villages have more than one well and the depths of the wells tested may be a factor in the arsenic concentration of the water. Additionally, it was found that in 15 villages where two separate tests were made, the reported arsenic concentrations for the same well differed significantly in two instances (Village 0-G: 0.03ppm and 0.437 ppm; Village 3-N, 0.032 ppm and 0.059 ppm); in both of these villages, only one well was tested in each instance. The cancer mortality rates in both of these villages were much more similar to the rates observed at the higher of the arsenic measurements, so those two villages were classified according to the higher of their two test values. In most villages, however, arsenic tests were made on several wells so the possibility for misclassification of the median concentration may be low.

Villages with mortality rates significantly different from others with similar arsenic concentrations may be candidates for possible misclassification by exposure. Two of the 10 villages at low exposure levels below 0.05ppm have exceptionally high rates of bladder cancer mortality (Village 3-5 in level 4 and Village 3-H in level 2). An assessment of risk below 0.05 ppm could hinge on whether these two villages are appropriately classified by exposure. We reached that conclusion by calculating the standardised mortality ratio (SMR) for bladder cancer at each of the 11 exposure categories relative to the whole country of Taiwan. These calculations and the associated significance levels are shown in Table 1 (an SMR value greater (less) than 1 indicates a higher (lower) mortality rate than Taiwan). The changes if Villages 3-H and 3-5 are omitted are also shown in Table 1. When the two villages are omitted, there is not only no significant increase in bladder cancer in the first five exposure levels, there is virtually no cancer mortality at all (it will be shown, however, that the sample sizes are inadequate to preclude the possibility of an undetected effect). Plots of the standardized mortality rates with the two villages omitted are shown in Figure 3 for comparison with Figure 2 which contains the two villages.

It is apparent from Figure 3 that without Villages 3-H and 3-5 the standardized mortality ratios are similar for exposure levels 1-5, 6-7, 8-9, and 10-11. Consequently we combined the data for levels 1-5, 6-7, 8-9, and 10-11 to narrow the confidence intervals on the standardized mortality ratios and to increase the power to detect a departure, in either direction, from the mortality rate for bladder cancer in Taiwan. The results are in Table 2, which shows the outcome with and without Villages 3-H and 3-5 included. It is readily apparent from Table 2 that conclusions about the risk of bladder cancer in the endemic region at arsenic concentrations

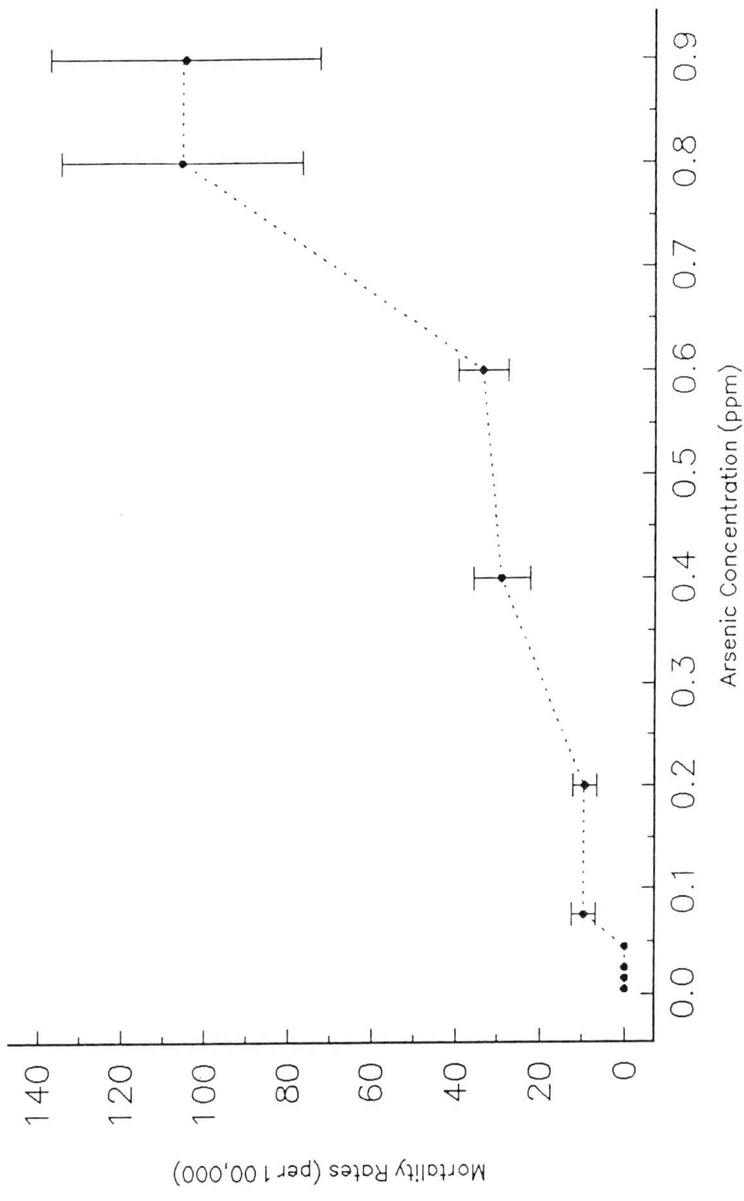

Figure 3a *How Figure 2a would appear if Villages 3-H and 3-5 were omitted.*

below 0.05 ppm could depend quite strongly on whether Villages 3-H and 3-5 are included or excluded. If those two villages are accurately classified below 0.05 ppm, then of course they should be included; if not, they should be excluded.

With inclusion of Villages 3-H and 3-5, the true risk for females at exposures

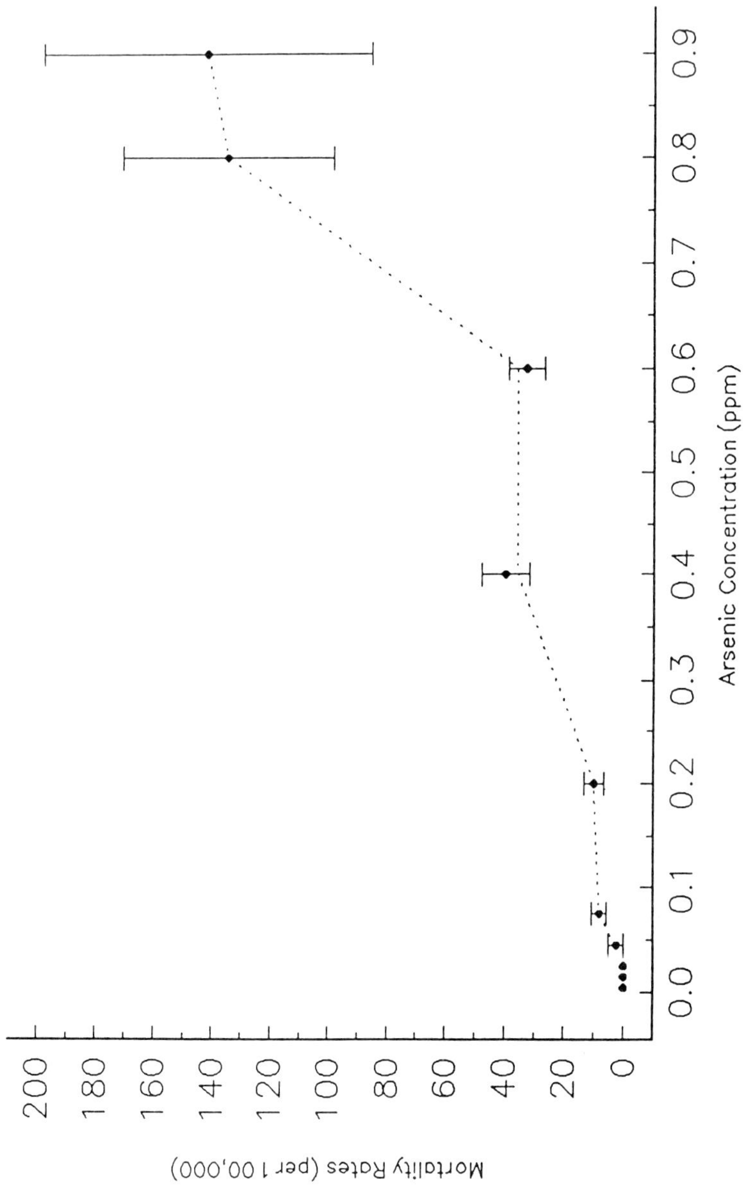

Figure 3b *How Figure 2b would appear if Villages 3-H and 3-5 were omitted.*

below 0.05 ppm (using the combined data at exposures below 0.05 ppm) is unquestionably much higher than for Taiwan; the significance level is far smaller than 0.0001 and the mortality rate is about 8 times that of Taiwan, with a 95% confidence interval of about 3.8 to 14.3. In terms of excess mortality rate, these figures correspond to an estimate of 9.3 with confidence interval of 3.8 to 18.4 (per 100,000 population per year, standardised to the world population of 1976 for age 20 and above). For males, the SMR is a more modest 1.8, which is not statistically significant (p>0.2 for a two-sided test), and the 95% confidence interval is 0.5 to 4.7. The excess mortality rate for males is estimated at 1.6 with a confidence interval of -1.5 to 10.6 (note: -1.5 means the lower confidence interval is 1.4 for the endemic region, 1.5 less than the 2.9 rate for males in Taiwan).

If Villages 3-H and 3-5 are excluded, the outcome is quiet different. For females, the SMR is 0.0 (there were 10 deaths due to bladder cancer previously, but all 10 were in Villages 3-H and 3-5). The 95% confidence interval is 0.0 to 3.7. For excess mortality, the estimate is -1.4 with confidence interval -1.4 to 3.3 (note: the lower confidence bound cannot go below -1.4 because 1.4 is the rate for Taiwan). For males, the SMR is 0.5 with confidence interval 0.0 to 3.0. For excess mortality, the estimate is -1.5, with confidence interval -2.9 to 5.7.

Discussion

Results of an assessment of the dose-response behavior of the Taiwan study at low exposure levels, *e.g.* <0.05ppm, may be sensitive to the surrogate value used for exposure to arsenic in drinking water, namely the single value taken as representative of arsenic concentration in the drinking water of a village. Using bladder cancer as an example, we have shown that the evidence of increased mortality in the ten villages of the endemic region with exposure measures below 0.05ppm is attributable to two of the ten villages (Villages 3-H and 3-5). Those two villages are also the only ones from the total of ten with a recorded prevalence of blackfoot disease in the previous study (Tseng, 1977). When the two villages (3-H and 3-5) are omitted from the analysis, the estimated mortality rate for male bladder cancer in the remaining eight villages is about half the rate for Taiwan as a whole; for women, the estimated rate is zero. While these data suggest a reduced mortality rate, however, they would not be conclusive even if it were known for certain that Villages 3-H and 3-5 should be reclassified at exposure levels above 0.05 ppm. Based on the 95% confidence intervals, the rate for bladder cancer mortality could be higher in the endemic region in these ten villages by as much as about 3 for females and 6 for males (per 100,000 population per year). It also should be kept in mind that the arsenic concentrations in the ten villages for which the data are combined are all less than 0.05 ppm, but cover a range of values. Consequently, the results apply to a mix of exposure values, not to 0.05 ppm itself.

If some villages may be misclassified at arsenic concentrations that are too low, or too high, it should become more apparent through conducting sensitivity analysis (with and without selected villages) and comparing results across cancer sites. For this purpose it may be useful to make each village the unit of observation. This could be done by classifying each village at its own arsenic concentration, only combining data across villages that have identical values. This approach, which would provide dose-response data at about 50 distinct exposure levels, should make discrepant village outcomes more readily apparent since they would not be averaged in with other villages. Additionally, ungrouped data is more informative for assessing behavior of the dose-response curve.

Although quite large, the Taiwan study is probably not large enough to exclude unequivocally the possibility that there may be some increased risk in cancer mortality arsenic concentration, even if Villages 3-H and 3-5 are excluded. But epidemiologic studies are designed and implemented to detect harmful effects, usually at high exposures where the power to detect an effect is greatest, not to verify that there is no possibility of an effect at any magnitude in situations where there is no evidence of an effect. For these reasons, and for others related to questions that arise in the context of extrapolating risk estimates from Taiwan or elsewhere to the US (which is outside the scope of this article), analysis of the Taiwan study data need to be supplemented with analysis of other data sets that may contain relevant epidemiologic evidence. When all sources of information are taken into account, including what is known about the pharmacokinetics and mechanisms of cancer or toxicity as well as observational data on humans from different countries and lifestyles, possibly with different modes and patterns of exposure that may include medical treatment with arsenic, for example, then one should have a more informed basis for assessing the total weight of evidence.

Acknowledgements

Funded by the American Water Works Association and the National Science Council, Republic of China.

References

Bates, M. N., Smith, A. H. and Hopenhayn-Rich, C. 1992. Arsenic ingestion and internal cancers: a review. *Amer. J. Epidem.*, **135(5)**, 462-476.

Brown, K. G., Boyle, K. E., Chen, C. W.and Gibb, H. J. 1989. A dose-response analysis of skin cancer from inorganic arsenic in drinking water. *Risk Analysis*, **9(4)**, 519-528.

Chen, C.-J., Chuang, Y.-C., Lin, T.-M.and Wu, H.-Y. 1985. Malignant neoplasms among residents of a Blackfoot disease-endemic area in Taiwan: high-arsenic artesian well water and cancers. *Cancer Res.*, **45**, 5895-5899.

Chen, C.-J.; Chuang, Y.-C.; You, S.-L.; Lin, T.-M and Wu, H.-Y. 1986. A retrospective study on malignant neoplasms of bladder, lung, and liver in Blackfoot disease endemic area in Taiwan. *Br. J. Cancer,* **53,** 399-405.

Chen, C.-J., Kuo, T.-L., Wu, M.-M. 1988. Arsenic and cancers (letter). *Lancet,* **II**, 414-415.

Chen, C.-J., Chen, C.-W., Wu, M.-M and ; Kuo, T-L. 1992. Cancer potential in liver, lung, bladder and kidney due to ingested inorganic arsenic in drinking water. *Br. J. Cancer,* **66**, 888-892.

Cuzick, J., Harris, R. and Mortimer, P. S. 1984. Palmar keratoses and cancers of the bladder and lung. *Lancet,* 530-533.

Cuzick, J., Sasieni, P.and Evans, S. 1992. Ingested arsenic, keratoses, and bladder cancer. *Amer. J. Epidem.,* **136(4)**, 417-421.

Feinglass, E. J. 1973. Arsenic intoxication from well water in the United States. *N. Engl. J. Med.,* **288(16)**, 828-830.

Fielder, R. J., Dale, E. A. and Williams, S. D. 1986. *Inorganic arsenic compounds.* Toxicity review 16. HMSO, London.

Hendricks, S. K. 1992. Arsenic in drinking water. Master's report, University of Colorado at Boulder.

Petito, C. T. and Beck, B. D. 1990. Evaluation of evidence of nonlinearities in the dose-response curve for arsenic carcinogenesis. *Trace Substances in Environmental Health.* **XXIV**, 143-176.

Smith, A. H., Hopenhayn-Rich, C., Bates, M. N., Goeden, H. M., Hertz-Picciotto, I., Duggan, H. M., Wood, R., Kosnett, M. J. and Smith, M. T. 1992. Cancer risks from arsenic in drinking water. *Environ. Health Perspect.,* **97**, 259-267.

Tseng, W.-P., Chu, H.-M., How, S.-W., Fong, J.-M., Lin, C.-S. and Yen, S. 1968. Prevalence of skin cancer in an endemic area of chronic arsenicism in Taiwan. *J. Natl. Cancer Inst.,* **40(3),** 453-463.

Tseng, W.-P. 1977. Effects and dose-response relationships of skin cancer and Blackfoot disease with arsenic. *Environ. Health Perspect.,* **19,** 109-119.

US Environmental Protection Agency (EPA). 1988. *Special Report on Ingested Inorganic Arsenic. Skin Cancer; Nutritional Essentiality.* EPA/625/3-87/013.

US Environmental Protection Agency (EPA). 1976. *National Interim Primary Drinking Water Regulations.* EPA-570/9-76-003.

US Environmental Protection Agency (EPA). 1984. *Health assessment document for inorganic arsenic.* Final report. EPA-600/8-83-021F.

US Environmental Protection Agency (EPA). 1993. *Draft drinking water criteria document on arsenic.* Submitted to EPA Science Advisory Board Drinking Water Committee, March 10, 1993, by Human Risk Assessment Branch, Health and Ecological Criteria Division, Office of Science and Technology, Office of Water, US EPA.

Wu, M.-M., Kuo, T.-L., Hwang, Y.-H.and Chen, C.-J. 1989. Dose-response relation between arsenic concentration in well water and mortality from cancers and vascular diseases. *Amer. J. Epidem.,* **130(6)**, 1123-1132.

16 Species differences in the metabolism of arsenic

Marie Vahter

Institute of Environmental Medicine, Karolinska Institutet, Box 210, 171 77 Stockholm, Sweden

Abstract

Biological methylation of arsenic has been known since the 1930s, when it was shown that the volatile, garlic-smelling gas, produced by micro-organisms growing in the presence of various arsenic compounds, was trimethylarsine. Evidence for methylation of arsenic in mammals was presented first in the early 1970s. Since then, a great number of studies have demonstrated that inorganic arsenic is methylated to methylarsonic acid (MMA) and dimethylarsinic acid (DMA) in most mammalian species. Pentavalent arsenic (As(V)) has to be reduced to trivalent arsenic (As(III)) before methyl groups can be attached. In fact, as much as 50–70% of absorbed As(V) is rapidly reduced to As(III), a reaction which seems to be common for most species. There are marked species differences in the methylation of inorganic arsenic. In most animal species studied so far, DMA is the main metabolite. Compared to human subjects, very little MMA is produced. However, the rate of methylation varies considerably between species. The marmoset monkey is the only species which has been shown to be unable to methylate inorganic arsenic. In contrast to other species, the rat shows a marked binding of DMA to the red blood cells, which results in a low rate of urinary excretion. In both human subjects and rodents exposed to DMA, about 5% of the dose is excreted in the urine as trimethylarsine oxide.

Introduction

Already in 1933 Challenger and co-workers at the University of Leeds reported that the volatile, garlic-smelling gas, which was produced by the mold *Scopulariopsis brevicaulis*, grown in the presence of various arsenic compounds, was trimethylarsine (TMA). Since it was known that mammals were able to produce dimethyl telluride from tellurium and dimethyl selenide from selenium, as well as several reports in the literature describing a garlic-smelling breath of animals exposed to cacodylic acid (dimethylarsinic acid; DMA) or its sodium salt, Challenger (1945) proposed that arsenic, at least cacodylic acid, could be methylated in the animal body and exhaled as trimethylarsine. However, as shown in Table 1, it was not until the 1970s that more direct evidence of methylation of inorganic arsenic in mammals were presented. In

$$AsO_4^{3-} \xrightarrow{2e^-} AsO_3^{3-} \xrightarrow{CH_3^+} CH_3AsO_3^{2-} \xrightarrow{2e^-} CH_3AsO_2^{2-} \xrightarrow{CH_3^+}$$

$$(CH_3)_2AsO_2^- \xrightarrow{2e^-} (CH_3)_2AsO^- \xrightarrow{CH_3^+} (CH_3)_3AsO \xrightarrow{2e^-} (CH_3)_3As$$

Figure 1 *Suggested mechanism for the methylation of arsenic*

Table 1 *Discovery of arsenic methylation in mammals.*

Year	Species	Exposure form	Methylated metabolites	Reference
1975	Dog, cow	As(III), As(V)	Methyl-As	Lakso & Peoples
1977	Man	As(III), As(V)	DMA, MMA	Crecelius Smith *et al.*
1978	Dog	As(V)	DMA, MMA	Tam *et al.*
1980	Hamster	As(V)	DMA, MMA	Charbonneau *et al.*
1980	Rat, Rabbit	As(III)	DMA, MMA	Marafante *et al.*
1980	Mouse	As(III), As(V)	DMA, MMA	Vahter & Gustafsson
1982	Marmoset	As(III)		Vahter *et al.*
1983	Cynomologus Rhesus	As(V)	DMA, MMA	Charbonneau *et al.* *c.f.* Vahter 1983

1973, Braman and Foreback reported that arsenic in human urine consisted of about 66% DMA, 8% methylarsonic acid (MMA) and 25% inorganic arsenic. Actual *in vivo* methylation of arsenic was demonstrated a couple of years later, when Lakso and Peoples (1975) reported high concentrations of methylated arsenic compounds, although not further specified, in urine of dogs and cows fed arsenite or arsenate. Since then, it has been well documented that DMA is the main metabolite of inorganic arsenic in many mammalian species, MMA being produced and excreted in less amounts.

Mechanism of arsenic methylation in mammals

The mechanism for arsenic methylation in mammals essentially follows the mechanism for arsenic methylation in microorganisms suggested by Challenger in 1945 (Figure 1). Arsenic in its pentavalent oxidation form (As(V)) is initially reduced to trivalent arsenic (As(III)), which is methylated to MMA and DMA. All the different steps in the arsenic biotransformation in mammals have not yet been elucidated, but it seems likely that the methylation takes place mainly in the liver by transfer of methyl groups from S-adenosylmethionine to arsenic in its trivalent oxidation state (Marafante and Vahter, 1984; Buchet and Lauwerys, 1985; Marafante *et al.*, 1985). It has been demonstrated both *in vitro* and *in vivo*, that the arsenic methylation requires reduced glutathione (GSH) (Buchet and Lauwerys, 1985; 1987; Hirata *et al.*, 1988). Depletion of hepatic GSH decreases the methylation. Studies on mice (Vahter and Gustafsson, 1980) and rabbits (Marafante *et al.*, 1985), as well as *in vitro* studies with blood and tissue homogenates of rats (Buchet and Lauwerys, 1985) have shown that the methylating capacity of red blood cells, brain, lung, intestine and kidneys is insignificant compared to that of the liver.

While most microorganisms produce TMA from inorganic arsenic, DMA is the endpoint of the methylation of inorganic arsenic in mammals. However, following administration of DMA to mice, hamsters and a human subject, about 5% of the dose was excreted in the urine as trimethylarsine oxide (TMAO) (Marafante *et al.*, 1987). There was no significant difference between species. There are some recent indications of the presence of trace amounts of trimethylarsenic compounds in the urine of mammals exposed to inorganic arsenic (Odanaka *et al.*, 1983; Yamauchi and Yamamura, 1985). However, it seems unlikely that significant amounts of TMAO are formed following exposure to inorganic arsenic, since the rate limiting factor in the synthesis of TMAO will be the hepatic concentration of DMA. This will not be very high, since DMA is rapidly eliminated from the liver and excreted in the urine.

Toxicological significance of arsenic metabolism

The methylation of inorganic arsenic in mammals functions as a detoxification mechanism. The methylated metabolites are less acutely toxic than inorganic arsenic (Tatken and Lewis, 1983; Yamauchi *et al.*, 1989; 1990). They are also considerably less reactive with tissue components (Vahter and Marafante, 1983) and faster excreted in urine (Buchet *et al.*, 1981 a, b; Vahter *et al.*, 1984), than is inorganic arsenic. The inorganic arsenic, especially As(III), is the main form of arsenic interacting with tissue constituents (Vahter and Marafante, 1983). This means that factors influencing the methylation (see below) may influence arsenic toxicity.

It should be noted that the proposed mechanism for biological methylation of inorganic arsenic requires that As(V) is reduced to the more toxic As(III) prior to the complexation and the biomethylation. This has been confirmed in a number of experimental studies (Vahter and Envall, 1983; Bertolero *et al.*, 1987). In rabbits exposed to ^{74}As-arsenate, ^{74}As-As(III) was detected in the plasma after only 15 minutes, while there was a lag time of about one hour before ^{74}As-DMA appeared in the plasma (Marafante *et al.*, 1985). It can be estimated based on studies on rabbits and marmoset monkeys (Vahter and Envall, 1983; Vahter and Marafante, 1985), that as much as 50-70% of a single dose of As(V) is rapidly reduced to As(III). There are no reports on species differences in the reduction of As(V).

Factors influencing the methylation of arsenic

When evaluating species differences in the metabolism of arsenic, it is important to recognize other factors which may influence the methylation. Such factors include exposure form of arsenic, dose level, route of exposure and type of diet.

Since the reduction of As(V) is not quantitative, exposure to As(III) results in a higher degree of methylation and more DMA in the urine than does exposure to As(V) (Vahter, 1981; Vahter and Marafante, 1983). In spite of this, the total amount of arsenic excreted in urine is higher following exposure to As(V) than to As(III), since As(V), being less reactive with tissue constituents, is excreted as such fairly rapidly in the urine.

The rate of methylation of inorganic arsenic decreases with increasing dose level (Vahter, 1981; Foa *et al.*, 1984). This can be explained by an overloading of the methylation capacity, and/or inhibition of the methylation by high tissue concentrations of As(III), which has been shown *in vitro* to inhibit the production of DMA (Buchet and Lauwerys, 1985). Also the route of exposure may influence the methylation. Since the liver is the main site of arsenic methylation, ingestion of arsenic gives a higher degree of methylation than does exposure via inhalation or parenteral administration (Vahter, 1981; 1983).

Chemicals that are capable of inhibiting the S-adenosylmethionine dependent

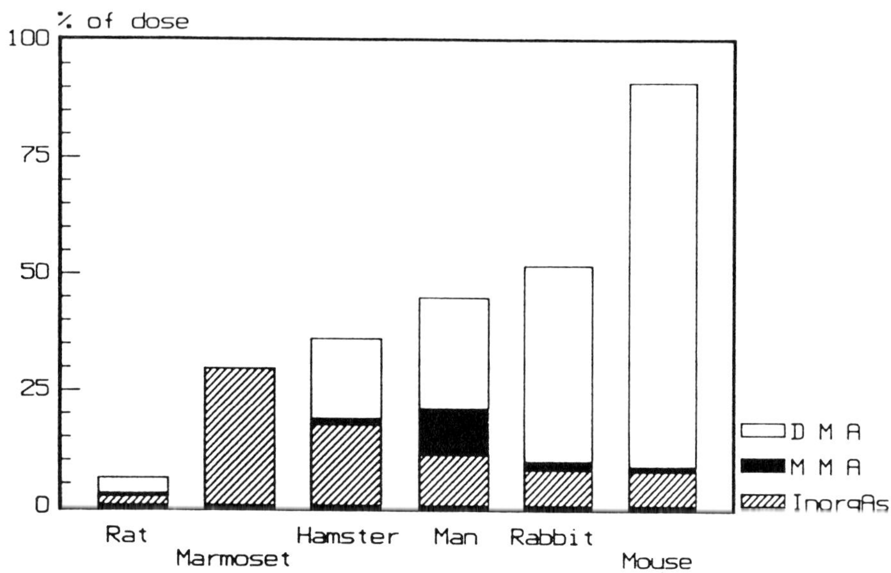

Figure 2 *Urinary excretion of arsenic metabolites in different species 2-4 days after a single dose of arsenite. Rat: 0.4 mg As kg^{-1} b.wt., p.o. (Vahter, 1981); Marmoset: 0.4 mg As kg^{-1} b.wt., i.p. (Vahter et al., 1982); Man: 0.007 mg As kg^{-1} b.wt., p.o. (Buchet et al, 1981b); Hamster: 2 mg As kg^{-1} b.wt., p.o. (Marafante and Vahter, 1987); Rabbit: 0.04 mg As kg^{-1} b.wt., i.v. (Vahter and Marafante, 1983); Mouse: 0.4 mg As kg^{-1} b.wt., p.o.(Vahter, 1981)*

methylation reactions, *e.g.* periodate-oxidized adenosine (PAD) which is a potent inhibitor of S-adenosylhomocysteine hydrolase, decreases the methylation of inorganic arsenic (Marafante and Vahter, 1984; Marafante *et al.*, 1985). A similar effect has also been observed in rabbits fed a diet deficient in methionine, choline or protein (Vahter and Marafante, 1987).

Species differences in the metabolism of inorganic arsenic

There is a large variation in the methylation of inorganic arsenic between animal species. Figure 2 shows the urinary excretion of arsenic metabolites in some different species 2-4 days after a single dose of arsenite. Since, in most species, MMA and DMA are rapidly excreted in the urine, the amounts of methylated arsenic metabolites in the urine gives a good indication of the formation of the methylated metabolites in the liver.

It is obvious from the figure that mice are very efficient in methylating arsenic. This results in a very fast excretion of DMA in the urine. A similar high degree of methylation has been reported for beagle dogs exposed to ^{74}As-arsenate (Charbonneau *et al.*, 1978; Tam *et al.*, 1978). Rats are also very efficient in methylating arsenic. The low urinary excretion of methylated arsenic metabolites in the rat shown in Figure 2 is not an indication of a low methylating capacity, but due to a specific retention of DMA in the erythrocytes. More than 95% of the arsenic in the erythrocytes of rats exposed to inorganic arsenic is in the form of DMA (Odanaka *et al.*, 1980; Lerman and Clarkson, 1983). The accumulation and long-term retention of DMA in the blood of rats causes a significantly longer retention of arsenic in rats compared to most other species and a different distribution pattern (Vahter, 1983; Vahter *et al.*, 1984; Stevens *et al.*, 1977).

The specific accumulation of arsenic in the rat erythrocytes has been known for many years. In 1942, Hunter and co-workers reported that arsenic, injected as potassium arsenite to rats, accumulated in the erythrocytes apparently bound to the hemoglobin. Other species studied showed much lower concentrations of arsenic in the blood. A few years later Ducoff *et al.*, (1948) and Lanz *et al.*, (1950) confirmed the results of Hunter et al (1942). One interesting finding of Lanz *et al.*, (1950) was that also cats accumulated arsenic in the blood following intramuscular injection of sodium arsenate, although not to the same degree as rats.

The accumulation of DMA in the red blood cells of rats has to be considered when evaluating the results of studies on arsenic metabolism in this species. An interesting effect of the specific binding of DMA in rats is that depletion of hepatic GSH gives rise to an increase in the urinary excretion of arsenic (Buchet and Lauwerys, 1987). Since the GSH depletion decreases the production of DMA, less arsenic in the form of DMA can be trapped in the rat erythrocytes, and more inorganic arsenic can be excreted in the urine. In other species, GSH depletion would lead to a decrease in the urinary excretion of arsenic.

The rat differs from other species also with respect to the biliary excretion of arsenic. Klaassen (1974) showed that following injection of 1 mg As kg^{-1} as ^{74}As-labelled arsenic trichloride, the rate of biliary excretion of arsenic in rats was

800-fold that in dogs and 37-fold that in rabbits. As much as 25% of the arsenic administered to rats was excreted in the bile within 2 hours. However, most of it was reabsorbed in the gut. It has been reported that arsenic in bile of isolated rat liver perfused with arsenite is associated with glutathione (Anundi *et al.*, 1982). No MMA or DMA was detected in bile following exposure to arsenite or arsenate (Vahter, 1983). Therefore, it is likely that the high biliary excretion of arsenic in rats is due to the high biliary excretion of GSH and related thiols and disulfides compared to most other species, (Stein *et al.*, 1988). Furthermore, GSH is the main thiol in rat bile, whereas bile of *e.g.* rabbits contains mainly cysteinylglycine and its disulfide.

The marmoset monkey is the only species which, so far, has been shown to be unable to methylate inorganic arsenic. We have carried out a series of studies on marmoset monkeys, given arsenite or arsenate, but never been able to detect any methylated arsenic metabolites in urine or tissues (Vahter *et al.*, 1982; Vahter and Marafante, 1985). The lack or methylation results in a long biological half-time of arsenic in the marmoset. In marmoset monkeys administered arsenate (As(V)), almost 60% of the dose was retained in the body after 3 days. About 10% of the dose was retained in the liver, where most of the arsenic was found in the rough microsomes, which is in contrast to findings in other species.

The hamster and the rabbit seem to have a metabolism of arsenic similar to that of human subjects, except for the lower urinary excretion of MMA. However, the gastrointestinal absorption of both inorganic arsenic and the methylated forms is somewhat lower in the hamster than in most other species (Charbonneau *et al.*, 1980; Yamauchi *et al.*, 1990; Marafante *et al.*, 1987), soluble arsenic compounds are almost completely absorbed. Thus, a perfect animal model for studies on arsenic toxicokinetics and toxicity has not yet been identified.

Conclusions

Inorganic As(III) is methylated to MMA and DMA in the liver of most mammals. As(V) is reduced in the blood to As(III), which then is methylated. DMA is the main metabolite in most mammals. Only human subjects excrete significant amounts of MMA in urine.

The rate of arsenic methylation varies considerably between animal species. Only the marmoset monkey is known not to methylate inorganic arsenic. In rats, most of the DMA produced is bound to the red blood cells. Rats also have a more pronounced biliary excretion of arsenic than other species.

There is no perfect animal model for studies on arsenic metabolism and toxicity. The rabbit and the hamster are most similar to man with regard to arsenic metabolism.

References

Anundi, I., Högberg, J. and Vahter, M. 1982. GSH release in bile as influenced by arsenite. *FEBS Lett.*, **145**, 285-288.

Bertolero, F., Pozzi, G., Sabbioni, E. and Saffiotti, U. 1987. Cellular uptake and metabolic

reduction of pentavalent to trivalent arsenic as determinant of cytotoxicity and morphological transformation. *Carcinogenesis,* **8**, 803-808.
Braman, R.S. and Foreback, C.C. 1973. Methylated forms of arsenic in the environment. *Science*, **182**, 1247-1249.
Buchet, J.P., Lauwerys, R. and Roels, H. 1981a. Urinary excretion of inorganic arsenic and its metabolites after repeated ingestion of sodium metaarsenite by volunteers. *Int. Arch. Occup. Environ. Health*, **48**, 111-118.
Buchet, J.P., Lauwerys, R. and Roels, H. 1981b. Comparison of the urinary excretion of arsenic metabolites after a single dose of sodium arsenite, monomethylarsonate or dimethylarsinate in man. *Int. Arch. Occup. Environ. Health*, **48**, 71-79.
Buchet, J. P. and Lauwerys, R. 1985. Study of inorganic arsenic methylation by rat liver in vitro: Relevance for the interpretation of observations in man. *Arch. Toxicol.*, **57**, 125-129.
Buchet, J.P. and Lauwerys, R. 1987. Study of factors influencing the *in vivo* methylation of inorganic arsenic in rats. *Toxicol. Appl. Pharmacol.*, **91**, 65-74.
Challenger, F., Higginbottom, C. and Ellis, L. 1933. The formation of organo-metalloidal compounds by microorganisms. I. Trimethylarsine and dimethylarsine. *J. Chem. Soc. Trans.*, 95-101.
Challenger, F. 1945. Biological methylation. *Chem. Rev.*, **36**, 315-361.
Charbonneau, S.M., Tam, G.K.H. and Bryce, F. 1978. Pharmacokinetics and metabolism of inorganic arsenic in the dog. In: Hemphill, D.D. (ed.), *Trace Substances in Environmental Health-XII. A symposium.*Un iversity of Missouri, Columbia.
Charbonneau, S.M., Hollins, J.G., Tam, G.K.H., Bryce, F., Ridgeway, J.M. and Willes, R.F. 1980. Whole-body retention, excretion and metabolism of [^{74}As]arsenic in the hamster. *Toxicol. Lett.*, **5**, 175-182.
Charbonneau, S.M., Bryce, F., Tam, G.K.H., and Sandi, E. 1983, personal communication. Cited from Vahter M., 1983.
Crecelius, E.A. 1977. Changes in the chemical speciation of arsenic following ingestion by man. *Environ. Health Perspect.*, **19**, 147-150.
Ducoff, H.S., Neal, W.B., Straube, R.L., Jacobson, L.O. and Brues, A.M. 1948. Biological studies with arsenic^{76}. II. Excretion and tissue localization. *Proc. Soc. Exp. Biol. Med.*, **69**, 548-554.
Foa, V., Colombi, A., Maroni, M., Buratti, M. and Calzaferri, G. 1984. The speciation of the chemical forms of arsenic in the biological monitoring of exposure to inorganic arsenic. *Sci. Total Environ.*, **34**, 241-259.
Hirata, M., Hisanaga, A., Tanaka, A. and Ishinishi, N. 1988. Glutathione and methylation of inorganic arsenic in hamsters. *Appl. Organomet. Chem.*, **2**, 315-321.
Hunter, F.T., Kip, A.F. and Irvine, J.W. 1942. Radioactive tracer studies on arsenic injected as potassium arsenite. *J. Pharmacol. Exp. Ther.*, **76**, 207-220.
Klaassen, C.D. 1974. Biliary excretion of arsenic in rats, rabbits and dogs. *Toxicol. Appl. Pharmacol.*, **29**, 447-457.
Lakso, J.U. and Peoples, S.A. 1975. Methylation of inorganic arsenic by mammals. *J. Agric. Food Chem.*, **23**, 674-676.
Lanz, H Jr, Wallance, P.C. and Hamilton, J.G. 1950. The metabolism of arsenic in laboratory animals using ^{74}As as a tracer. *Univ. California Publ. Pharmacol.*, **2**, 263-282.
Lerman, S. and Clarkson, T.W. 1983. The metabolism of arsenite and arsenate by the rat. *Fundam. Appl. Toxicol.*, **3**, 309-314.
Marafante, E. and Vahter, M. 1984. The effect of methyltransferase in hibition on the metabolism of (^{74}As)arsenite in mice and rabbits. *Chem. Biol. Interact.*, **50**, 49-57.

Marafante, E. and Vahter, M. 1987. Solubility, retention, and metabolism of intratracheally and orally administered inorganic arsenic compounds in the hamster. *Environ. Res.*, **42**, 72-82.

Marafante, E, Rade, J, Pietra, R, Sabbioni, E, and Bertolero, F. 1980. Metabolic fate of inorganic arsenic in laboratory animals. In: Anke, M., Schneider, H.J., Bruckner, C. (eds). *Proc. 3rd Symposium on Trace Elements. Arsenic. July 7-11, 1980, Jena. pp.49-55.* Abteilung Wissenschaftliche Publikationen der Friedrich-Schiller-Universität Jena.

Marafante, E., Vahter, M. and Envall, J. 1985. The role of the methylation in the detoxication of arsenate in the rabbit. *Chem. Biol. Interact.*, **56**, 225-238.

Marafante, E., Vahter, M., Norin, H., Envall, J., Sandström, M., Christakopoulos, A. and Ryhage, R. 1987. Biotransformation of dimethylarsinic acid in mouse, hamster and man. *J. Appl. Toxicol.*, **7(2)**, 111-117.

Odanaka, Y, Matano, O. and Goto, S. 1980. Biomethylation of inorganic arsenic by the rat and some laboratory animals. *Bull. Environm. Contam. Toxicol.*, **24**, 452-459.

Odanaka, Y., Tsuchiya, N., Matano, O. and Goto, S. 1983. Determination of inorganic arsenic and methylarsenic compounds by gas chromatography and multiple ion detection mass spectrometry after hydride generation - heptane cold trap. *Anal. Chem.*, **55**, 929-932.

Smith, T.J., Crecelius, E. A. and Reading, J.C. 1977. Airborne arsenic exposure and excretion of methylated arsenic compounds. *Environ. Health Perspect.*, **19**, 89-93.

Stein, A.F., Gregus, Z. and Klaassen, C.D. 1988. Species variations in biliary excretion of glutathione-related thiols and methylmercury. *Toxicol. Appl. Pharmacol.*, **93**, 351-359.

Stevens, J.T., Halle, L.L., Farmer, J. D., DiPasquale, L. C., Chernoff, N. and Durham, W. F. 1977. Disposition of ^{14}C and/or ^{74}As-cacodylic acid in rats after intravenous, intratracheal or peroral administration. *Environ. Health Perspect.*, **19**, 151-157.

Tam, G.K.H., Charbonneau, S.M., Bryce, F. and Lacroix, G. 1978. Separation of arsenic metabolites in dog plasma and urine following intravenous injection of ^{74}As. *Anal. Biochem.*, **86**, 505-511.

Tatken, R.L. and Lewis, R.J. (eds.) 1983. *Registry of Toxic effects of Chemical Substances*, 1981-82 edn., US Department of Health and Human Services, Cincinnati, Ohio.

Vahter, M. 1981. Biotransformation of trivalent and pentavalent inorganic arsenic in mice and rats. *Environ. Res.*, **25**, 286-293.

Vahter, M. 1983. Metabolism of inorganic arsenic in relation to chemical form and animal species. Doctoral thesis. Departments of Toxicology and Environmental Hygiene, Karolinska Institute, and National Institute of Environmental Medicine, Stockholm, 62 pp.

Vahter, M. and Envall, J. 1983. *In vivo* reduction of arsenate in mice and rabbits. *Environ. Res.*, **32**, 14-24.

Vahter, M. and Gustafsson, B. 1980. Biotransformation of arsenic in germfree and conventional mice. In: Anke, M., Schneider, H.-J., Bruckner, C. (eds) *Proc. 3rd Symposium on Trace Elements. Arsenic. July 7-11, 1980, Jena, pp.123-129.* Abteilung Wissenschaftliche Publikationen der Friedrich-Schiller-Universität Jena.

Vahter, M. and Marafante, E. 1983. Intracellular interaction and metabolic fate of arsenite and arsenate in mice and rabbits. *Chem. Biol. Interact.*, **47**, 29-44.

Vahter, M. and Marafante, E. 1985. Reduction and binding of arsenate in marmoset monkeys. *Arch. Toxicol.*, **57**, 119-124.

Vahter, M. and Marafante, E. 1987. Effects of low dietary intake of methionine, choline or proteins on the biotransformation of arsenite in the rabbit. *Toxicology Letters*, **37**, 41-46.

Vahter, M., Marafante, E., Lindgren, A. and Dencker, L. 1982. Tissue distribution and subcellular binding of arsenic in marmoset monkeys after injection of ^{74}As-arsenite. *Arch. Toxicol.*, **51**, 65-77.

Vahter, M., Marafante, E. and Dencker, L. 1984. Tissue distribution and retention of ^{74}As-dimethylarsinic acid in mice and rats. *Arch. Environ. Contam. Toxicol.*, **13**, 259-264.

Yamauchi, H. and Yamamura, Y. 1985 Metabolism and excretion of orally administrated arsenic triocide in the hamster. *Toxicology*, **34**, 11-121.

Yamauchi, H., Takahashi, K. and Yamamura, Y. 1989. Metabolism and excretion of orally and intraperitoneally administered trimethylarsine oxide in the hamster. *Toxicol. Environ. Chem.*, **22**, 69-76.

Yamauchi, H., Kaise, T., Takahashi, K. and Yamamura, Y. 1990. Toxicity and metabolism of trimethylarsine in mice and hamsters. *Fund. Appl. Toxicol.*, **14**, 399-407.

17 Inorganic arsenic metabolism in humans

J.P. Buchet* and R. Lauwerys

Catholic University of Louvain, Industrial Toxicology and Occupational Medicine Unit, Clos Chapelle-aux-Champs, 30.54, B-1200 Brussels, Belgium

Abstract
For the general population, the toxicological risks of arsenic mainly result from the possible presence of some inorganic derivatives in food, ambiant air and drinking water. Soluble inorganic arsenic (As_i) compounds are usually well absorbed by the oral and pulmonary routes mainly when the element is in the trivalent state. Arsenic is rapidly eliminated through the kidney either unchanged or after methylation to less biologically active metabolites. A fraction of arsenic is incorporated in hair and nails. As_i can also be excreted in bile as a glutathione complex which may be partly reabsorbed.
The only arsenic species excreted in urine when exposure has been to As_i are As_i itself, particulary during the first hours after the start of exposure, monomethylarsonic (MMA) and dimethylarsinic (DMA) acids, the latter being the preponderant species after a few hours. A high uptake of As can transiently inhibit the production of DMA but so far in human this phenomenon has mainly been observed after acute intoxication. Like the organoarsenicals present in marine organisms, DMA is excreted unchanged while MMA is slightly methylated into DMA before excretion.
The study of As_i metabolism by rat tissues in vitro and observations made in humans exposed to As_i suggest that it involves two consecutive methylation steps both requiring S adenosylmethionine but catalysed by enzymatic systems differing in their resistance to inhibition by As_i and their requirement for reduced glutathione.

Introduction

Environmental pollution may be responsible for exposure of the general population to inorganic arsenic (As_i) through inhaled air (*e.g.* surroundings of non ferrous metal smelters) or drinking water (geochemical contamination). The As_i content of food may also contribute to the daily absorbed dose but the methylated As derivatives, particularly the trimethylated forms sometimes present in high amount in some

*To whom correspondence should be addressed.

marine organisms are not considered to represent an important health risk. In microorganisms as well as in mammals, As_i is a substrate for methylating enzymes which produce less toxic metabolites. As_i is a human carcinogen and the extent of its methylation may modulate its carcinogenic activity. We have studied some aspects of arsenic toxicokinetics in healthy volunteers, in workers exposed to As_i, in patients with liver diseases and in subjects acutely intoxicated by As_i; in each case, the urinary concentration of several arsenicals (total As, inorganic forms-As_i, monomethylarsonic acid-MMA and dimethylarsinic acid-DMA) was followed up during known periods of time.

Methodology

Arsenic analysis

As_i (tri-and pentavalent forms) and its methylated derivatives MMA and DMA were measured by atomic absorption spectrometry of the corresponding arsines generated at low pH with sodium borohydride following a modification of the Braman and Foreback (1973) procedure (Buchet and Lauwerys, 1981). For the measurement of the total amount of As, the sample was submitted to a drastic ashing step using Mg $(NO_3)_2$ in a MgO matrix at 650^0 C (Buchet *et al.*, 1980).

Experimental protocols

Members of the laboratory staff and patients from the university hospital volunteered to the studies the design of which was approved by the ethical committee of the UCL Faculty of Medicine. Prior to the administration of As, urine was collected for 24 h to obtain background excretion values. Administration of As occurred just after the volunteers had voided their bladder; doses were adjusted so that the subject had to drink 150 mL of water enriched with the test compound or be injected IV with 0.2 mL/10 kg body weight of a sterile aqueous solution containing 0.62 mg $NaAsO_2$ and 0.5 mg ascorbic acid per milliliter. Depending upon the experiment urine was collected at least for 24 h or up to 14 days in one or several samples corresponding to 4, 12 or 24 h collection periods. In cases of severe acute intoxication (3 accidental poisonings and 2 suicide attempts) urine collection started at the admission in the intensive care unit (between 5 to 18 h after the As intake).

Results

Chemical forms of As in urine

The excretion of high quantities of As in urine after seafood consumption is well documented, but the exact chemical form is generally unknown. Trimethylated arsenicals reducible to trimethylarsine has been sometimes found, but frequently seafood arsenic in urine can only be measured after efficient mineralisation. We have determined the chemical forms of arsenic present in urine of control subjects and workers occupationally exposed to As_2O_3; both groups of subjects had refrained from eating seafood for several days before urine sampling. Most of the time a very close agreement was observed between the total amount of urinary As (obtained after complete ashing of the sample) and the sum of As_i (tri and pentavalent), MMA and

DMA measured by the hydride generation technique, suggesting that no other metabolite was excreted after exposure to As_i; in the range of urinary concentrations between 2 and 1,600 μg As L^{-1}, the correlation coefficient between the results

Table 1 *Changes in the chemical speciation of arsenic in urine following acute intoxication by* As_2O_3.

Subject Age	Sex	Day	Total As excreted μg/24 h	% excreted as As_i	MMA	DMA
44	M	1	1,900	64	17	19
		2	9,400	73	13	14
		3	17,600	83	6	11
		6	12,700	43	17	40
		9	7,400	3	2	95
40	F	1	16,700	76	11	13
		3	10,700	62	18	20
		8	4,600	2	13	87

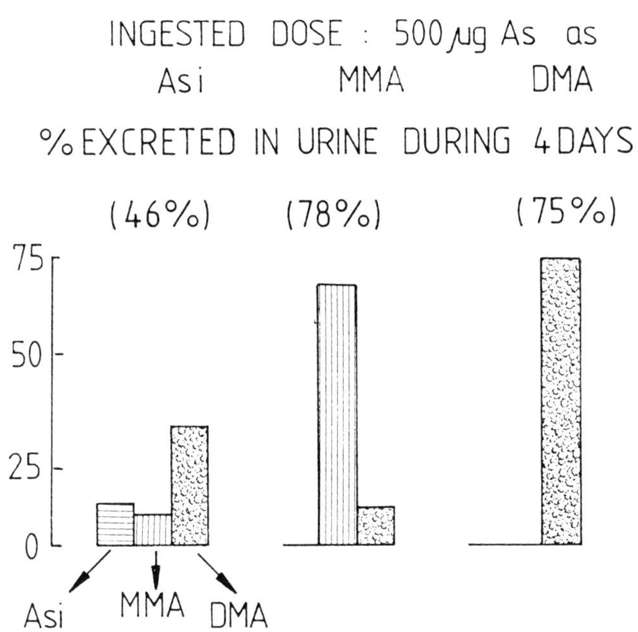

Figure 1 *Comparison of the urinary excretion of* As_i *and its methylated metabolites in 4 days after ingestion of 500* μg *As as* As_i, *MMA or DMA.*

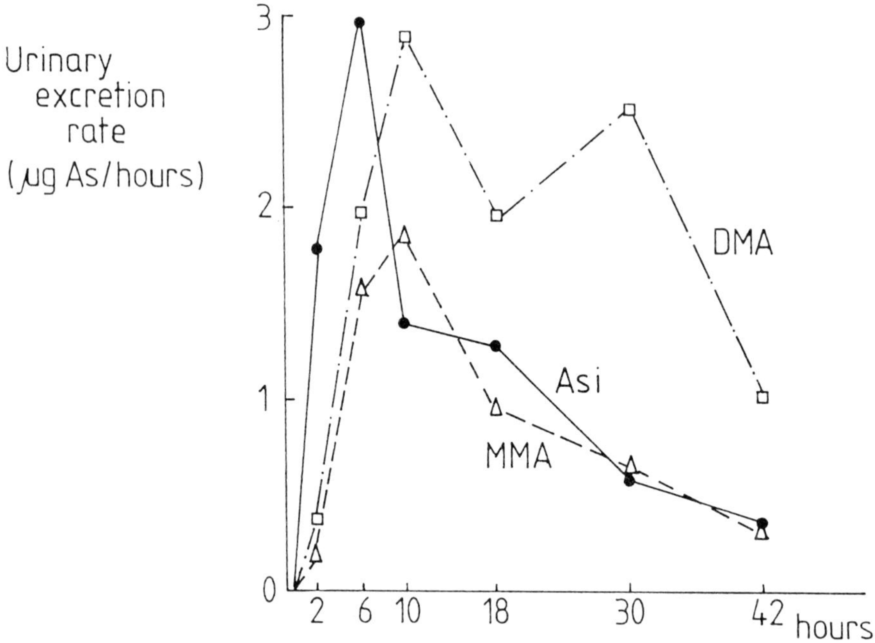

Figure 2 *As_i, MMA and DMA excretion rates following an oral dose of 500 µg As_i ($NaAsO_2$).*

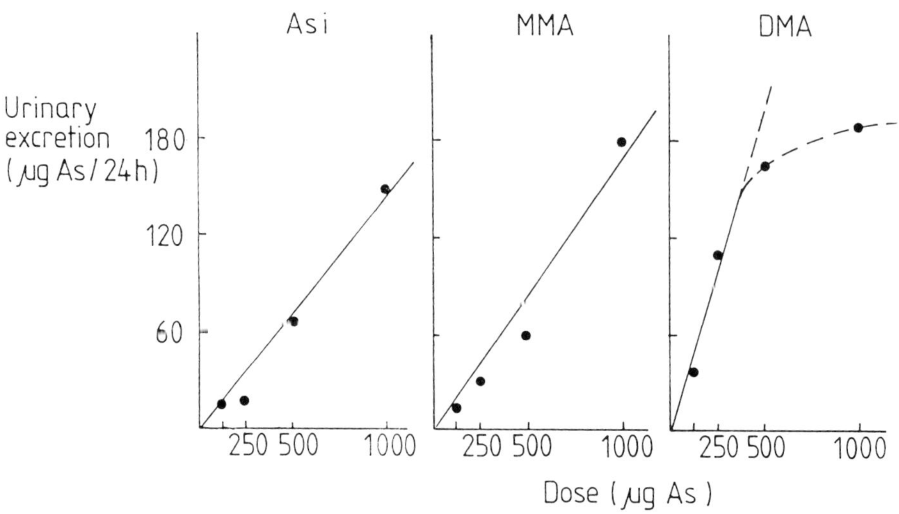

Figure 3 *Urinary excretion of As_i and its metabolites during 24 h following 5 daily administration of As_i orally.*

Table 2 *Methylation of As_i injected IV in control subjects and in patients with liver disease.*

Subjects	*n*	Total As excretion within 24 h % of the injected dose	Proportions of metabolites (% of total excretion) As_i	MMA	DMA
Controls					
Women	6	29.9 ± 2.8	62.2 ± 3.7	12.6 ± 1.2	25.2 ± 3.4
Men	7	29.8 ± 1.5	63.5 ± 1.8	12.9 ± 0.9	23.5 ± 1.1
Patients					
Alcoholic cirrhosis	9	27.5 ± 2.7	53.9 ± 3.3	5.1 ± 0.7	40.6 ± 3.3
Chronic hepatitis	5	39.4 ± 3.0	57.3 ± 2.8	9.2 ± 1.4	33.5 ± 3.4
Hemochromatosis	3	32.3 ± 3.0	45.1 ± 9.3	4.1 ± 2.0	50.9 ± 7.6
Postnecotric cirrhosis	9	28.8 ± 3.0	57.2 ± 3.1	6.5 ± 1.6	36.3 ± 2.2

obtained with or without sample ashing amounts to 0.993 in a group of 34 workers and control subjects.

Comparison of the urinary excretion of arsenic metabolites after a single oral dose of As_i, MMA or DMA (Buchet et al., 1981 a)

The urinary excretion of metabolites (unchanged Asi, MMA and DMA) after the administration of a single oral dose of 500 μg As as Asi, MMA or DMA, indicates that within 4 days (Figure 1) : 1. 75 % of the ingested dose of DMA is excreted unchanged; 2. 68 % of the ingested dose of MMA is excreted unchanged, but a small amount (10 %) is excreted as DMA; 3. 46 % only of the ingested dose of As_i is excreted in 4 days; and 4. the proportions of As_i, MMA and DMA amount to 11, 8.3 and 26.7 % respectively. Results obtained after mineralisation of the samples confirmed that no chemical forms other than As_i, MMA and DMA are present in urine.

After an oral dose of MMA or DMA, the substances are excreted with an apparent half-time of 4 and 11 h respectively, while the time required to reduce the As urinary excretion by 50 % after an oral dose of As_i amounts to 28 h. As seen in Figure 2, As_i is preponderant during the first hours following ingestion and the decrease of As_i excretion is associated with an increased proportion of the methylated forms; the increase of the MMA fraction is transient and that of DMA lasts until it reaches the percentage found in pre-exposure urine samples.

Urinary excretion of inorganic arsenic and its metabolites after repeated ingestion of trivalent As_i by volunteers (Buchet et al., 1981 b).

Arsenic (125, 250, 500 or 1,000 μg As as $NaAsO_2$) was administered orally once a day for five consecutive days to 4 volunteers who refrained from eating seafood during the experiment. Urine was collected during 24 h periods starting one day before the first administration and up to 14 days later. The speciation of As_i and its methylated metabolites MMA and DMA shows that a steady state in the urinary

Table 3 *Correlation coefficients and parameters of regression equations ($y = a + bx$) between the concentration of As metabolites in urine (y) and the concentration of As_2O_3 (x) in air measured with personal sampler (log distribution; $n = 82$).*

Y	r	*p*	a	b
Post shift sample				
log As_i	0.53	0.0001	0.443	0.401
log MMA	0.24	0.0315	0.589	0.173
log DMA	0.43	0.0001	0.843	0.376
log As total	0.48	0.0001	1.130	0.353
Before next shift sample				
log As_i	0.40	0.0002	0.350	0.273
log MMA	0.32	0.0035	0.542	0.240
log DMA	0.45	0.0001	0.812	0.414
log As total	0.46	0.0001	1.098	0.353

excretion of arsenic is reached within 5 days and that the total amount of As excreted in urine per day amounts to 60 % of the ingested dose. However measurements made on the samples collected during the day following the last administration show that while the excretion of As_i and MMA are linearly retated to the dose administered, the excretion of DMA levels off at the highest dose, indicating a possible saturation of the methylation capacity (Figure 3).

Pattern of urinary excretion of As_i and its metabolites after acute intoxication (Mahieu et al., 1981).

The urinary excretion of As_i and its metabolites MMA and DMA has been followed-up in several cases of acute oral intoxication by As_2O_3. The proportion of the three species of urinary As changes markedly with time (Table 1). During the first 2 to 4 days after intoxication, As is excreted mainly as its unmetabolized inorganic form but this is rapidly followed by a progressive increase of the proportion excreted as MMA and DMA The time at which the majority of arsenic is excreted as its organic metabolites depends on the severity of the intoxication but in all the cases, more than 95 % of the excreted As is in the organic form after 9 days, DMA being preponderant.

Influence of liver disease on the methylation of arsenite in humans (Buchet et al., 1984).

To avoid misleading interpretations possibly due to variable rates of gastrointestinal absorption of As_i in controls and patients with liver disease, the IV route of administration was selected. A comparison of the toxicokinetics after its IV and PO administration was also performed in three control subjects. After IV injection of 500 μg As_i / 70 kg b.w., the percentage of the administered dose excreted in urine within 24 h is higher (34 %) than that observed after oral ingestion of the same dose (20 %). However, the proportion of the methylated derivatives is smaller after IV injection (37 % IV versus 60 % PO).

The influence of liver diseases on the methylation of arsenite is presented in Table 2. When compared to controls, the percent of the injected dose excreted within 24 h by patients is not significantly different, but the proportion of As_i excreted as DMA is significantly increased whereas that excreted as MMA is significantly decreased.

Relationship between airborne arsenic trioxide and the urinary excretion of inorganic arsenic and its methylated metabolites (Offergelt et al. 1992)
During five consecutive days, the concentration of As in the breathing zone of 18 workers from a sulphuric acid producing plant was measured using a personal air sampling system with a collection efficiency higher than 95 %. Each worker provided a urine sample before and after each shift for As_i, MMA and DMA excretion measurements. As concentrations in these spot samples were standardized for a creatinine concentration of 1 g L^{-1}. The time weighted average exposure concentrations of As_2O_3 ranged from 6 to 502 μg As m^{-3} with a lognormal distribution (SD = 2.4). Despite the fact that exposure to As_2O_3 is likely to occur not only by inhalation but also by ingestion, low but statistically significant correlation coefficients are found between the concentration of each As metabolite in urine collected before or after the shift and the airborne concentration of As_i, they vary between 0.4-0.5 for As_i and DMA and are lower (0.2-0.3) for MMA (Table 3). Only for As_i is the mean concentration in urine collected after the shift significantly higher than that found the next morning. The proportion of As_i in urine falls from 25 % of the total after the shift to 14 % before the next shift.

Discussion and conclusions

Our studies of As_i metabolism in volunteers indicate that the measurement of the urinary excretion of the element and its metabolites MMA and DMA accounts for an important proportion of the absorbed dose of As_i. The methylation of As_i in human is characterised by the synthesis of only mono- and dimethylated derivatives which are efficiently excreted in urine along with unchanged inorganic form : no trimethylation is detected. A close examination of the excretion rates of As_i and its metabolites indicates that after an oral dose of As_i several hours are required before the methylated metabolites (DMA particularly) become the preponderant arsenical excreted. This latency period is prolonged when the absorbed dose of As_i is increased (*e.g.* when repeated and elevated doses are ingested or in the case of acute intoxication). Another conclusion derived from our studies relates to the importance of the liver in the As_i metabolism : significant differences in the pattern of excretion of arsenic metabolites are observed depending on the route of administration or on the presence of liver insufficiency. These observations in humans and other in vitro or in vivo studies in the rat (Buchet and Lauwerys, 1985; Georis *et al.*, 1990) are consistent with a metabolic scheme made up of two consecutive reactions catalysed by two different methyltransferases. These enzymes use S adenosylmethionine to methylate trivalent As_i into a derivative which is either transformed into MMA or is used as substrate for the formation of a dimethylated derivative giving rise to DMA. The rate limiting step is the first methylation reaction which is facilitated by the presence of reduced glutathione (GSH); the second methylation reaction appears very efficient but is easily inhibited by an excess of As_i, at the opposite of the first methylation reaction which is more sensitive to inhibition by the product MMA. The existence of two consecutive methylating reactions may explain the changes with time in the proportions of As_i, MMA and DMA found in urine after absorption of small As_i dose. The high sensitivity of the second methylation step to inhibition by As_i may help explain the reduced or even the abolished production of DMA in case of exposure to high amount of As_i. Eventually the change in excretion pattern observed between healthy controls and patients with liver diseases is probably due to the reduced GSH liver content in case of liver insufficiency. Indeed we found in animals a reduced uptake of As_i by hepatocytes when hepatic GSH is low, and a decreased accumulation of MMA in the liver due to a higher efficiency of its transformation into DMA caused by a decreased inhibition of the second methylation step by As_i : as a result, the amount of As excreted in urine as MMA decreases while that of DMA increases (Buchet and Lauwerys, 1987).

In workers exposed to As_2O_3, the measurement of the urinary excretion of As_i, MMA and DMA allows a better evaluation of the health risk because it takes into account all absorption routes. However the relative proportions of metabolites strongly depend upon the sampling time. In addidtion, according to recent observations, the biological monitoring of exposure to As_i should preferably be performed on subjects who refrained from eating seafood for at least 48 hours because organoarsenicals from marine origin can release a small amount of DMA *in vivo*.

References

Braman, R.S. and Foreback, C.C. 1973. Methylated forms of arsenic in the environment. *Science*, **182**, 1247-1249.

Buchet, J.P. and Lauwerys R. 1981. Evaluation of exposure to inorganic arsenic in man. In: Facchetti, S. (ed.), *Analytical techniques for heavy metals in biological fluids*, pp.75-90. Elsevier, Amsterdam

Buchet, J.P., Lauwerys, R. and Roels, H. 1980. Comparison of several methods for the determination of arsenic compounds in water and in urine. *Int.Arch.Occup.Environ.Health*, **46**, 11-29.

Buchet, J.P., Lauwerys, R. and Roels, H. 1981 a. Comparison of the urinary excretion of arsenic metabolites after a single oral dose of sodium arsenite, monomethylarsonate, or dimethylarsinate in man. *Int. Arch. Occup. Environ. Health*, **48**, 71-79.

Buchet, J.P., Lauwerys, R. and Roels, H. 1981 b. Urinary excretion of inorganic arsenic and its metabolites after repeated ingestion of sodium metaarsenite by volunteers. *Int.Arch.Occup.Environ.Health*, **48**, 111-118.

Buchet, J.P., Geubel, A., Pauwels, S., Mahieu, P. and Lauwerys, R. 1984. The influence of liver disease on the methylation of arsenite in humans. *Arch. Toxicol.*, **55**, 151-154.

Buchet, J.P. and Lauwerys, R. 1985. Study of inorganic arsenic methylation by rat liver *in vitro* : relevance for the interpretation of observations in man. *Arch. Toxicol.*, **57**, 125-129.

Buchet, J.P. and Lauwerys R. 1987. Study of factors influencing the *in vivo* methylation of inorganic arsenic in rats. *Tox. Appl. Pharmacol.*, **91**, 65-74.

Georis, B., Cardenas, A., Buchet, J.P. and Lauwerys, R. 1990. Inorganic arsenic methylation by rat tissue slices. *Toxicology,* **63**, 73-84.

Mahieu, P., Buchet, J.P., Roels, H. and Lauwerys, R. 1981. The metabolism of arsenic in humans acutely intoxicated by As_2O_3. Its significance for the duration of BAL therapy. Clin.Toxicol, **18,** 1067-1075.

Offergelt, J.A., Roels, H., Buchet, J.P., Boeckx, M. and Lauwerys, R. 1992. Relation between airborne arsenic trioxide and urinary excretion of inorganic arsenic and its methylated metabolites. *Brit.J.Ind.Med.*, **49**, 387-393.

18 Genotoxicity of arsenic and its methylated metabolites

Martha M. Moore,[1] Karen Harrington-Brock[2] and Carolyn L. Doerr[2]

[1] *Mutagenesis and Cellular Toxicology Branch, Genetic Toxicology Division, Health Effects Research Laboratory, U.S. Environmental Protection Agency, Research Triangle Park, NC 27711,USA*
[2] *Environmental Health Research and Testing, Inc., Research Triangle Park , NC 27709, USA*

Abstract

Arsenic is one of the few identified human carcinogens that does not cause cancer in rodents. The sensitivity of humans to arsenic-induced cancer may be related to the limited ability of humans to detoxify arsenic. Methylation is potentially an important detoxification pathway for arsenic. If methylation detoxifies arsenic, one would predict that the methylated arsenicals might be less genotoxic than the inorganic arsenicals. To test this hypothesis, we have evaluated four arsenicals using an in vitro gene mutation and gross aberration assay. Sodium arsenite and sodium arsenate were active at concentrations of 1-2 g mL^{-1} and 10-14 g mL^{-1}, respectively. Monomethylarsonic acid (MMA) and dimethylarsinic acid (DMA) were significantly less potent than the inorganic arsenicals. MMA was active between 2,500-5,000 g mL^{-1} while, DMA required almost 10,000 g mL^{-1} to induce a genotoxic response.

A review of the literature and our current study indicate that the inorganic arsenicals are capable of inducing chromosomal mutations but not point mutations. The current study demonstrates that the methylated arsenicals (although much less potent than the inorganic arsenicals) also act by mechanisms that cause chromosomal mutations.

In selecting a risk assessment model for arsenic, it is important to understand both the metabolism (detoxification) and the genotoxicity of the arsenicals. Because arsenic exerts its genotoxicity by causing chromosomal mutations, it may well act in one of the latter stages in cancer development. This is consistent with the

---

This manuscript has been reviewed by the Health Effects Research Laboratory, US Environmental Protection Agency, and approved for publication. Approval does not signify that the contents necessarily reflect the views and policies of the Agency, nor does mention of trade names or commercial products constitute endorsement or recommendation for use.

epidemiologic evidence that suggests that arsenic is a progressor rather than a classic initiator or promotor.

Introduction

Arsenic is one of the few identified human carcinogens that has yet to be shown to cause cancer in rodents (IARC, 1980, 1990). Because it is somewhat unique as a human but not rodent carcinogen, considerable attention has been placed in attempting to understand the mechanisms by which arsenic is metabolized by humans and rodents and the mechanisms by which arsenic exerts its carcinogenicity in humans.

A considerable data base exists for the genotoxicity of both the 3+ and 5+ valence states of arsenic. Most of the available genotoxicity data has been generated using inorganic arsenicals. Jacobson-Kram and Montalbano reviewed the mutagenicity data for inorganic arsenic in 1985. A recent article by Mass (1992) summarizes the genotoxicity data base for arsenicals and includes the Genetic Activity Profiles developed by Dr. Michael Waters and coworkers. The Profiles graphically depict the available data from a number of possible test systems and show either the lowest effective or highest ineffective concentration for each individual test evaluation. The complete details of the profiles can be found in the IARC Monographs on the Evaluation of Carcinogenic Risks to Humans, genetic and related effects: an updating of selected IARC monographs from Vols 1 to 42. Supplement 6. Lyon, France, 1987. Briefly, the available data indicates that arsenic does not induce point mutations. Arsenic does, however, induce both sister chromatid exchange and chromosome aberrations.

Methylation is potentially an important detoxification pathway for arsenic (Vahter 1981). If methylation detoxifies arsenic, one would predict that the methylated arsenicals might be less genotoxic than the inorganic arsenicals. To test this hypothesis, we have evaluated four arsenicals (sodium arsenite, sodium arsenate, MMA and DMA) for their relative mutagenic potency. For these analyses, we selected the L5178Y $TK^{+/-}$-3.7.2C mouse lymphoma assay. Although this is a specific locus assay, it has been shown that the thymidine kinase (tk) locus used in this assay is capable of undergoing a broad spectrum of genetic damage that can be detected as loss of function of the tk gene (Moore *et al.*, 1985b; Applegate *et al.*, 1990). Mutational events that may be detected by this assay include point mutation, deletion (including multilocus deletion), translocations, somatic mitotic recombination, and gene conversion (see Applegate *et al.*, 1990, for a discussion of the genetic alterations detected.) One feature of this assay that is particularly useful for evaluating potential mutagenicity is the detection of mutants that fall into two basic classes. These classes can be separately quantitated based on the colony size of the mutants and provide insight into the mechanism (single gene mutation vs chromosomal mutation) of induced damage (Moore *et al.*, 1985a). Small colony mutants contain cells that grow more slowly than the normal cell population and have been associated with the induction of chromosomal mutations [and are induced by chemicals that have clastogenic activity (Moore and Doerr, 1990)]. Large colony mutants contain cells that grow at normal population rates and have been associated

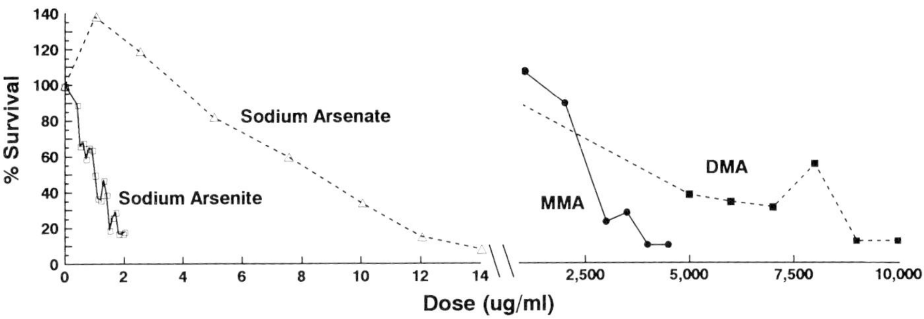

Figure 1 *Cytotoxicity observed following treatment of mouse lymphoma cells with either sodium arsenite, sodium arsenate, MMA or DMA. The % Survival is calculated based on the methods of Clive and Spector, 1975.*

with the induction of relatively small scale (single gene) damage (Applegate *et al.*, 1990).

In the present study, we evaluated the four arsenicals for their ability to induce thymidine kinase deficient mutants and also for their ability to induce gross chromosome aberrations.

Methodology

Chemicals

Sodium arsenite (CAS no. 7784-46-5), sodium arsenate (Cas no. 7631-89-2), and dimethylarsinic acid (Cas no. 124-65-2) were obtained from Sigma Chemical Company (St. Louis, MO), and monomethylarsonic acid (Cas no. 2163-80-6) was obtained from ISK-Biotech Corporation (Mentor, Ohio).

Cells

The $TK^{+/-}$-3.7.2C heterozygote of L5178Y mouse lymphoma cells was used. Cells were maintained in culture using Fischer's Medium for Leukemic Cells of Mice supplemented with Sodium pyruvate, pluronic, penicillin, streptomycin, and horse serum according to the procedures described by Turner *et al.* (1984).

Experimental methods

Duplicate cultures of L5178Y/$TK^{+/-}$-3.7.2C cells were treated without exogenous activation for 4 h according to the procedures of Turner *et al.* (1984). The mutagenicity assay and the cytogenetic analysis were performed according to the procedures described by Doerr *et al.* (1989).

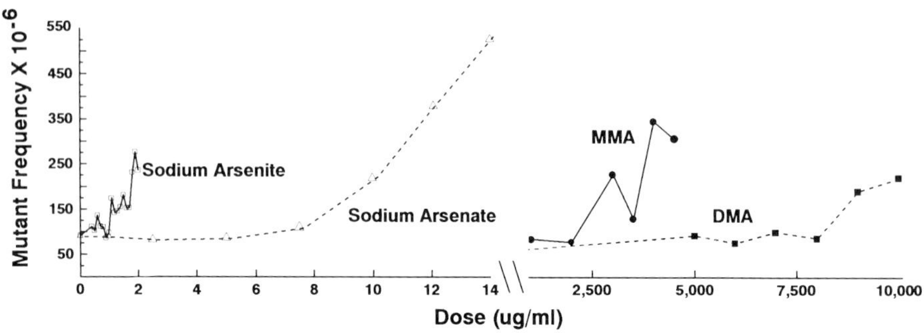

Figure 2 *Mutagenicity of sodium arsenite, sodium arsenate, MMA and DMA to the thymidine kinase gene of mouse lymphoma cells.*

Results

In the course of performing the mutagenesis assay, it is necessary to define the cytotoxicity of the test chemicals. This cytotoxicity is expressed as the relative total growth or percent survival. It is calculated according to the method of Clive and Spector (1975) and includes the relative (to the untreated control) suspension cell growth and the relative plating efficiency during the mutant selection. Test chemicals are evaluated over a concentration range that yields little cytotoxicity (close to 100% survival) to substantial cytotoxicity (approx. 10-20% survival). The relative cytotoxicity of the four arsenicals is shown in Figure 1. Sodium arsenite is the most cytotoxic followed by sodium arsenate, MMA and DMA. The relative mutagenicity (at the tk locus) of the four arsenicals is shown in Figure 2. All four chemicals are mutagenic with the following relative potency: sodium arsenite>sodium arsenate>>MMA>DMA. The majority of mutants were small colony mutants (data not shown). Very few if any large colony mutants were induced. This is indicative of the induction of chromosomal mutations but not single gene mutations. The clastogenicity (analyzed by the induction of microscopically visible gross chromosomal aberrations) of the four arsenicals is shown in Figure 3. Sodium arsenite, sodium arsenate, and MMA are clastogenic. DMA induced a slight increase in gross aberration frequency, but it was insufficient to be declared clastogenic.

Discussion

Organic arsenicals are substantially less cytotoxic, mutagenic, and clastogenic than the inorganic arsenicals. It should be noted that DMA demonstrated such little cytotoxicity that had it not been extremely soluble we would not have continued with the evaluation. Most investigators do not continue the evaluation of chemicals that are not cytotoxic at 5,000 g mL $^{-1}$. This practice is based on the feeling that chemicals requiring more than 5,000 g mL $^{-1}$ to induce cytotoxicity or mutagenicity are not of significant health concern.

The mutagenic activity of the arsenicals based both on the available literature and on the current evaluations is due to their ability to induce chromosomal mutations. Sodium arsenite and sodium arsenate demonstrate substantial potency as mutagenic agents. They exert their effects over dose ranges of 1-2 and 10-14 g mL $^{-1}$, respectively.

The comutagenicity and coclastogenicity of inorganic arsenic (primarily sodium arsenite) has been evaluated by several investigators (Li and Rossman, 1989, 1991,

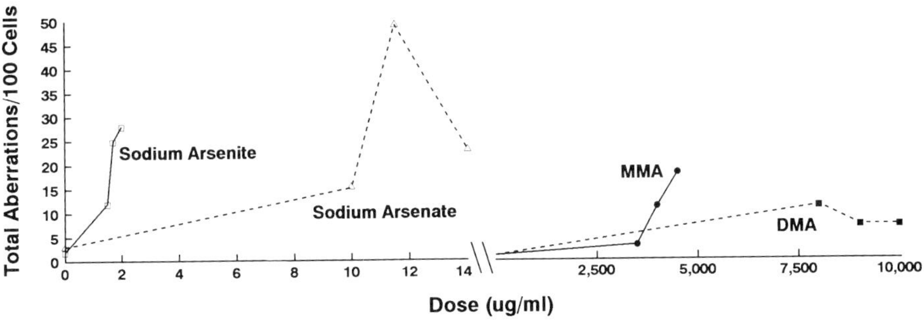

Figure 3 *Induction of microscopically visible chromosome aberrations following treatment of mouse lymphoma cells with sodium arsenite, sodium arsenate, MMA or DMA.*

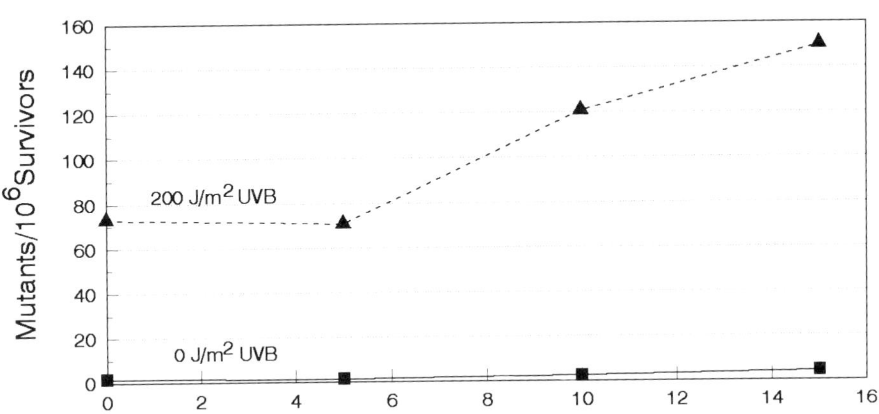

Figure 4 *Comutagenicity of sodium arsenite with ultraviolet light. Data from Li and Rossman (1991) has been graphically expressed. Four cultures of V79 cells were either treated with (200 J m^{-2}) or not treated with ultraviolet light. Six of the eight cultures received either 5, 10 or 15 μM sodium arsenite. The hprt mutant frequency for each culture was determined.*

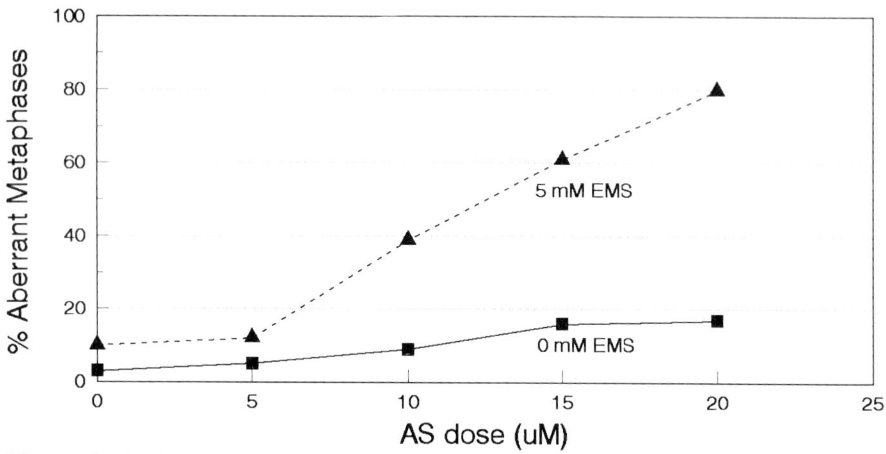

Figure 5 *Coclastogenicity of sodium arsenite and ethyl methanesulfonate. Data from Jan* et al. *(1986) has been graphically expressed. Ten cultures of Chinese hamster ovary cells were either treated with (5mM) or not treated with ethyl methanesulfonate. Eight of the ten cultures received either 5, 10, 15 or 20 µM sodium arsenite. The percent aberrant metaphases were determined for each culture.*

Jan *et al.*, 1986, Wiencke and Yager, 1992). Of particular interest and significance (in the potential role of arsenic in the induction of skin cancer) is the research of Li and Rossman (1991). These investigators demonstrated the comutagenicity of sodium arsenite with ultraviolet radiation (see Figure 4 for a graphical representation of their data). For these studies the hypoxanthine guanine phosphoribosyltransferase (hprt) locus of V79 cells was used. The hprt locus does not detect the ability of chemicals to induce chromosomal mutations (see Moore *et al.*, 1989, for a comparison of the hprt and tk loci). Thus sodium arsenite is negative in the V79 hprt assay. Ultraviolet light induces point mutations and is thus detected using the hprt locus as the marker gene. When sodium arsenite and ultraviolet light are combined there is an increase (over that seen with ultraviolet light alone) in the frequency of hprt mutants.

Sodium arsenite demonstrates a dramatic coclastogenicity. Jan *et al.* (1986) found that sodium arsenite greatly enhances the clastogenicity of ethyl methanesulfonate. Their data is shown in graphic form in Figure 5.

The comutagenic/coclastogenic data is particularly significant. Humans are obviously exposed to complex mixtures of chemicals rather than to single chemicals. These mixtures will likely contain chemicals that can be point mutagens as well as clastogens. If arsenic is included in the mixture, the human risk from that arsenic exposure may well be substantially higher than would be predicted from the arsenic exposure alone.

Conclusions

In selecting a risk assessment model for arsenic, it is important to understand both the metabolism (detoxification) and the genotoxicity of the arsenicals. Because arsenic exerts its genotoxicity by causing chromosomal mutations, it may well act in one of the latter stages in cancer development. This is consistent with the epidemiologic evidence that suggests that arsenic is a progressor rather than a classic initiator or promotor. The comutagenic/coclastogenic potential of arsenic is of particular importance in developing the quantitative risk assessment for human exposure to arsenic.

Acknowledgements

We acknowledge the expert technical assistance of Lisa P. Burnette in the conduct of these experiments and of Ms. Velva Milholland in the preparation of the graphics.

References

Applegate, M.L., Moore, M.M., Broder, C.B., Burrell, A., Juhn, G., Kasweck, K.L., Lin, P.-F., Wadhams, A.and Hozier, J.C. 1990. Molecular dissection of mutations at the heterozygous thymidine kinase locus in mouse lymphoma cells. *Proc. Natl. Acad. Sci.*, **87**, 51-55.

Clive, D and Spector, J.F.S. 1975. Laboratory procedure for assessing specific locus mutations at the TK locus in cultured L5178Y mouse lymphoma cells. *Mutat. Res.*, **31**, 17-29.

Doerr, C.L., Harrington-Brock, K., and Moore, M.M. 1989. Micronucleus, chromosome aberration, and small-colony TK mutant analysis to quantitate chromosomal damage in L5178Y mouse lymphoma cells. *Mutat. Res.*, **222,** 191-203.

IARC (International Agency for Research on Cancer). 1980. In: IARC Monographs on the Evaluation of the Carcinogenic Risk of Chemicals to Humans, 20, *Some Metals and Metallic Compounds,* pp.39-140. IARC, Lyon.

IARC (International Agency for Research on Cancer). 1990. In: IARC Monographs on the Evaluation of the Carcinogenic Risk of Chemicals to Humans, **50**, *Pharmaceutical Drugs, pp.,* 47-63. IARC, Lyon.

Jacobson-Kram, D. and Montalbano, D. 1985. The reproductive effects assessment group's report on the mutagenicity of inorganic arsenic. *Environ. Mutagen.* **7,** 787-804.

Jan, K.Y., Huang, R.Y., and Lee, T.C. 1986. Different modes of action of sodium arsenite, 3-aminobenzamide, and caffeine on the enhancement of ethyl methanesulfonate clastogenicity. *Cytogenet. Cell Genet.,* **41**, 202-208.

Li, J-H. and Rossman, T.G. 1989. Mechanism of comutagenesis of sodium arsenite with n-methyl-n-nitrosourea. *Biol. Trace Element Res.,* **21**, 373-381.

Li, J-H. and Rossman, T.G. 1991. Comutagenesis of sodium arsenite with ultraviolet radiation in Chinese hamster V79 cells. **Biol. Metals, 4**, 197-200.

Mass, M.J. 1992. Human carcinogenesis by arsenic. *Environ. Geochem. Health,* **14,** 49-54.

Moore, M.M., Clive, D., Howard, B.E., Batson, A.G., and Turner, N.T. 1985a. *In situ* analysis of trifluorothymidine-resistant (TFT^r) mutants of $L5178Y/TK^{+/-}$ mouse lymphoma cells. *Mutat. Res.*, **151**, 147-159.

Moore, M.M., Clive, D., Hozier, J.C., Howard, B.E., Batson, A.G., Turner, N.T., and

Sawyer, J. 1985b. Analysis of trifluorothymidine-resistant (TFT^r) mutants of L5178Y/$TK^{+/-}$ mouse lymphoma cells. *Mutat. Res.*, **151**, 161-174.

Moore, M.M. and Doerr, C.L. 1990. Comparison of chromosome aberration frequency and small-colony TK-deficient mutant frequency in L5178Y/$TK^{+/-}$ -3.7.2C mouse lymphoma cells. *Mutagenesis*, **5**, 609-614.

Moore, M.M., Harrington-Brock, K., Doerr, C.L. and Dearfield, K.L. 1989. Differential mutant quantitation at the mouse lymphoma *tk* and CHO *hgprt* loci. *Mutagenesis,* **4**, 394-403.

Turner, N.T., Batson, A.G., and Clive, D. 1984. Procedures for the L5178Y/$TK^{-/-}$Mouse Lymphoma Cell Mutagenicity Assay. *Handbook of Mutagenicity Test Procedures,* pp.239-268.

Vahter, M. 1981. Biotransformation of trivalent and pentavalent inorganic arsenic in mice and rats. *Environ. Res.*, **25**, 286-293.

Wiencke, J.K. and Yager J.W. 1992. Specificity of arsenite in potentiating cytogenetic damage induced by the DNA crosslinking agent diepoxybutane. *Environ. Mol. Mutagen.* **19,** 195-200.

19 Arsenic essentiality and factors affecting its importance

Eric O. Uthus

United States Department of Agriculture Agricultural Research Service, Grand Forks Human Nutrition Research Center, PO Box 9034 , Grand Forks, ND, 58202-9034 USA

Abstract

Homeostatic mechanisms and the diversity in sources and types of foods consumed make the occurrence of overt, simple, or uncomplicated deficiencies of trace or ultratrace elements in humans relatively uncommon. There are, however, situations that make trace elements of practical nutritional importance. These situations either affect homeostasis or cause a metabolic stress, resulting in signs of deficiency at concentrations of the element not normally associated with deficiency. The first strong evidence supporting the essentiality of arsenic for animals became available in the 1970s. Although notable effects of arsenic deprivation were found in pregnant rats, goats, and minipigs and their offspring, the most striking effects were seen with lactating goats. Some of these animals died, apparently from myocardial dysfunction. Later studies, using rats, chicks, and hamsters, have not used pregnancy but other stressors, such as those that alter methionine metabolism, to exacerbate arsenic deprivation. Authors of a recent study in humans concluded that arsenic homeostasis was altered by hemodialysis treatment and certain diseases and, therefore, maintaining desirable arsenic concentrations in the body seemed reasonable. Thus, arsenic may be similar to selenium in that a metabolic stressor (such as vitamin E deficiency for selenium) is necessary to produce marked signs of deficiency in laboratory animals or in humans. These studies have suggested that arsenic has a physiological role that affects methionine metabolism.

Factors influencing signs of trace or ultratrace element deficiency

As stated by Frieden (1984), "The simplest definition of an essential element is that it is an element required for the maintenance of life; its absence results in death or a severe malfunction of the organism". He further states "Experimentally, this rigorous criterion cannot always be satisfied and this has led to a broader definition of essentiality. An element is considered essential when a deficient intake produces an

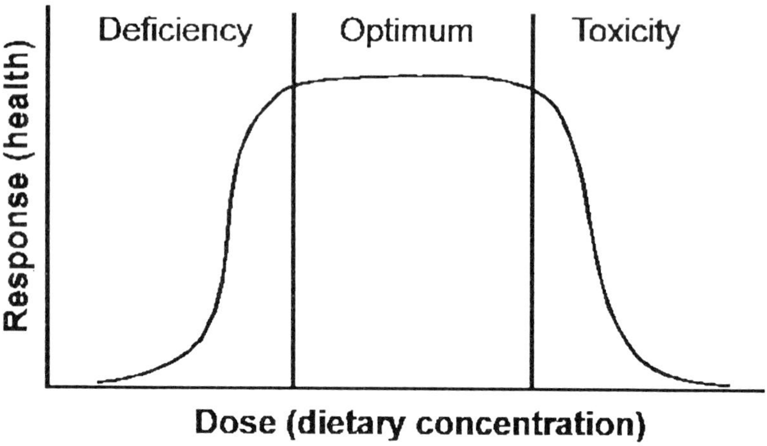

Figure 1 *Dose-response curve.*

impairment of function and when restoration of physiological levels of that element prevents or relieves the deficiency". This can be illustrated by the dose-response curve (Figure 1) as first proposed by Bertrand (1912) and further elucidated by Frieden (1984). The dose response curve illustrates that there is a range (concentration of the element in the diet, for example) at which optimal health is maintained (the center portion of the graph between the 2 vertical lines). As deficiency increases (decrease in the dietary intake of the element, for example) or toxicity increases, the curve declines from optimal (to the left for deficiency; right for toxicity) indicating detrimental effects on health. As the curve falls lower and lower, the effects of deficiency or toxicity become more and more severe until death occurs. Thus, as the deficiency of an element becomes more and more severe, the detrimental effects on health increase and the nutritional importance of the element becomes more and more evident. Nielsen (1991) described two factors affecting the nutritional importance of trace and ultratrace elements. Because of these factors, the occurrence of overt, simple, or uncomplicated deficiencies of trace or ultratrace elements in humans is relatively uncommon. The first factor involves the homeostatic mechanisms that regulate absorption, storage, and excretion of the trace and ultratrace elements. The extent or importance of each of these three homeostatic processes is dependent on the specific trace element. Basically, the homeostatic mechanisms shift the dose-response curve to the left. As a result, a smaller amount or dose of the element is necessary to achieve optimal health. For example, absorption may be increased or excretion decreased by suboptimal intake of an element. The second factor, especially noteworthy in industrialized nations, is the diversity in sources and types of foods consumed. This effectively results in providing the necessary amounts and forms of trace elements to humans. Thus, as stated, overt or unconfounded trace or ultratrace element deficiencies are uncommon. There are, however, situations that make trace elements of practical nutritional importance (Nielsen, 1990; 1991). These are: (1) Inborn errors of metabolism that affect the homeostatic mechanisms of the element; (2) Alterations in metabolism as a secondary consequence to malnutrition, disease, or injury; (3) Inadvertent omission from a total parenteral nutrition solution or baby formula; prolonged hemodialysis; (4) Marginal deficiencies induced by various dietary manipulations or by direct or indirect interaction with another nutrient or drug; and (5) The enhanced requirement caused by some nutritional, metabolic, hormonal, or physiologic stressors.

Because the statements are general in nature, there is, no doubt, some overlap between or among these situations. These five situations can be summarized by the statement "insufficient intake of a specific trace element becomes obvious only when the body is stressed in some way that enhances the need, or interferes with the utilization, of that element" (Nielsen, 1990).

Tapp and Natelson (1988) studied the effect of stress on hamsters with inherited heart disease. Their conclusions were summarized by the formula:

Pathological effects = stress x organic vulnerability

This formula can be adapted to trace elements accordingly:

Pathological effects (signs of deficiency) = stress x degree of deficiency

The formula implies that with a severe deficiency, little or no stress is needed to produce pathological or deficiency symptoms. In the presence of one or more of the five situations, either the degree of deficiency is exacerbated (situation 1 or 3, for example) or the effect of the stressor (situation 2, 4, or 5, for example) causes a shift in the dose response curve to the right. This shift would result in deficiency symptoms appearing at a dose higher than would otherwise be expected to produce adverse affects. In other words, in the presence of stressors, the degree of deficiency need not be great to induce deficiency signs. Excluding iron and iodine, most human trace element deficiencies are probably the result of the presence of one of the five situations. For example, signs of selenium deficiency have not been observed in free-living animals adequate in vitamin E (Levander and Burk, 1990) and in humans, apparently only a confounded selenium deficiency occurs (Keshan disease) (Whanger, 1989). Thus, arsenic may be similar to selenium in that a simple or non-confounded deficiency is difficult to produce.

Arsenic deprivation studies

The first strong evidence that arsenic is essential came from studies by Nielsen *et al.*, (1975). This, and the following studies headed by Anke, used pregnancy and lactation as stressors. Rats were fed a diet that contained approximately 30 ng arsenic g^{-1} on an air-dried basis. Controls were supplemented with 4.5 μg arsenic (4.0 μg arsenate and 0.5 μg arsenite) g^{-1} diet. The offspring of dams placed on the purified diet prior to being bred displayed a rougher coat and a slower growth rate than controls. Arsenic deprivation affected males more than females. At 12 to 15 weeks, the arsenic-deprived males exhibited significantly depressed hematocrits and enlarged, blackened spleens containing 50% more iron on a per gram basis. The spleens of arsenic-deprived females were also enlarged. Both sexes exhibited an elevated erythrocyte osmotic fragility. Although the amount of arsenic (4.5 μg g^{-1}) fed the control animals was not physiologic, changes in the experimental animals were indicative of a deficiency, rather than a pharmacologic effect. That is, parameters obtained from the arsenic-deprived animals were the abnormal findings; the findings obtained from the arsenic-supplemented animals were normal. Thus, this was probably a study of arsenic deprivation rather than a study of pharmacologic actions of arsenic.

Anke *et al.*, (1976) presented evidence for the essentiality of arsenic for goats and minipigs. They fed a semi-synthetic diet containing less than 50 ng arsenic g^{-1} to growing, pregnant, and lactating goats and minipigs, and to their first and second generation offspring. Controls received 350 ng arsenic g^{-1} diet. Only 58% of the deficient goats and 62% of the deficient minipigs produced offspring compared to 92% and 100% of the respective controls. The deficient offspring exhibited depressed growth and an elevated mortality rate. The slower growth was most marked in the second generation deficient goats. Some lactating arsenic-deficient goats died apparently from myocardial abnormalities. Ultrastructurally, there was an increase in the formation of mitochondrial membrane contours. A fine granular and electron dense material (possibly insoluble calcium phosphates) was found in the membranes; this material would lie free in the cytoplasm upon rupture of the mitochondrial membrane. These mitochondrial changes were found in skeletal and heart muscle and to a lesser extent, in liver (Schmidt *et al.*, 1984). Arsenic deficiency in goats also affected organ trace element concentration and lipid metabolism (Anke, 1986).

Nielsen and Shuler (1978) reported findings indicating an essential physiologic role for arsenic in growing chicks. Day-old cockerel chicks were fed a diet containing about 1.3% arginine and approximately 15-25 ng arsenic g^{-1}. Controls received the basal diet supplemented with 1 μg arsenic g^{-1}. Some of the arsenic-deprived and arsenic-supplemented chicks were supplemented with an additional 20 g arginine kg^{-1} diet. During the first three weeks, there was no obvious difference in growth or appearance between the arsenic-deprived and control chicks. In week four, the arsenic-deprived chicks drank and excreted more water and exhibited slower growth. At four weeks, the arginine-supplemented, arsenic-deprived chicks weighed significantly less than the comparable controls; the difference in weight at four weeks was not as marked in the groups not fed additional arginine. Other findings in the arginine-supplemented, arsenic-deprived chicks included: an elevated liver weight/body weight ratio and an elevated concentration of zinc and depressed concentrations of arsenic, iron, and manganese in liver. In the groups not supplemented with arginine, similar changes, but of lesser magnitude, occurred except the arsenic-deprived chicks had a significantly lower liver weight/body weight ratio than controls.

This was one of the first studies on arsenic essentiality in which animals were intentionally stressed through dietary means to induce a more severe arsenic deprivation. In this case the stressor was the additional dietary arginine. The requirement of arginine for the chick is 1.2%; the arginine-supplemented chicks received about 3.3% arginine. As indicated (Uthus, 1992), the effect on arsenic

Table 1 *Effects in rats of arsenic, diethyl maleate, gender, and their interaction on body weight, whole blood arsenic concentration, kidney weight/body weight ratio, and the concentration of calcium and iron in kidney*

Treatment As[a] (μg g^{-1})	DEM[a] (%)	Body Weight (g)	Blood As (μg g^{-1})	Kidney weight/ body weight x 100	Kidney Ca (μg g^{-1})	Fe (μg g^{-1})
Male						
0	0	346 ± 6[b]	0.084 ± 0.016	0.280 ± 0.006	245 ± 10[b]	237 ± 15
0.5	0	369 ± 12	13.7 ± 1.4	0.280 ± 0.005	257 ± 5	248 ± 7
0	0.25	283 ± 11	0.093 ± 0.014	0.336 ± 0.012	241 ± 19	231 ± 9
0.5	0.25	276 ± 10	16.3 ± 0.7	0.325 ± 0.011	224 ± 7	261 ± 20
Female						
0	0	210 ± 6	0.111± 0.007	0.308 ±0.013	4,850± 1570	544 ± 154
0.5	0	195 ± 5	21.3± 1.3	0.326 ± 0.008	1,520± 870	303 ± 13
0	0.25	190 ± 4	0.113± 0.011	0.339 ± 0.011	19,500± 7,270	551 ± 141
0.5	0.25	190 ± 8	24.4 ± 0.7	0.309 ± 0.012	5,080± 1,940	329 ± 52
Analysis of variance - *p* Values						
As		NS	0.0001	NS	0.01	0.04
DEM		0.0001	0.01	0.0002	0.006	NS
As x DEM		NS	0.01	0.05	NS	NS
G[c]		0.0001	0.0001	0.04	0.0001	0.0004
As x G		NS	0.0001	NS	0.01	0.01
DEM x G		0.0001	NS	0.004	0.002	NS
As x DEM x G		0.06	NS	NS	NS	NS
EMS[d]		419.9	3.506	0.0006	0.605[e]	26,355.8

[a]Amounts of arsenic (as $Na_2HAsO_4{\cdot}7H_2O$) and diethyl maleate (DEM) supplemented to the diet.
[b]Mean ± SEM.
[c]G = gender.
[d]EMS = error mean square.
[e]Data for kidney calcium were transformed by using the natural logarithm before running analysis of variance. Data (mean ± SEM) for kidney calcium are reported as non-transformed values.
N = 6 per group.

deprivation of the additional arginine was probably caused indirectly by the effect of arginine on methyl and, consequently, methionine metabolism. Other studies ascertaining the effect of stressing methionine metabolism on signs of arsenic deprivation in rats, chicks, and hamsters were recently reviewed (Uthus, 1992). In the studies reviewed, methionine metabolism was stressed directly by feeding guanidoacetic acid, diets marginally or severely deficient in methionine, or diets deficient in pyridoxal phosphate, or indirectly by using pregnancy and lactation.

Some of the effects of arsenic deprivation included decreases in growth, plasma taurine concentration, liver polyamine concentration, specific activity of liver S-adenosylmethionine decarboxylase, and in the activities of several enzymes necessary for phosphatidylcholine biosynthesis. Effects also were seen in tissue trace elements, other plasma and liver amino acids, and certain blood variables (hematocrit, hemoglobin, mean corpuscular volume, and mean corpuscular hemoglobin). Because many of the effects of arsenic deprivation became manifest in methionine metabolism and because alteration in methionine status or metabolism affects signs of arsenic deprivation, the hypothesis was developed that arsenic has a physiologic role that affects methionine or methyl metabolism.

Recent experiments have focused on determining where in methionine metabolism (methionine cycling or transsulfuration) arsenic is having its effect. Methionine cycling involves the donation of a methyl group via S-adenosylmethionine and the reactions that regenerate methionine by methylation of homocysteine utilizing methyl sources from choline (betaine homocysteine methyltransferase) or methyltetrahydrofolate (methionine synthase). Transsulfuration involves the irreversible catabolism of methionine through homocysteine and cystathionine ß-synthase. Transsulfuration results in the formation of the methionine metabolites cyst(e)ine and taurine.

The general site (recycling or transsulfuration) at which arsenic physiologically affects methionine metabolism is unknown. As stated above, studies with rats, chicks, and hamsters have shown that the nature and severity of the signs of arsenic deprivation are affected by dietary manipulations which impact or stress methionine metabolism and that many signs of arsenic deprivation include apparent perturbations in methionine metabolism. For the most part, the dietary manipulations (methionine deprivation, for example) which have been tried impact mainly on methionine cycling (Eloranta *et al.*, 1989). Thus, a study was designed to determine the effects of manipulation of the transsulfuration pathway on arsenic deprivation. This was accomplished by feeding diethyl maleate (DEM) which can enzymatically conjugate with reduced glutathione (GSH) (Plummer *et al.*, 1981), a cysteine-containing tripeptide. GSH is thought to be a reservoir for cysteine (Soon *et al.*, 1984); depletion of cysteine by DEM would cause an increase in the flux of methionine through the transsulfuration pathway. Also, it is known that males and females handle methionine/cysteine differently. For example, De La Rosa and Stipanuk (1985) found that the total amount of cysteine oxidized in eight hours to taurine and pyruvate differed between male and female rats. Also, the specific activity of cystinesulfinate decarboxylase, the rate-limiting enzyme in taurine biosynthesis, is decreased in female, compared to male, rats (De La Rosa and Stipanuk, 1985). Limited work has been done on quantifying the recycling of homocysteine to methionine in rats (Barak *et al.*, 1982; Eloranta *et al.*, 1989; Xue and Snoswell, 1985) and humans (Mudd and Poole, 1975) and even less work has been done studying the effect of gender on this recycling. In a human study that used young adult males and females, Mudd and Poole (1975) determined that the average homocysteinyl moiety recycled between methionine and homocysteine at least 1.9 times in males and 1.5 times in females before entering the transsulfuration pathway; when labile methyl

Table 2 *Effects in female rats of arsenic on liver S-adenosylhomocysteine (SAH), S-adenosylmethionine (SAM), and their ratio*

Treatment As[a] (μg g^{-1})	SAH (nmol g^{-1})	SAM (nmol g^{-1})	SAM/SAH
0	26.7 ± 2.2[b]	86.7± 6.2	3.29 0.22
0.5	26.7 ± 1.5	104.4 ± 4.3	3.99 0.38
	NS	*p*<0.05	NS

[a]Amount of arsenic (as As_2O_3) supplemented to the basal diet.
[b]Mean ± SEM.
N = 5 per group.

intake was limited, the number rose to 3.9 for males and 3.0 for females. Thus, because of the gender differences in methionine/cysteine metabolism, it was of interest to determine the effects of DEM feeding on arsenic deprivation in both male and female rats. Some of the results from this experiment are reported in Table 1.

Perhaps the most notable findings of this experiment were the effects of arsenic and gender on kidney iron and the effects of arsenic, gender, and DEM on kidney calcium. Males were not affected; iron and calcium in the kidneys of females were markedly elevated by arsenic deprivation. This, in part, can be explained by the diet used in this experiment. The diet contained 0.5% calcium, 0.5% phosphorus (0.4% added by mineral mix and 0.1% added by casein), and 0.04% magnesium. These are similar to the amounts of calcium, phosphorus, and magnesium found in the AIN-76 diet. The molar ratio of Ca:P is 0.78. Numerous studies have indicated that female rats fed diets with a low Ca:P molar ratio, such as the AIN-76 diet, are prone to nephrocalcinosis (Adams *et al.*, 1989; Reeves, 1993; Shah and Belonje, 1991; Shah *et al.*, 1986). Reeves *et al.*, (1993) has shown that supplementation of cystine, in lieu of methionine, to an AIN-76 type diet decreased the accumulation of calcium in the kidneys of female rats. Because there are differences in methionine/cysteine metabolism between male and female rats, females may be affected by arsenic deprivation differently, as reflected by kidney calcium and iron. Also, Reeves *et al.*, (1993) has shown that ovariectomized rats, when fed AIN-76 type diets, are not prone to nephrocalcinosis. Supplementation of estrogen to these animals, however, allows for nephrocalcinosis (Reeves *et al.*, 1993). Thus, the effects found in this experiment (Table 1) may be hormonally related.

It is not known if the effects of arsenic deprivation on kidney calcium in female rats are related to the insoluble calcium phosphate material that was found in mitochondrial membranes in lactating goats as reported by Schmidt and co-workers (1984).

In another recent experiment, a basal diet having a molar ratio of Ca:P of 1.3 was fed to female rats for 10 weeks. This diet does not promote nephrocalcinosis in female rats. The basal diet contained about 10 ng arsenic g^{-1}; controls were supplemented with 0.5 μg arsenic g^{-1}. This was not a pregnancy study and methionine metabolism was not stressed by any means other than through arsenic deprivation. Livers were analyzed for S-adenosylmethionine (SAM) and S-adenosylhomocysteine; the concentration of SAM was decreased by arsenic deprivation. These findings add to the growing evidence that arsenic has a physiologic role related to or affecting methionine metabolism. Results are shown in Table 2.

Previous work on arsenic essentiality has only been done in animals. Recently, however, a report has been published correlating low serum arsenic concentrations in hemodialysis patients with central nervous system disorders, vascular diseases, and cancer (Mayer *et al.*, 1993). Mean serum arsenic concentration in the hemodialysis patients was 8.5 ±1.8 ng mL^{-1} , in controls the concentration was 10.6 ± 1.3 ng mL^{-1}. The authors concluded that arsenic homeostasis was altered by hemodialysis treatment and certain diseases and, therefore, maintaining desirable arsenic concentrations in the body seemed reasonable. Furthermore, the authors suggested that arsenic deficiency may impose an additional risk to hemodialysis patients and that supplementation may be appropriate. These may be the first reported cases of arsenic deficiency in humans. As stated, finding a simple or non-confounded arsenic deficiency, especially in humans, is probably unlikely. That is, the loss of arsenic through hemodialysis (situation 3), resulted in serum arsenic concentrations low enough to produce affects attributable to arsenic deficiency.

Conclusions

There are two factors (homeostasis and the diversity in sources and types of foods consumed) that make the occurrence of overt, simple, or uncomplicated deficiencies of trace or ultratrace elements in humans relatively uncommon. There are, however, situations that make trace elements of practical nutritional importance. These situations either affect homeostasis (in effect lowering the amount absorbed, decreasing bioavailability, or increasing the amount excreted) or cause a metabolic stress; the stress can lead to signs of deficiency at concentrations of the element not normally associated with deficiency. Perturbations of methionine metabolism can alter signs of arsenic deprivation. Also, many of the signs of arsenic deprivation are manifest in methionine or methyl metabolism. Because of this, arsenic is thought to have a role in methionine metabolism. Evidence of arsenic essentiality has been shown in studies done with rats, hamster, chicks, goats, minipigs, and humans.

References

Adams, C., Ritskes-Hoitinga, J., Lemmens, A.G. and Beynen, A.C. 1989. Dietary restriction of phosphorus and nephrocalcinosis in two inbred strains of rats. *Nutr. Rep. Internat.*, **40**, 923-929.

Anke, M. 1986. Arsenic. In: Mertz, W. (ed.), *Trace Elements in Human and Animal Nutrition*, pp. 347-372. Academic Press, Orlando.

Anke, M., Grün, M. and Partschefeld, M. 1976. The essentiality of arsenic for animals. In: Hemphill, D.D. (ed.), *Trace Substances in Environmental Health*, pp. 403-409. University of Missouri, Columbia.

Barak, A.J., Beckenhauer, H.C. and Tuma, D.J. 1982. Use of S-adenosylmethionine as an index of methionine recycling. *Anal. Biochem.*, **127**, 372-375.

Bertrand, G. 1912. Sur le role des infiniment petits en agriculture. *Proc. Int. Congr. Appl. Chem., 8th*, 30-49.

De La Rosa, J. and Stipanuk, M.H. 1985. Evidence for a rate-limiting role of cysteinesulfinate decarboxylase activity in taurine biosynthesis *in vivo*. *Comparative Biochem. Physiol. B*, **81**, 565-571.

Eloranta, T.O., Martikainen, V. and Smith, T.K. 1989. Adaptation of adenosylmethionine metabolism and methionine recycling to variations in dietary methionine in the rat. *Proc. Soc. Exp. Biol. Med.*, **194**, 364-371.

Frieden, E. 1984. A survey of the essential biochemical elements. In: Frieden, E. (ed.), *Biochemistry of the Essential Ultratrace Elements*, pp. 1-15. Plenum Press, New York.

Levander, O.A. and Burk, R.F. 1990. Selenium. In: Brown, M.L. (ed.), *Present Knowledge in Nutrition,* 6th edn., pp. 268-273. International Life Sciences Institute, Washington, DC.

Mayer, D.R., Kosmus, W., Pogglitsch, H., Mayer, D. and Beyer, W. 1993. Essential trace elements in humans: serum arsenic concentrations in hemodialysis patients in comparison to healthy controls. *Biol. Trace Elem. Res.*, **37**, 27-38.

Mudd, S.H. and Poole, J.R. 1975. Labile methyl balances for normal humans on various dietary regimens. *Metabolism*, **24**, 721-735.

Nielsen, F.H. 1990. New essential trace elements for the life sciences. *Biol. Trace Elem. Res.*, **26-27**, 599-611.

Nielsen, F.H. 1991. Trace elements (nutrition). In: Dulbecco, R. (ed.), *Encyclopedia of Human Biology,* vol. 7, pp. 603-613. Academic Press, San Diego.

Nielsen, F.H., Givand, S.H. and Myron, D.R. 1975. Evidence of a possible requirement for arsenic by the rat. *Fed. Proc., Fed. Am. Soc. Exp. Biol.*, **34**, 923 (Abs).

Nielsen, F.H. and Shuler, T.R. 1978. Arsenic deprivation studies in chicks. *Fed. Proc., Fed. Am. Soc. Exp. Biol.*, **37**, 893 (abstract).

Plummer, J.L., Smith, B.R., Sies, H. and Bend, J.R. 1981. Chemical depletion of glutathione *in vivo*. In: Jakoby, W.B. (ed.), *Methods in Enzymology,* vol. 77, pp. 50-59. Academic Press, Inc., New York.

Reeves, P.G., Rossow, K.L. and Lindlauf, J. 1993. Development and testing of the AIN-93 purified diets for rodents: results on growth, kidney calcification and bone mineralization in rats and mice. *J. Nutr.*, **123**, 1923-1931.

Schmidt, A., Anke, M., Groppel, B. and Kronemann, H. 1984. Effects of As-deficiency on skeletal muscle, myocardium and liver. A histochemical and ultrastructural study. *Exp. Pathol.*, **25**, 195-197.

Shah, B.G. and Belonje, B. 1991. Different dietary calcium levels required to prevent nephrocalcinosis in male and female rats. *Nutr. Res.*, **11**, 385-390.

Shah, B.G., Trick, K.D. and Belonje, B. 1986. Factors affecting nephrocalcinosis in male and female rats fed AIN-76 salt mixture. *Nutr. Res.*, **6**, 559-570.

Soon, E., Johnson, N. and Snider, B.C.F. 1984. Tissue glutathione as a cyst(e)ine reservoir during cystine depletion in growing rats. *J. Nutr.*, **114**, 1853-1862.

Tapp, W.N. and Natelson, B.H. 1988. Consequences of stress: a multiplicative function of health status. *FASEB J.*, **2**, 2268-2271.

Uthus, E.O. 1992. Evidence for arsenic essentiality. *Environ. Geochem. Health*, **14**, 55-58.
Whanger, P.D. 1989. China, a country with both selenium deficiency and toxicity: some thoughts and impressions. *J. Nutr.* **119**, 1236-1239.
Xue, G.-P. and Snoswell, A.M. 1985. Comparative studies on the methionine synthesis in sheep and rat tissues. *Comparative Biochem. Physiol. B*, **80**, 489-494.

20 A physiologically based pharmacokinetic model for ingested arsenic

Daniel B. Menzel[1], Matthew Ross[1], Stephen V. Oddo[1], Paul Bergstrom[2], Howard Greene[3] and Randy N. Roth[3]

[1] *Department of Community and Environmental Medicine, University of California, Irvine, CA 92717, Atlantic Richfield Company,* [2] *Denver, CO, and* [3] *Los Angeles, CA , USA*

Abstract

Mammals metabolise arsenate (As(V)) to arsenite (As(III)), monomethylarsenic (MMA) and dimethylarsonic acids (DMA). As(V) is thought to be a human skin carcinogen. Because the mechanism of carcinogenesis for As is not known, definition and estimation of target organ doses from As exposure is critical to cancer risk assessment. Physiologically based pharmacokinetic (PB-PK) models can account for steps of metabolism and distribution to calculate target organ doses of one or more arsenic compounds. We present here a preliminary version of a PB-PK model which accounts for metabolism and distribution in mice, rats, and humans. Model uncertainties are in the extraction coefficient for organs, the site of reduction of As(V) to As(III), the site of methylation to MMA and DMA, and the tissue binding of As(III). Current research is being conducted to fill these data gaps by an approach to steady state study in mice and chemical speciation of As in treated mouse organs, blood and urine. Genetic variations in reduction and methylation and of diet on excretion of As in human populations will be studied in the completed and validated model.

Introduction

Arsenic (As) is one of the metalloid group of elements. Arsenic is thought to be a human skin carcinogen (EPA, 1984). More recent work suggests an association of arsenic ingestion and cancer of other organs, but as yet no mechanism of action has been identified or an experimental animal model found. Because As is widely distributed in the earth's crust, avoidance of exposure to As is impossible. The occurrence of As in food, air, and drinking water and its potential carcinogenicity poses a major public health dilemma of whether to eliminate As from these sources of exposure.

The precise dose of As at any target organ in humans is unknown. Lacking this information, cancer risk estimates have been based on the simplifying assumption that the cancer risk is proportional to the total As dose. Cancer risk has been assumed to be linear with As dose, and the same for all forms of As. The acute toxicity of As, however, varies widely with its valence state and chemical form. Humans and experimental animals metabolize arsenate (As(V)) changing both the valence state of ingested As and its chemical form.

This paper is a progress report of our efforts towards the development of a mathematical model predicting As internal organ burdens during specific As exposures and accounting for metabolism, distribution to organs and binding to organs.

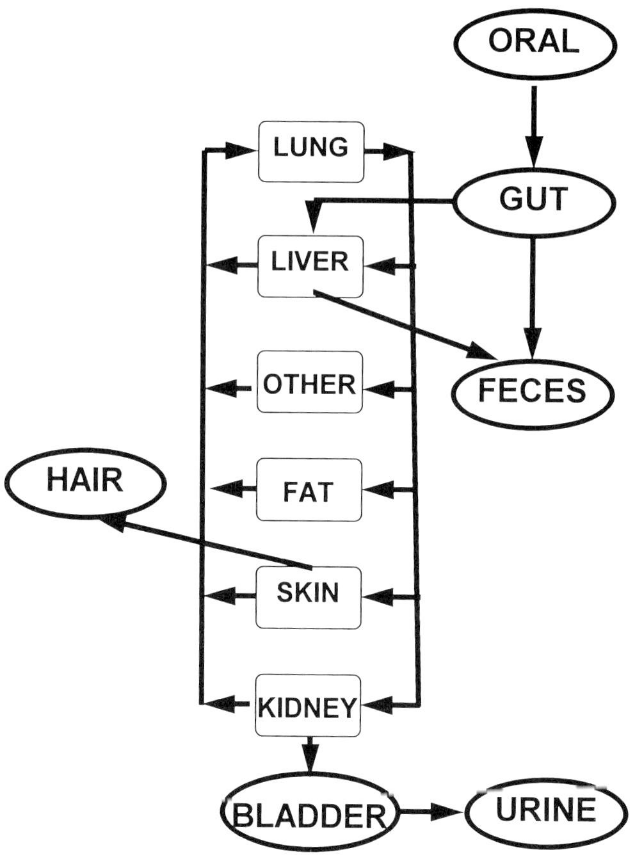

Figure 1 *Schematic representation of the physiologically based pharmacokinetic model for arsenate ingestion. The ovals represent pools of arsenic species not influenced by blood perfusion. Rectangles with rounded corners represent organs perfused by the blood. Blood flow from one organ to another is indicated by the arrows in the central parallel series. The interrelations between organs and metabolism of arsenic is described in the text.*

Methodology

Model structure

We have chosen the physiologically based pharmacokinetic modeling (PB-PK) approach articulated most recently by Andersen and his coworkers and the Safe Drinking Water Committee of the National Research Council (Andersen, 1987). The PB-PK modeling approach utilizes the organ volumes and blood flow to construct a multicompartmental model of the distribution of chemicals in the body with time. The overall structure of the model is shown schematically in Figure 1.

Our model follows that of M.A. Andersen (personal communication) for dioxin. The model is implemented through a set of simultaneous differential equations describing blood flow to the organ, chemical partitioning, metabolism and binding to the tissue (Ramsey and Andersen, 1984).The model is being developed using the commercial simulation control and optimization program SIMU-SOLV,® which runs on a DEC5000 workstation under ULTRIX.

Separation and identification of arsenate and its metabolites

In our animal experiments we are using sodium (^{73}As) arsenate as the radiotracer. Arsenic species were separated by ion chromatography using a Dionex AS7 anion exchange column with the separation and regeneration conditions of Roehl, *et al.*, (1992). Arsenic species were detected by a flow-through liquid scintillation detector (Berthold, Germany). The individual As species were identified by relative retention times as described by Roehl *et al.* (1992).

Treatment of experimental animals

Male Swiss Webster white mice were obtained as specific respiratory pathogen free adults weighing 18–23g. Mice were injected interperitoneally or intravenously with a saline solution or given a solution of ^{73}As-containing drinking water *ad libitum*. Animals are housed, cared for and euthanised according to the NIH guidelines. All animals were maintained and treated according to a protocol approved and supervised by the UCI Institutional Animal Care Committee and overseen by the University Veterinarian.

Results

Arsenic exposures considered

In our model As enters the body from drinking water and food. Inhalation of As in the particulate phase or as arsine gas is not considered . The chemical species of As in drinking water and food are unclear. Identification of As species in drinking water is now in progress. We assume that As is present as As(V) (see other presentations dealing with the national survey of drinking water supplies in this volume). Chlorination of drinking water and air oxidize As to As(V). We assume that As(V) is transported across the gut as As(V) and is distributed to the organs according to organ uptake. Such a model is shown schematically in Figure 1.

The ratio of blood As(V) to tissue As(V) is the extraction coefficient, and is difficult to estimate from the literature. We are in the process of confirming these coefficients experimentally. The organ burden of As species in the preliminary model is estimated from the extraction coefficient, blood As concentration and the portion of the cardiac output which the organ receives. We have limited the organs treated in the model to those shown in Figure 1, because these organs appear either important in As metabolism or because they are potential targets for As carcinogenicity.

Arsenic uptake
It is unclear from the literature if As (V) is taken up exclusively in the stomach. So far we have found that the velocity of As(V) uptake is so rapid that uptake beyond the stomach and the upper small intestine seems minimal. Uptake by the lower small intestine could lead to a direct transport via the portal circulation to the liver. Benchmark experiments, however, indicate little difference in the conversion of As(V) to DMA in mice given As(V) orally, i.v. or i.p. suggesting that As(V) does not exhibit a first pass effect. Biliary excretion of about 5-10% of the total dose appears to occur.

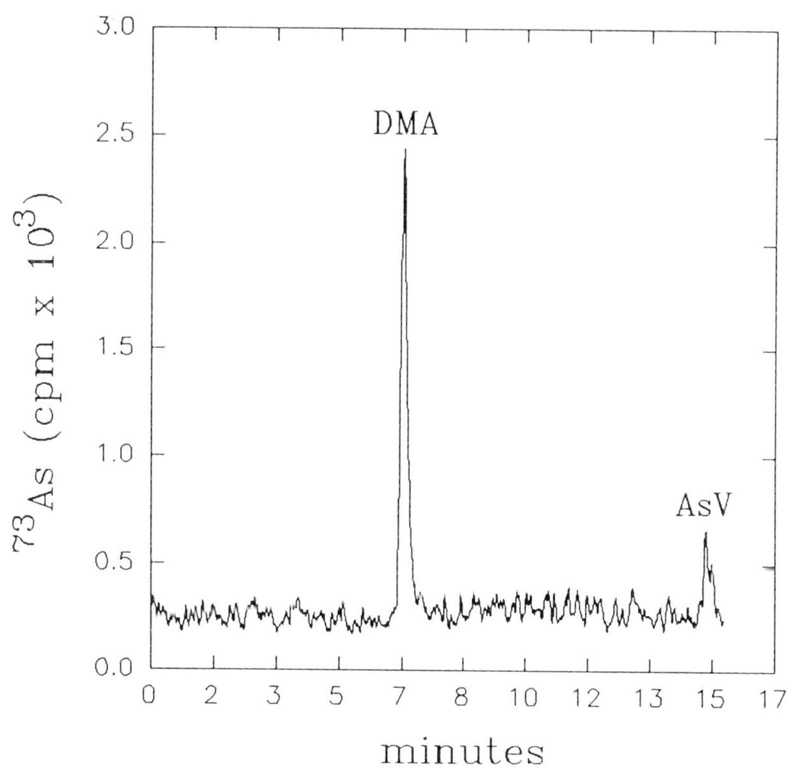

Figure 2 *Ion chromatogram of ^{73}As compounds found in the urine of mice treated with sodium (^{73}As) arsenate. DMA is dimethylarsonic acid, while As(V) is arsenate.*

Arsenate excretion

All species examined, including humans, rapidly excrete As(V) in the urine following ingestion or inhalation of As(V) containing compounds (Buchet *et al.*, 1981). In our model reabsorption via the kidney is not considered.

Concerns have been expressed that bladder tumors might arise from As excreted into the urine. To provide for this potential occurrence and to help guide epidemiological studies in the future, As(V) excretion from the kidney is accumulated in a bladder compartment. The bladder is not considered to be a perfused organ in this preliminary model to simplify the description of As(V) distribution in the body. The dose of As(V) to the bladder is calculated as the time weighted average concentration of As(V) in urine assuming arbitrary urine volumes in the bladder and periodic emptying of the bladder. In human experiments it is possible to determine the mixed urinary As(V) concentration and precise emptying times, while it is not possible to control bladder emptying in unrestrained mice or rats.

Biliary excretion may occur as a by-product of the affinity of As species for sulfur containing chemicals, especially glutathione (GSH). GSH conjugates of larger molecular weight xenobiotic chemicals are known to be excreted in the bile. The biliary excretion products and unabsorbed As(V) appear in the feces. Because little is known about the chemical composition of As species in food aside from the well known occurrence of arsenobetaine and similar As analogs of phospholipid metabolism in shellfish and shrimp, the fecal As content may represent both unabsorbed As(V) and complex organic As species yet to be identified in food.

Non-Enzymatic

GSH + AsV → GS- AsIII-SG (Eq.1)

Glutathione S-Transferase Enzymatic

GS- AsIII-SG → MMA + GSH

MMA → DMA (Eq. 2)

AsV + GSH + SAM → AsIII + MMA + DMA (Eq.3)

Protein-SH + AsIII + GSH → Protein-S-AsIII-SG (Eq.4)

Figure 3 *Metabolism of arsenate.*

Arsenic metabolism

Arsenic, as a metalloid, forms covalent bonds with the biologically important elements sulfur, carbon and phosphorus. Metabolism of As into different oxidation states and metabolites alters both the chemical and physical properties of As. In order to construct our PB-PK model, data on the As content of tissues and fluids must be quantitated by chemical species. Figure 2 shows the ion chromatogram from a typical mouse urine sample illustrating the separation of As(V) from DMA, the principal excretory metabolite of As(V) and the feasibility of quantitating As metabolites by this method.

Methylation is known to occur in all species. A mixture of As(III), MMA and DMA has been reported in the blood, feces and urine of human and experimental animals. Chemical speciation of As by wet chemical means (as carried out in the older literature, for example Norin and Vahter, 1981) is no simple process so that the apportionment of As content into the four potential species may have been in error or variable from experiment to experiment compared to ion chromatography. As oxidation states are freely interconvertible so that storage or extraction conditions may also have led to analytical errors.

Very recently, the formation of As(V) and As(III) complexes with GSH has been observed with nuclear magnetic resonance spectra (Scott, *et al.*, 1993). Equation (1) (Figure 3) illustrates a theoretical equation for the reduction of As(V) by GSH. In our laboratory As(V) is reduced by GSH at physiological conditions non-enzymatically and probably by this reaction. Enzymatic reduction could take place by several mechanisms including that in equation (1). So far no single enzyme or preparation has been reported which catalyzes the reduction of As(V) to As(III). The velocity of these reactions with GSH in crude tissue extracts and with purified glutathione S-transferases (GSTs) is under investigation.

GSTs are potential candidates for As(V) reductases because GSTs catalyse similar reductions with organic peroxides, hydrogen peroxide and ozonides (Vos *et al.*, 1987; and Mannervik and Danielson, 1988).

Conversion of As(V) to MMA and DMA probably occurs in two steps since MMA is converted to DMA (equation (2) Figure 3). The methyl group appears to arise from S-adenosyl methionine (SAM), but the As(III) transmethylase has not been isolated. DNA methylases are not likely candidates, but DMA does react with the DNA nucleotide specific binding site of the Eco DNA methylase isolated from E. coli (Lam *et al.*, 1992). Non-specific methylases are potential candidates for this conversion.

For purposes of modeling the overall metabolism and distribution of As(V), the enzymatic reaction can be written as equation (3) (Figure 3), in which GSH and SAM are cosubstrates, but the velocity of the reaction is described by the As(III) and As(V) content. Equation (3) describes the organ enzymatic conversion so that the apparent K_m and V_{max} calculated from whole animal experiments or human accidental poisoning may not be identical to the kinetic parameters of isolated purified enzymes. Because of the conversion of As(V) to at least three metabolites, we have chosen to maintain mass balance for all of the metabolites as well as As(V) in our model.

Ion chromatography determinations of As species in freshly isolated mouse blood does not show any As(III) or MMA. Thompson *et al.* (1993) report the appearance of MMA and As(III) at higher doses of As(V) than so far examined by us. The occurrence of MMA and As(III) in the blood of mice may represent the saturation of As metabolism in the liver.

To account for As(III), MMA and DMA distribution in the body we assume a scheme similar to that shown in Figure 1 for each metabolite. No oral uptake for these metabolites is assumed so this portion of the As(V) model is eliminated from the model describing As(III), MMA and DMA distribution. For simplicity the conversion of As(III) to MMA and DMA is assumed to follow the same scheme as in equation 3 (Figure 3).

Arsenic accumulation in skin, hair and nails

The skin has been the focus of As carcinogen risk assessment because of the association between As ingestion and skin tumors (US EPA, 1984). Arsenic accumulation in skin, nails and hair has not been a particularly useful indicator of long term As exposures. If As distribution to hair and nails could be adequately described, existing data on As content of hair and nails could be reinterpreted and new studies of long term exposure to As initiated.

Vahter and Marafante (1989) found As bound to a variety of unidentified proteins isolated from the liver of As treated mice and rabbits by gel filtration chromatography Mixed disulfides between GSH and protein SH groups readily occur in tissues (Reed, 1990). By analogy with equation 1, equation 4 (Figure 3) represents the binding of As(III) to tissue proteins with cantenation with GSH. Protein SH groups are incorporated into permanent storage sites in hair and nails as these tissues grow outwards from the body. As(III) could be bound to these proteins, but not be an exchangeable As store as far as the rest of the body is concerned. Our model treats this step by assuming that hair and nail growth are constant with time and that As(III) reacts directly with hair and nail SH groups before hair and nail cells die; *e.g.*, As is irreversibly bound by cell death and transport to hair and nails.

Discussion

PB-PK modeling of arsenate

We have laid out a preliminary version of a PB-PK model based on the current literature on the metabolism and distribution of As in rats, mice, rabbits, and humans. The literature has not provided sufficient data to quantitate specific parameters sufficiently to finalize the model. These data gaps occur because most toxicity and human exposure studies are directed at mechanisms and provide little kinetic data. We are in the process of filling these data gaps. Validation of theoretical models is a necessary step before the models can be applied confidently to human dose estimates. Our aim is to validate the model by independent experiments as described as a methodological aim in PB-PK modeling by Menzel (1988).

Inorganic toxicants such as As(V) represent difficult subjects for investigation because the transport into tissues may be by normal ion channels utilized to maintain the intracellular milieu. To overcome this difficulty we are using an approach to steady state method of estimating the blood to organ ratio of As(V) and its metabolites. Mice are provided As(V) in graded concentrations in drinking water to attain doses of 0.4 to 5 mg $kg^{-1} day^{-1}$. When the urinary output of As reaches a steady state, we assume that the organ distribution of As(V) and its metabolites in blood and tissue is also at steady state. Since we do not plan life term exposures, the disruption of this steady state by accumulation in irreversible stores such as the hair and nails should not be a limiting factor.

What are the uncertainties in the preliminary model?
Several uncertainties are evident from the current literature. We list these and some comments of how they might affect human risk estimation below.

Arsenate reduction
The enzyme(s) responsible for reduction of As(V) to As(III) is (are) very poorly described because experiments have been undertaken in tissue slices. GSH is thought to function in this reaction from two points of evidence: First, addition of GSH to homogenates increases the conversion of As(V) to DMA and MMA. Second, As(V) can be reduced by thiols. The propensity of As compounds to react with thiols is well recognized. The enzyme(s) may not be a specific As(V) reductase but may be akin to the GSH dependent peroxidase reactions with hydrogen peroxide, organic peroxides, ozonides and oxorane compounds of GSTs. The GSTs possess peroxidase activity probably as a secondary property associated with their S-transferase active site.

GSH can react non-enzymatically with As(V) to produce As(III). The rate of this reaction under physiological conditions is not clear. We are investigating this reaction in order to determine if the conversion of As(V) to As(III) in tissues by non-enzymatic reactions with GSH is a significant pathway of conversion.

If the pseudo reductase activity is due to GST or some other GSH dependent enzyme, metabolism may be tissue specific. GSTs are dimers and are part of a supergene family with some 8 monomers so far identified (Mannervik and Danielson, 1988). GSTs form homo- and hetero-dimers with maximum enzymatic activity, although some of the monomers appear to be enzymatically active. These dimers form the three major classes of GSTs in humans and rodents, alpha, pi and mu. Although the genes for the major subunits are present in most tissue, the tissue distribution of GST expression varies between species. In humans about 50% of European white males have a gene deletion for the pi class (Daly *et al.*, 1993). As(III) formation from circulating As(V) may be variable and tissue specific if GSTs are responsible for the reduction. Depending on the second order rate constant for the non-enzymatic reduction of As(V) with GSH some tissue specificity may also exist due to GSH content. GSH levels vary between tissue and even between cell types in an organ, but the magnitude of these differences in GSH concentrations may be too small to have a major effect on As(V) metabolism.

Are there heritable difference in arsenic metabolism and do they represent different risk factors for arsenic carcinogenicity?
Drugs are methylated with varying velocities in humans. This variance leads to well known adverse reactions to drugs when individuals lack the ability to methylate a drug as a detoxification step (Daly *et al.*, 1993). At present nothing is known about

the diversity of As methylation in the human population since too few subjects have been studied with this objective in mind. We suggest above that the reduction of As(V) to As(III) could well be a point for metabolic diversity. Diversity in methylation could also result from more than one methyl transferase and differences in inheritance.

Are there dietary factors in As cancer risk?
Folic acid, vitamin B12, and methionine are well known factors in the metabolism of carbon at the C1 level. Folic acid antagonists inhibit the transfer of methyl groups from dietary sources, mostly methionine, to a variety of pathways. Methylation is a particularly important means by which the phosphatidyl choline series of phospholipids is maintained. Lack of dietary sources of the nutrients needed to carry out methyl metabolism could well affect As(III) conversion to MMA and DMA. If As(III) is the ultimate carcinogenic form of As, any step which alters the As(III) content at the target site will affect the incidence of cancer.

Do the different ratios of arsenic metabolite reflect qualitative differences in arsenic metabolism between species?
Much frustration has been expressed over the lack of an experimental animal model of arsenic carcinogenicity. Since we lack an isolated specific enzyme(s) to which to ascribe the metabolic conversion of As, uncertainties have been raised about studies in different animal species and humans. The ratio of MMA to DMA found in the urine of rats, mice, rabbits, and humans could represent true qualitative differences in metabolic pathways. Alternatively, the differences may be quantitative. The pool of free MMA in mice, for example, may be small so that MMA is converted efficiently to DMA, the only As metabolite found in mouse urine at low As levels. As the As dose is increased, more MMA may be formed increasing the free MMA pool, thus MMA diffuses into the blood and appears in urine. In humans MMA and DMA occur in urine even at low doses, but this may be because of a higher MMA pool.

Some conversion of As(V) in tissues other than the liver must occur because tissue slices of lung convert As(V) to MMA and DMA. The organ activity may vary between species quantitatively leading to apparent qualitative differences in As metabolism. All of these differences in quantitative conversion are easily accommodated by PB-PK models so that the apparent discordant data can be unified into a concurrent hypothesis.

Summary

A PB-PK model of oral ingested As(V) has been conceptualised and preliminary steps taken to implement a quantitative model. The model provides estimates of As dose to lung, liver, skin and bladder as potential target sites of As action. While quantitative data are being accumulated on the extraction of As(V) and its metabolites by several organs, the model offers both a conceptual and quantitative means of testing several hypotheses. By the addition of a second biological effects model, an overall prediction of the biological effects of As ingestion should be possible. Hopefully the PB-PK modeling approach will provide a better estimate of the risk of As to humans so that rational schemes of intervention can be devised.

Acknowledgements

We thank Mrs Sally Menzel, Messrs. S. Owh, E. Lee and P. Lee for their able assistance. Dr. Raymon Roehl provided invaluable assistance in the analysis of As species. Dr. M. A. Andersen provided invaluable help and outstanding constructive criticism of our PB-PK modeling approach. This work was supported by a contract from ARCO. M. Ross is a National Institute of Environmental Health Sciences pre-doctoral fellow.

References

Andersen, M.E. 1987. Physiological Based Pharmacokinetic and the Risk Assessment Process for Methylene Chloride, *Toxicol. Appl. Pharm.,* **87**, 185-205.

Andersen, M.E. 1993. A physiological based pharmacokinetic model of dioxin (personal communication.)

Buchet J.P, Lauwerys, R.and Roels, H.1981. Urinary Excretion of Inorganic Arsenic and its Metabolites after Repeated Ingestion of Sodium Metaarsenite by Volunteers. *Int. Archv. Occup. Envir. Health,* **48,**111-118.

Daly, A.K., Cholerton, S., Gregory W. and J. R. Idle, 1993. Metabolic Polymorphisms *Pharmac. Ther.,* **57,**129-160.

Lam, W.C., Tsao, D.H., Maki, A.H., Maegley, K.A. and Reich, N.O. 1992. Spectroscopic Studies of Arsenic (III) Binding to Escherichia coli RI Methyltransferase and to Two Mutants, C223S and W183F. *Biochemistry,* **31,** 10438-10442.

Mannervik, B. and Danielson, U.H. 1988 Glutathione S-Transferases--Structure and Catalytic Activity.*CRC Critical Reviews in Biochemistry,***23,** 283-339.

Menzel, D.B. 1988. Planning and Using PB-PK Models: An Integrated Inhalation and Distribution Model for Nickel, *Toxicol. Lett.,* **43,** 67-83.

Norin, H. and Vahter, M. 1981. A Rapid Method for the Selective Analysis of Total Urinary Metabolites of Inorganic Arsenic. *Scand. J. Work, Envir. Health,* **7,** 38-44.

Ramsey, J.C. and Andersen, M.E. 1984. A Physiological Based Description of the Inhalation Pharmacokinetics of Styrene in Rats and Human, *Toxicol. Appl, Pharm.,* **73,** 159-175.

Reed, D.J. 1990. Glutathione and Its Toxicological Implications. *Ann. Rev. Pharm. Toxicol.,* **10,** 603-631.

Roehl, R., Alforque, M. M. and J. Riviello. 1992. Arsenic Speciation in Biological and Environmental Samples by Liquid Chromatography Combined With On-Line Generation and Inductively Coupled Plasma Mass Spectrometry. Summary of a Paper Presented At the 1992 Winter Conference on Plasma Spectrometry, San Diego, CA, January 6-11.

Scott, N., Hatlelid, K.M., MacKenzie, N.D. and D.E. Carter. 1993]. Reactions of Arsenic (III) and Arsenic (V) Species With Glutathione. *Chem. Res. Toxicol.,* **6,** 102-106.

US EPA 1984. *"Health Assessment Document for Inorganic Arsenic" (Final Report). EPA/600/8083/021F,* Washington, D.C..

Vahter, M; and Marafante, E. 1989. Intracellular Distribution and Chemical Forms of Arsenic in Rabbits Exposed to Arsenate. *Biological Trace Element Res.,* **21,** 233-239.

Vos, R. M., Reitjens, I.M., Alink, G. M.and Van Bladeren , P. J. 1987. Methyl Linoleate Ozonide: A Substrate for Rat Glutathione S-Transferase.*Europ. J. Drug Metab. Pharmacokin.,***12,** 245-277

21 A physiologically based pharmacokinetic model for the four major arsenic species in mammals

S. Mann[1], P.O. Droz[1] and M. Vahter[2]

[1]*Institute of Occupational Health Sciences, Lausanne University, Avenue du Bugnon 19, 1005 Lausanne, Switzerland*
[2]*Karolinska Institute, Institute of Environmental Medicine, PO Box 210, Stockholm, Sweden*

Abstract

The general concepts of a physiologically based pharmacokinetic model (PB-PK) are presented. It describes the absorption, distribution, metabolism and excretion of arsenate (As(V)), arsenite (As(III)), methylarsonate (MMA) and dimethylarsinate (DMA), the four major metabolites of inorganic arsenic. Routes of intake considered are respiratory for dust and fumes, and oral for arsenic in water and food. Tissue distribution is calculated based on physiological and physico-chemical parameters: blood perfusion, tissue volumes, diffusion coefficients, membrane characteristics, and tissue affinities. Oxidation-reduction reactions between As(V) and As(III) are described by first order kinetic reactions in plasma, and methylation based on Michaelis-Menten kinetics in the liver. Renal excretion is calculated including glomerular filtration, tubular secretion and reabsorption. Numerical values are obtained from physiological references and by fitting the model to in vitro *and* in vivo *studies. Validation of the model is carried out by comparison of the results obtained to experimental data for rabbits and hamsters. In a second step validation is also carried out for human subjects exposed to arsenic orally or by inhalation.The PB-PK model allows prediction of concentrations in urine of As(III), As(V), MMA, and DMA, frequently used as indicators of exposure. It is also possible to estimate levels of the various arsenic metabolites in target organs such as the lungs, the liver and the kidneys. A few scenarios are presented for estimation of biological concentrations under various conditions: for example, exposure to aerosols of different mass median diameters and to different concentrations of arsenic. Future applications of the PB-PK model are also discussed.*

Introduction

Environmental exposure to inorganic arsenic compounds occurs via drinking water, food and air. In all these media, arsenic may occur in the trivalent (AsIII) or pentavalent (AsV) oxidation state. Chronic exposure to inorganic arsenic may give rise to several health effects, including effects on the gastrointestinal tract, respiratory tract, skin, liver, cardiovascular system, hematopoietic system, and nervous system (for review see WHO, 1981; US EPA, 1988; Pershagen and Vahter, 1991; Smith *et al,* 1992). The critical effect following ingestion of inorganic arsenic is skin cancer. Recently it has been reported that ingestion of inorganic arsenic via drinking water (probably both AsIII and AsV) is also associated with increased risk for cancer in lung, liver, bladder and kidney (Chen *et al,* 1988; Smith *et al,* 1992). The critical effect following inhalation of inorganic arsenic is lung cancer, with the most extensive evidence coming from studies on smelter workers exposed mainly to arsenic trioxide (AsIII) and workers exposed to both AsIII and AsV in pesticide production (IARC, 1980; US EPA, 1984; Järup *et al,* 1989; Lee-Feldstein, 1989). An increased lung cancer risk has also been reported in populations living in the vicinity of copper smelters (Pershagen, 1985).

Fly ash from coal-fired power plants may contain principally arsenic(V), although definite data are not yet available. No epidemiological study is available to describe the potential risk incurred, either by maintenance workers in these plants, or by populations living nearby. There is therefore a need for a tool to integrate knowledge on the toxicokinetics of trivalent and pentavalent arsenic in order to reduce uncertainties for risk assessment of pulmonary coal fly ash exposure.

Physiologically based pharmacokinetic models (PB-PK) are now more often used in risk assessment, since they allow the introduction of biological concepts in extrapolations. They are often used in dose extrapolations and in comparisons between different animal species and humans. In the present work, we propose to use a PB-PK model to also permit chemical species extrapolation. Such a model should allow an integration of our knowledge about arsenic pharmacokinetics, in order to compare and contrast AsV pulmonary exposure, with AsIII pulmonary exposure and AsV or AsIII oral exposure.

The objective of this work is to develop a PB-PK model for the four major arsenic metabolites (AsIII, AsV, methylarsonic acid (MMA) and dimethylarsinic acid (DMA)) in hamster, rabbit, marmoset monkey and human. This model includes oral and pulmonary intakes, and for development purposes also allows one to simulate intravenous injection and intratracheal instillation. Results of the PB-PK model include tissues and blood concentrations, and urinary excretion for the four chemical species. This presentation gives an overview of the development of the model, and discussion of its possible applications. More detailed descriptions are presented elsewhere (Mann *et al.* 1994a, 1994b).

Methodology

General structure

A simplified version of the arsenic model for the animal species and humans is presented in Figure 1. There are nine tissue compartments (red blood cells, RBCs not shown in Figure 1), chosen according to known affinities of arsenic metabolites, and health effects. Four arsenic metabolites are considered here: arsenic (V), arsenic (III), methylarsonic acid (MMA) and dimethylarsinic acid (DMA). Therefore the schema of Figure 1 is repeated four times in the model, once for each metabolite.

Absorption

Four routes of absorption are considered in the model: pulmonary absorption, oral absorption, intratracheal instillation and intravenous injection. The last two modes of exposure are included to make possible simulation of various experimental designs used in animal studies for fitting and validation purposes.

For exposure by inhalation, a polydispersed aerosol is assumed. In this case,

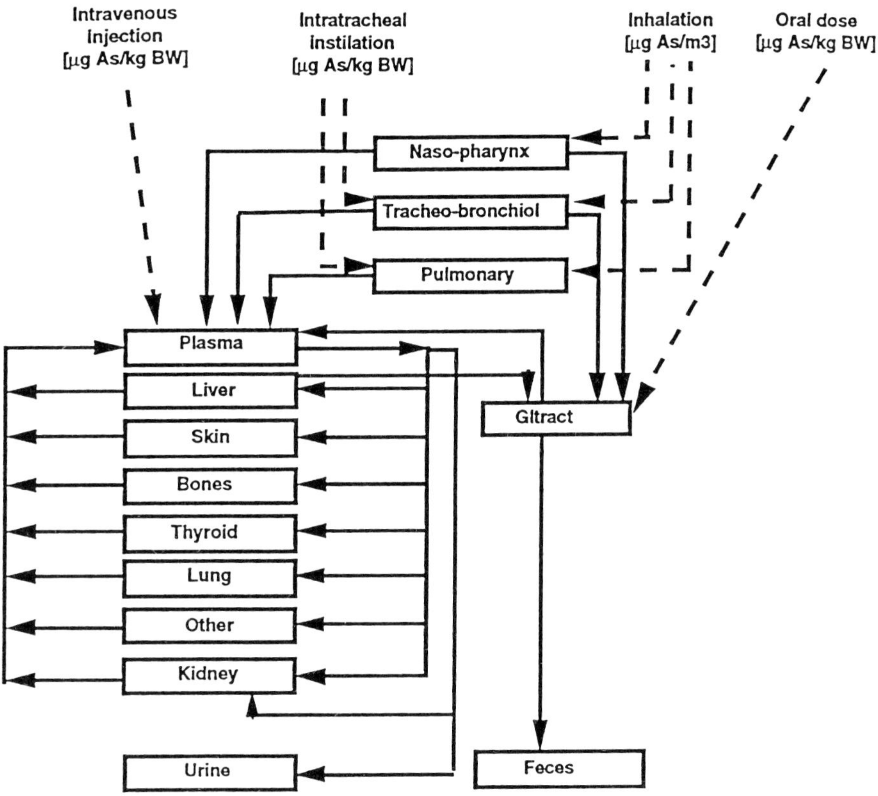

Figure 1 *Schematic presentation of the PB-PK model. This diagram is repeated 4 times, once for each of AsIII, AsV, MMA and DMA. The 4 diagrams are linked together by the metabolic reactions: oxidation-reduction and methylation.*

lungs are described by three parts, *i.e.* the naso-pharynx (N-P), tracheo-bronchial (T-B), and pulmonary regions (P) (ISO, 1991). Penetration to and deposition in them is calculated using the Weibel model A (Weibel, 1963), and empirical equations (James *et al.*, 1989; Persons *et al.*, 1987). The aerosol inhaled is described by the two parameter lognormal distribution (aerodynamic mass median diameter MMD, geometric standard deviation GSD) (Aitchison and Brown, 1957). The aerosol deposited can be transfered by muccocillary clearance (NP and T-B) to the gastro-intestinal (GI) tract (Johnson and Milencoff, 1989). Direct absorption from all three lung regions into the plasma occurs according to a first order process, the rate of which varies with the arsenic compound. To simulate intratracheal instillation, the dose is assumed to be introduced in equal proportions into the T-B and P regions. For intravenous injection, the dose is introduced as a single bolus into the plasma.

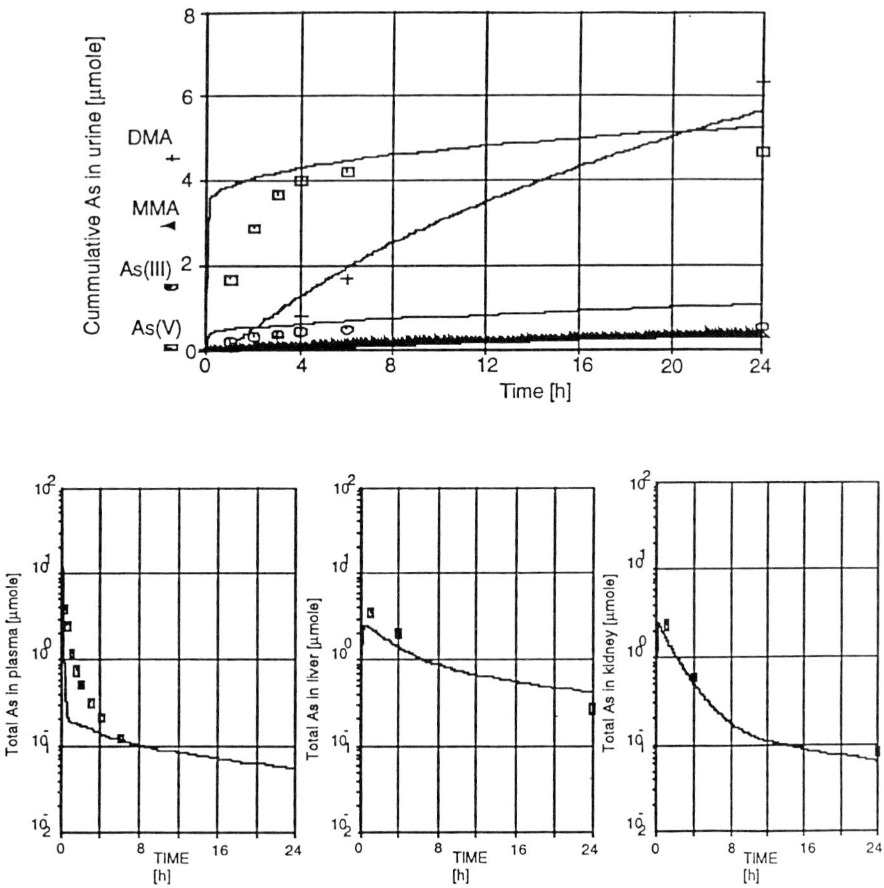

Figure 2 *Comparison of model data with results obtained after an intravenous injection of 400 mg As kg^{-1} in rabbits (Marafante* et al., *1985).*

For oral intake, arsenic is introduced directly into the GI tract, from which it is absorbed with a first order rate constant, depending on the chemical form.

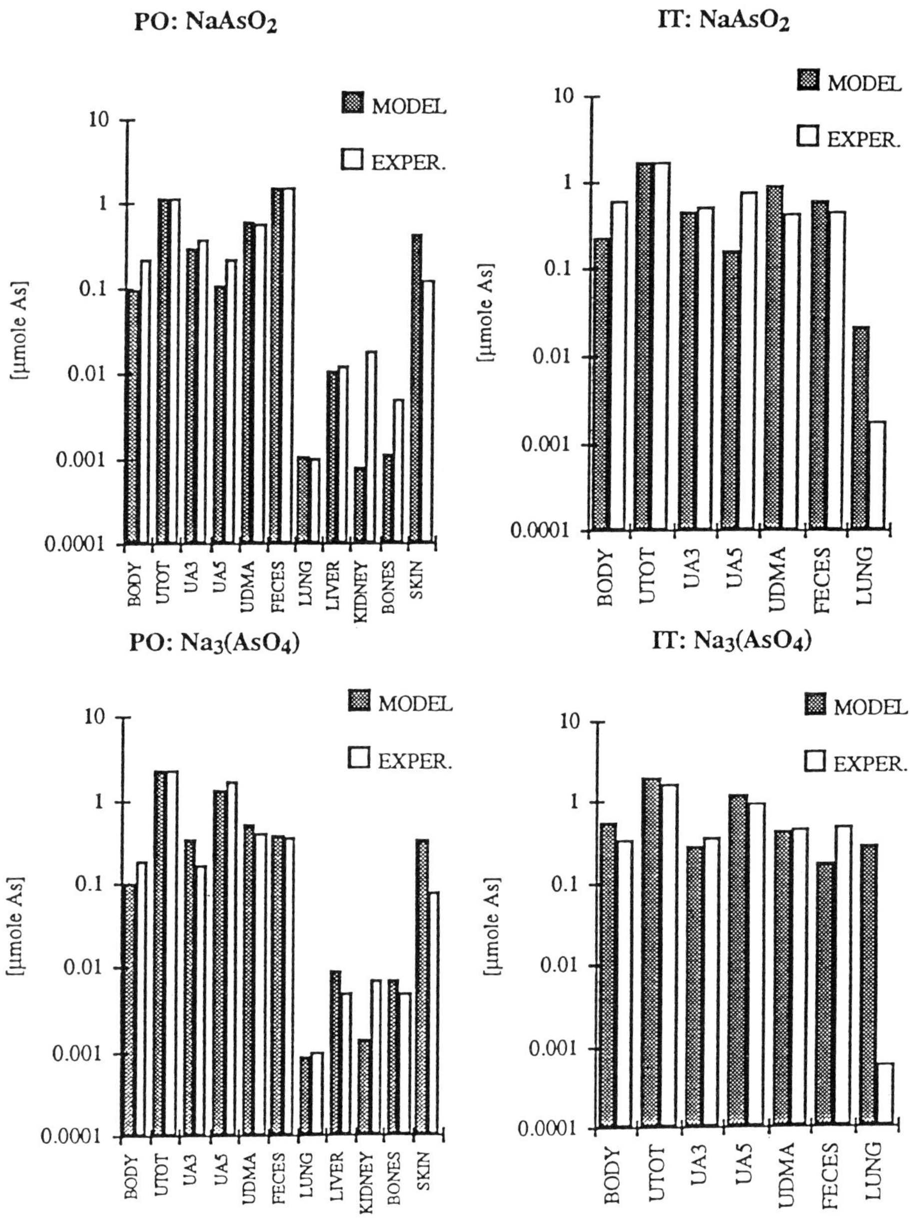

Figure 3 *Comparison of model data with results obtained after oral administration (PO) and intratracheal instillation (IT) of 2,000 mg As* kg^{-1} *in hamster (Marafante and Vahter, 1987).*

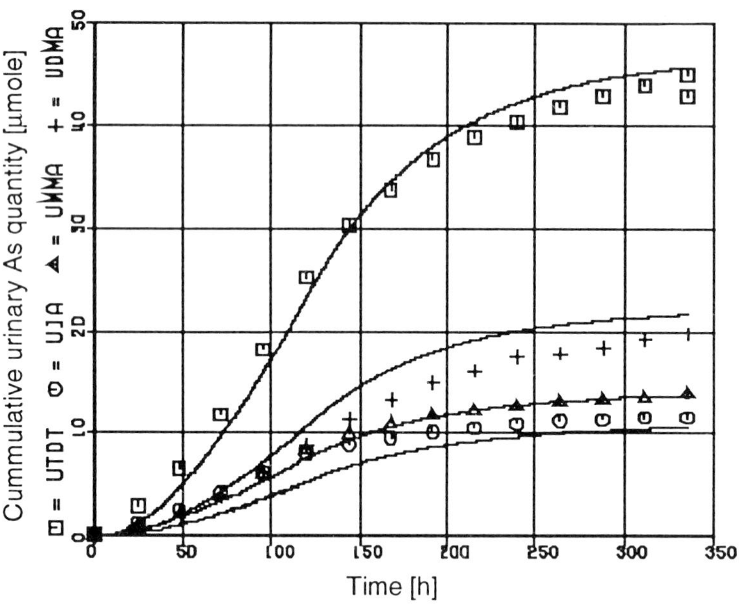

Figure 4 *Validation of the model with data on urinary excretion of arsenic in human volunteer ingesting 5 consecutive daily doses of 1,000 mg As (as $NaAsO_2$) (Buchet* et al., *1981b)*

Distribution

The distribution of arsenic compounds to the tissues depends on their blood perfusion, on the permeability of the capillary membranes, and on the affinity of the tissue for the arsenic metabolites under consideration. Blood flows, blood volumes, tissues volumes and capillary surface areas are taken from physiology reference data. These parameters are scaled to body weight in order to make possible simulation of different animal species (Lindstedt, 1992; Fiserova-Bergerova, 1983).

To simulate membrane transfers of arsenic compounds, it is considered that non-ionized species diffuse freely through capillary membranes, but that ionized species diffuse only through pores in the membranes (Berner and Cooper, 1985). The description of this process is based on: molecular weight and size, pore size and number, capillary membrane surface area and thickness. Affinity of arsenic metabolites for different tissues is considered to be linear, and it is described by an affinity constant K. This is obtained by fitting the model results to experimental data, as no direct measurements are as yet available.

Metabolism

In the model, two types of metabolic reactions are considered: (i) oxidation-reduction reactions between AsV and AsIII in the plasma, and (ii) consecutive methylation reactions in the liver. The first type of reactions is described by a first order reaction in the concentration range of interest. The second is simulated as a Michaelis-Menten enzymatic reaction. The maximum rate (V_{max} and the affinity constant between the enzyme and the substrate (K_m), as well as the first order reaction constants, are estimated by fitting the model results to experimental data. This is done for each animal species separetely, because of the well known metabolic difference between various animals and man (Vahter and Marafante, 1988).

Excretion

Elimination of arsenic occurs mainly via urine, and to some extent also via feces and skin. The difference in fecal elimination following oral and parenteral administration

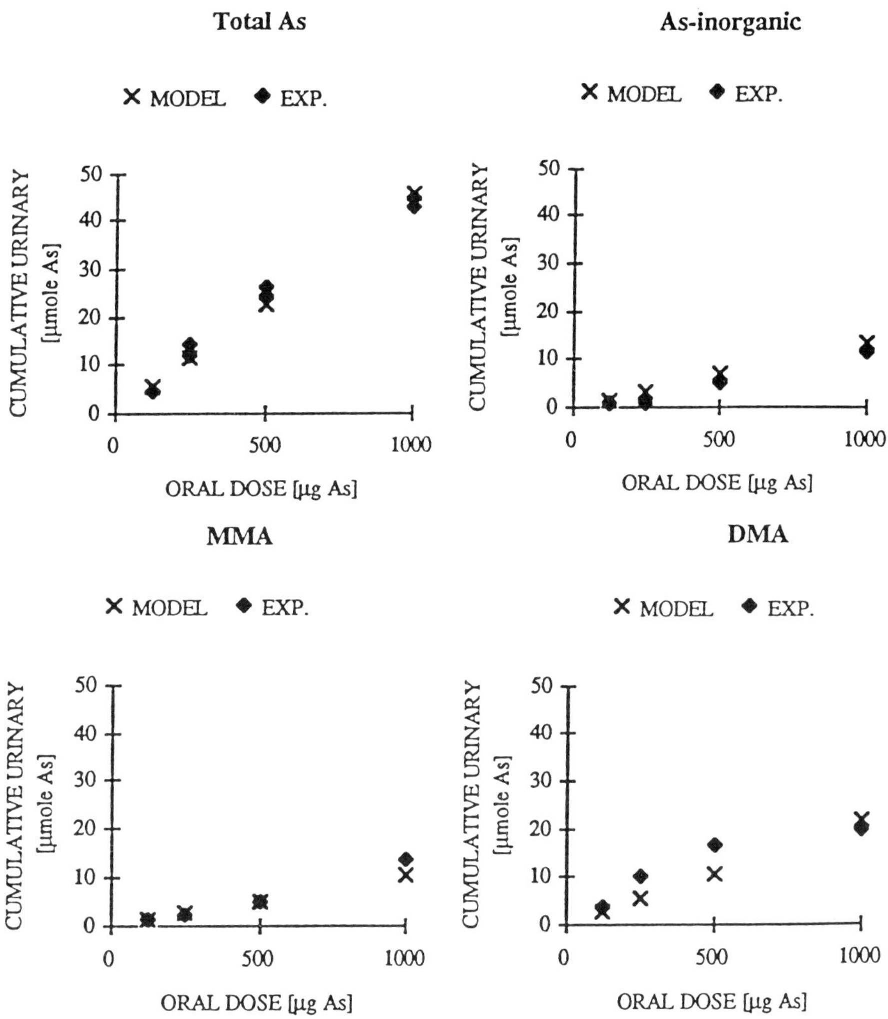

Figure 5 *Validation of the model with data on urinary excretion of arsenic in 4 human volunteers ingesting 4 different doses of $NaAsO_2$ over 5 days. Results represent the cumulative urinary excretion during 14 days (Buchet* et al.*, 1981b).*

may be considered as the fraction of arsenic not absorbed. Arsenic in the skin is transfered to a keratin compartment in order to simulate excretion via this route (hair, nails, skin cells).

It is assumed that all four arsenic metabolites are filtered at the glomerular filtration rate, which is scaled to body weight (Fiserova-Bergerova, 1983). Only AsV is reabsorbed in the tubules, from which there is a very small tubular secretion of AsIII (Ginsburg, 1965). Urinary excretion is calculated as an absolute amount or as a

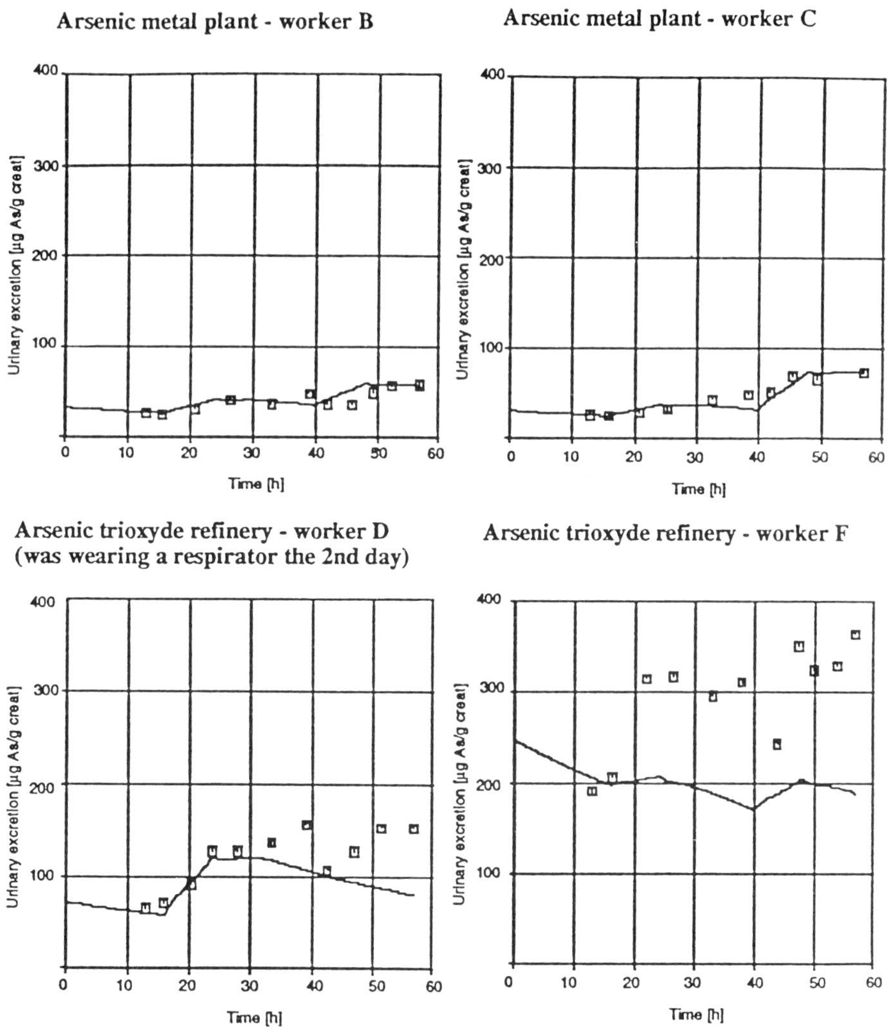

Figure 6 *Validation of the model with data on urinary excretion of arsenic in workers in an arsenic metal plant and an arsenic trioxide refinery (Vahter* et al., *1986).*

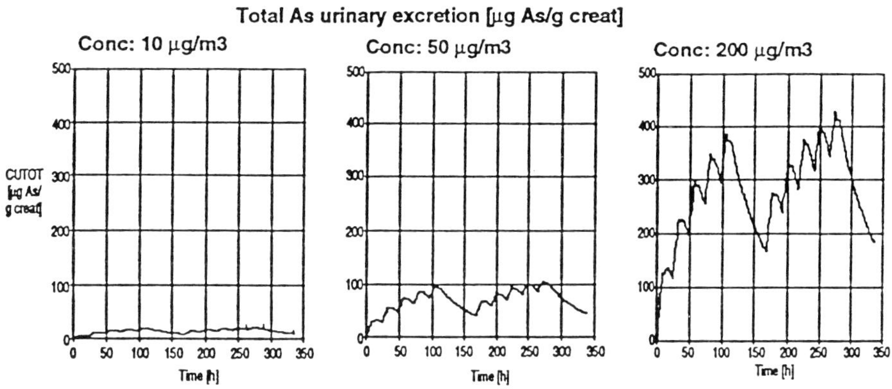

Figure 7 *Simulation of 3 scenarios of pulmonary exposure to arsenic containing aersosol at concentrations of 10, 50 and 200 mg As m^{-3}. Two weeks of exposure, 5 days a week, 8 hours a day, with a light physical workload (50 W).*

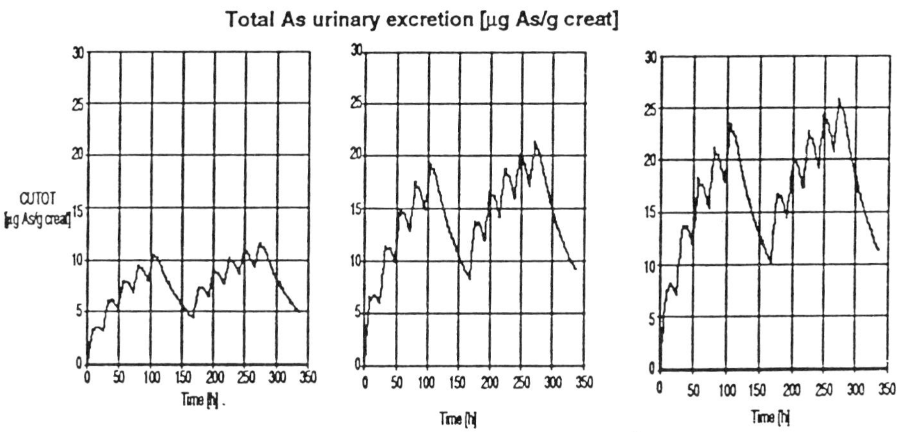

Figure 8 *Simulation of pulmonary exposure at 10 mg As m^{-3} with aerosols of MMD 2.0, 5.0 and 10 μm (GSD 2.0). Other parameters are the same as in Figure 7.*

concentration corrected for creatinine excretion rate, which is function of body weight.

Results and discussion

Tissue affinity and metabolism

In order to establish numerical values for tissue affinity and metabolism constants, data obtained in experimental studies on marmoset monkey, rabbit, hamster and human were used (Lindgren *et al.*, 1984; Vahter and Marafante, 1985; Marafante *et al.*, 1985; Marafante and Vahter, 1987; Buchet *et al.*, 1981a). They include three routes of intake: intratracheal instillation, oral intake and intravenous injection. Experimental data used were derived from plasma, RBCs, urine, feces, and several tissues. Fitting of the model was done by simulation of the experimental protocols, while varying affinity and metabolic constants, and visually comparing with experimental data.

Figure 2 gives an example of comparative data obtained in rabbit after intravenous injection of 400 mg As(V)/kg (Marafante *et al.*, 1985). The amounts in the tissues or excreted are presented as a function of time. Agreement is satisfactory except for As(V) in urine (and in plasma, of course). Figure 3 presents a summary of data obtained in hamster with 2,000 mg kg $^{-1}$of various arsenic compounds by oral intake and intratracheal instillation (Marafante and Vahter, 1987). The amounts of arsenic found after 72 hours are compared.

Validation

The model was validated by comparison with experimental data obtained in human volunteers and in workers. Figure 4 presents results obtained after ingestion of five daily doses of 1,000 mg As in the form of NaAsO2 over a 5 day period by one man (Buchet *et al.*, 1981b). The agreement between the model and the experimental data is good for all four arsenic metabolites in urine. Figure 5 presents the same experiment with data plotted for different arsenic doses (one subject per dose), as cummulative urinary excretion during 14 days.

Finally, Figure 6 presents validation of the model with data on urinary excretion of arsenic in four workers from an arsenic metal plant and a arsenic trioxide refinery (Vahter *et al.,* 1986). Plotted data are for 2 consecutive days, Monday and Tuesday. Worker D was wearing a respirator on the second day, and therefore exposure was set to zero in the model. The fact the model underestimated urinary levels might be an indication that complete respiratory protection was not achieved. There is a large discrepency in the results from worker F which probably reflects a substantial oral intake of arsenic via contaminated hands and mouth. Worker F was found to take snuff regularly during work.

Application

Preliminary results are shown in Figures 7 and 8. Figure 7 indicates the effect of repeated inhalation exposure at three different exposure levels on the urinary concentrations of total arsenic metabolites. Exposure consists of an arsenic trioxide aerosol with a MMD of 5.0 μm and a GSD of 2.0. Physical workload is light (50W). Figure 7 shows that the urinary concentration is close to a steady-state after 2 weeks of exposure. Morning and end of shift concentrations are slightly different, but this can probably not be detected when measured in workers. The concentration of exposure has an almost linear effect on urinary concentrations in this range of exposures (10 to 200 mg m^{-3}).

In Figure 8, the effect of a change in the particle size distribution is shown. In all 3 examples, exposure concentration to total As dust is kept constant (10mg As m^{-3}), as is the GSD, while the MMD is changed from 2.0 up to 10.0 μm. It is interesting to note that for a MMD of 2.0 μm the urinary concentrations are much lower than in the other two cases. This is due to the fact that, although smaller particles penetrate more efficiently to the deep region of the lungs, they do not deposit as well as the large particle. From MMD 5 to 10 mm, the effect of size distribution is not very important anymore.

Future work will focus on simulating various exposure scenarios for comparison of urinary and tissue arsenic levels under various conditions. It is expected that the results will be of importance for the evaluation of their relative risks. Situations to be studied are: drinking water, copper smelter worker, fly ash worker and community expoure to ambient aerosols.

Conclusions

A pharmacokinetic model has been developed for the description of the uptake, distribution, metabolism and elimination of AsV, AsIII, MMA and DMA. It was shown to be valid in at least four animal species: marmoset monkey, rabbit, hamster and human. It takes into account four different routes of exposure, namely inhalation, intratracheal instillation, oral and intravenous.

The model is principally a physiologically based pharmacokinetic model, especially for lung deposition, tissue distribution, capillary transfers, and renal excretion. However some quantitative data were missing and had to be estimated by fitting. These include tissue affinities, fecal and bilary secretion rates, and distribution within blood constituents. These parameters were assumed to be independent of the animal species. The metabolism of arsenic however, is known to differ between species. Specific metabolic constants were therefore found by fitting the model species by species. Application of the present model to different exposure scenarios will notably improve our understanding of arsenic pharmacokinetics. It will also contribute to reducing uncertainties in risk assessment for exposure to different arsenic compounds and by different routes.

Acknowledgments

We want to thank Prof. H.Savolainen for his helpful comments and advices. This work was supported by the Electric Power Research Institute, Palo Alto, CA 94303, USA (RP 3370-02).

References

Aitchison, J. and Brown, J.A.C. 1957. *The lognormal distribution.* Cambridge University Press.

Berner, B. and Cooper, E.R. 1985. Models of skin permeability. In: A.F. Kydoniev, B.Berner (ed.), *Transdermal delivery of drugs,* Vol. II. pp. 41-55. CRC Press, Florida.

Buchet, J.P., Lauwerys, R. and Roels, H. 1981a. Comparison of the urinary excretion of arsenic metabolites after a single oral dose of sodium arsenite, monomethylarsonate or dimethylarsinate in man. *Int. Arch. Occup. Environ. Health*, **48**, 71-79.

Buchet, J.P., Lauwerys, R. and Roels, H. 1981b. Urinary excretion of inorganic arsenic and its metabolites after repeated ingestion of sodium metaarsenite by volunteers. *Int. Arch. Occup. Environ. Health,* **48**, 111-118.

Chen, C.J., Kuo, T.L. and Wu, M.M. 1988. Arsenic and cancers. *Lancet*, **I**, 414-415.Fiserova-Bergerova, V. 1983. *Modeling of inhalation exposure to vapors: uptake, distribution and elimination.* CRC Press, Florida.

Ginsburg, J.P. 1963. Renal mechanism for excretion and transformation of arsenic in the dog. *Amer. J. Physiol*, **208**, 832-840.

IARC. 1980. Arsenic and arsenic compounds. In: I*ARC Monographs on the evaluation of the carcinogenic risk of chemicals to humans.* Some metals and metallic compounds, Vol. 23, pp. 39-141.

International Agency for Research on Cancer, Lyon. ISO. 1991. International Organization for Standardization (ISO). *Air quality - Particle size fraction definitions for health-related sampling.* Approved for publication as CD 7708. Geneva.

James, A.C., Birchall, A., Cross, F.T., Cuddihy, R.G. and Johnson, J.R. 1989. The current approach of the ICRP task group for modeling doses to respiratory tract tissues. *Health Physics*, **57**, 271-289.

Järup, L., Pershagen, G. and Wall, S. 1989. Cumulative arsenic exposure and lung cancer in smelter workers: dose-response study. *Amer. J. Ind. Med.,* **15**, 31-41.

Johnson, J.R. and Milencoff, S. 1989. A comparison of excretion and retention between the current ICRP lung model and a proposed new model. *Health Physics*, **57**, 263-270.

Lee-Feldstein, A. 1989. A comparison of several measures of exposure to arsenic. Matched case-control study of copper smelter employees. *Amer. J. Epidemiol.*, **129**, 112-124.

Lindgren, A., Danielsson, R.G., Dencher, L. and Vahter, M. 1984. Embryotoxicity of arsenite and arsenate in pregnant mice and monkeys and effects on embryonic cells *in vitro. Acta Pharmacol. Toxicol.*, **54**, 311-320.

Lindstedt, S.L. 1992. Allometric extrapolation. *International workshop on physiologically based pharmacokinetic modeling and risk assessment.* Fort Collins CO.

Mann, S., Droz. P.O. and Vahter. M. 1994. *A physiologically based pharmacokinetic model for arsenic exposure.* Part I. Development and validation in animals and human. (in preparation).

Mann, S., Droz, P.O. and Vahter, M. 1994. *A physiologically based pharmacokinetic model for arsenic exposure.* Part II. Application to oral and pulmonary exposure to various arsenic species. (in preparation).

Marafante, E., Vahter, M. and Envall, J. 1985. The role of the methylation in the detoxification of arsenate in the rabbit. *Chem. Biol. Interact.,* **56**, 225-238.

Marafante, E. and Vahter, M. 1987. Solubility, retention, and metabolims of intratracheally and orally administered inorganic arsenic compounds in the hamster. *Environ. Res.*, **42**, 72-82.

Pershagen, G. 1985. Lung cancer mortality among men living near an arsenic-emitting smelter. *Amer. J. Epidemiol.*, **122**, 684-694.

Pershagen, G. and Vahter, M. 1991. Inorganic arsenic. In: Beije, B., and Lundberg, P. (eds), *Arbete och Hälsa* No 50.Criteria documents from the Nordic Expert Group . .

Persons, D.D., Hess, G.D., Muller, W.J. and Scherer, P.W. 1987. Airway deposition of hygroscopic heterodispersed aerosols: results of a computer calculation. *J. Appl. Physiol.*, **63**, 1195-1204.

Smith, A.H., Hopenhayn_Rich, C., Bates, M.N., Goeden, H.M., HertzPicciotto, I., Duggan, H.M., Wood, R., Kosnett, M.J. and Smith, M.T. 1992. Cancer risks from arsenic in drinking water. *Environ. Health Perspect.,* **97,** 259-267.

Tam, G.K.H., Charbonneau, S.M., Bryce, F., Pomoroy, C. and Sandi, E. 1979. Metabolism of inorganic arsenic (74As) in humans following oral ingestion. *Toxicol. Appl. Pharmacol.*, **50,**319-322.

US EPA. 1984. *Health assessment document for inoragnic arsenic.* Final report. Research Triangle Park, NC: US. Environmental Protection Agency. Publication No. EPA-6008-83-021F.

US EPA. 1988. *Special report on ingested inoragnic arsenic: skin cancer; nutritional essentiality.* Research Triange Park, NC: US. Environmental Protection Agency. Publication No. EPA-6253-87-013.

Vahter, M. and Marafante, E. 1985. Reduction and binding of arsenate in the marmoset monkey. *Arch. Toxicol.,* **57**, 119-124.

Vahter, M., Friberg, L. Rahster, B., Nygren, A. and Nolinder, P. 1986. Airborne arsenic and urinary excretion of metabolites of inorganic arsenic among smelter workers. *Int. Arch. Occup. Environ. Health,* **57**, 79-91.

Vahter, M. and Marafante, E. 1988. In vivo methylation and detoxyfication of arsenic. In: P.J. Craig and F. Glockling (ed.), *The biological alkylation of heavy elements,* pp. 105-119. Royal Society of Chemistry, London.

Weibel, E.R. 1963. *Morphology of the human lung. Springer-Verlag,* New York.

WHO. 1981. *Arsenic - Environmental Health Criteria,* 18, World Health Organization, Geneva, 174 pp.

22 Water utility regulatory and research projects on arsenic

J. Alan Roberson

American Water Works Association, 1401 New York Ave NW, Suite 640, Washington, D.C. 20005, USA

Abstract

Water utilities will be greatly impacted by the US Environmental Protection Agency's (US EPA's) arsenic regulation. The final Maximum Contaminant Level (MCL) will determine how many utilities will have to install additional treatment for the removal of arsenic. US EPA is having difficulty in extrapolating the existing occurrence data, treatment technologies, and analytical methods to the levels that are being considered for the arsenic MCL.

The American Water Works Association (AWWA), through the Water Industry Technical Action Fund (WITAF), is developing several projects to fill in the data gaps for the development of this regulation. The first project is a study on the Practical Quantitation Level (PQL) for arsenic. The second project is a occurrence and treatment matrix, to determine how many utilities will have to install specific treatment technologies to reach different MCLs. The third project is a study on the affordability impacts from the arsenic regulation. Rate increases driven by the costs of the additional treatment will be determined, and analyzed against various income levels.

The AWWA Research Foundation (AWWARF) is starting a project to compare the analytical techniques to reach the levels that are being considered for the arsenic MCL. Potential improvements to the analytical techniques will be analyzed, and potential recommendations to improve the techniques will be developed.

Introduction

The American Water Works Association (AWWA) is an international non-profit, scientific and educational society dedicated to the improvement of drinking water quality and supply. Founded in 1881, the Association is the largest organization of water supply professionals in the world. Our over 55,000 members represent the full spectrum of the "drinking water community": treatment plant operators and managers, public health officials, scientists, academicians and others who hold a genuine interest in water supply and public health. Our membership includes over 3,700 utilities which supply roughly 75 percent of the nation's drinking water.

The AWWA Water Utility Council (WUC) directs the legislative and regulatory activities for the Association. The WUC is provided technical guidance on arsenic through the Inorganics and Corrosion By-Products (IOC/CBP) Technical Advisory Workgroup (TAW). The IOC/CBP TAW provides technical analysis for regulatory and legislative actions impacting inorganic contaminants in drinking water, including arsenic.

The IOC/CBP TAW reviews the available data on health effects, occurrence data, treatability, and treatment costs in order to assess, from the perspective of AWWA, the viability and feasibility of the proposed US EPA regulations. The IOC/CBP TAW prepares technical recommendations and expert testimony for the WUC. The TAW is also responsible for developing studies and research projects to provide data to develop and support the AWWA position on future regulations and/or to challenge the US EPA position on proposed or existing regulations. These research projects are managed by either the TAW or a Project Advisory Committee (PAC) from the AWWA sister organization, the AWWA Research Foundation (AWWARF).

The 1986 Amendments to the Safe Drinking Water Act (SDWA) listed 83 contaminants for regulation in drinking water. Arsenic, along with other inorganic contaminants, was included in the list of 83 contaminants. Arsenic, sulfate, aldicarb and its metabolites, and radionuclides are the remaining contaminants to be regulated from the list of 83 contaminants.

Based on the increasing number of contaminants being regulated in drinking water, the five major water industry associations formed a coalition to provide technical input to these regulations. The following five associations formed the Water Industry Coordinating Committee (WICC):

-American Water Works Association (AWWA)

-Association of Metropolitan Water Agencies (AMWA)

-National Association of Water Companies (NAWC)

-National Rural Water Association (NRWA)

-National Water Resources Association (NWRA)

In order to provide funding to develop technical input on the drinking water regulations, the WICC started the Water Industry Technical Action Fund (WITAF) by assessing the drinking water utilities. The WITAF is used for consultant contracts to assist in the technical development of the AWWA's comments to the proposed drinking water regulations.

Arsenic has recently emerged as a major contaminant of concern for the future. Through a settlement with the Bull Run Coalition, the Environmental Protection Agency (US EPA) is committed to proposing the arsenic regulation in November, 1995, and the final arsenic regulation in November, 1997. This paper will address the AWWA and AWWARF efforts on the upcoming US EPA's regulation for arsenic in drinking water.

WITAF arsenic projects

Through WITAF, the following three arsenic projects are currently underway:

-An independent cancer risk analysis

-A study of the arsenic Practical Quantitation Level (PQL)

-Co-sponsorship of this SEGH conference

The following three additional projects are in the planning stages:

-Development of a national occurrence database

-Development of the number of systems impacted, and the estimated national treatment cost

-Studying the affordability of the treatment costs

The drinking water regulation for arsenic raises many difficult scientific and policy questions. The goal of the above listed projects is to develop the best scientific data to answer these difficult questions.

How solid is the data that US EPA is using in their risk assessment? One project being sponsored by the Water Industry Technical Action Fund (WITAF) is an independent analysis of the cancer risk from arsenic in drinking water. This project will provide an independent evaluation of the Taiwan data. This project is currently underway, and consists of an independent review of the US EPA health criteria document, and all of the backup studies. This project began in May, 1993, and although scheduled for completion in November, 1993, may extend well beyond that in order to critically evaluate the real state of the existing data. The contractor is Dr Ken Brown.

Are the analytical methods advanced enough to provide practical and reliable quantitation at levels between 0.2 and 20 $\mu g\ L^{-1}$? Over the past decade, the method detection limits for arsenic have been in the range of 10-20 $\mu g\ L^{-1}$. In the past few years, improvements to the method have lowered the detection limit to the range of 5-10 $\mu g\ L^{-1}$. But how practical and reliable are the analytical methods at these low levels? What is the true quantitation level?

A WITAF project currently underway is a study on the Practical Quantitation Level (PQL) for arsenic. This project began in May, 1993, and will be completed in November, 1993. The contractor is Montgomery Laboratories, with Dr Andy Eaton being the principal investigator. Three drinking water matrices and reagent water will be spiked at arsenic levels of 0,1,2,4,6, and 10 $\mu g\ L^{-1}$. 78 samples will be run by each of 22 laboratories, for a total of 1,404 samples. 11 commercial laboratories will be used, along with 11 utility laboratories. Both graphite furnace atomic absorbance and hydride generation will be used in this PQL analysis. The results of these analyses will be compiled to determine the PQL for arsenic, using the different analytical methods.

How would these low levels for the MCL affect water utilities? For example, assume US EPA sets the arsenic MCL at 2 $\mu g\ L^{-1.}$ According to recent survey by Metropolitan Water District of Southern California of 144 utilities, 14% exceeded a 2 $\mu g\ L^{-1}$ MCL in their effluent. In a recent study of Wisconsin groundwaters, 14% of the groundwaters exceeded a 2 $\mu g\ L^{-1}$ MCL. However, a much higher percentage of utilities could exceed this level in the Western States. Further occurrence studies are needed to refine the estimates of the number of utilities that could exceed a proposed MCL.

Through the Water Industry Technical Action Fund (WITAF), AWWA is planning exactly such a project to gather all of the existing occurrence studies, and to further refine the number of utilities that will exceed the MCL at different levels. Additionally, this project will analyze the treatment technologies for arsenic. Arsenic removal at different points in the treatment process will be analyzed. This analysis will compare the raw water arsenic levels with proposed MCL levels, and determine how many utilities will be able to use each treatment technology to meet each of the proposed MCLs. The number of utilities using each treatment technology will be compiled, in order to determine the total treatment cost for the arsenic regulation, at different MCL levels. The goal of this project is an independent Regulatory Impact Analysis (RIA) that estimates the total cost of the arsenic regulation. This independent RIA will compared to the RIA that US EPA will develop as part of the arsenic regulation.

How will the drinking water consumers be impacted by increased rates to finance arsenic treatment? Another WITAF project will further analyze the total costs of this regulation, by looking at affordability. This project will analyze the impact on water rates of the proposed regulation, and then determine the impact on the low-income and fixed-income populations. The focus of this project is the impact on the lower end of the distribution of monthly income.

AWWARF arsenic projects

The AWWA Research Foundation (AWWARF) conducts basic research for the drinking water utilities. AWWARF's mission is "advancing the science of water to promote the quality of life." Funding for AWWARF's projects comes from a voluntary subscription program, primarily consisting of drinking water utilities, but recently expanded to include consulting engineers and equipment manufacturers. AWWARF currently has 260 projects completed or underway, and its 1993 research budget is \$5 million. The 1993 research budget, including in-kind contributions from the researchers and participating utilities, is \$7.9 million.

AWWARF has two arsenic research projects underway. The first is a study of the analytical chemistry of arsenic in drinking water. The principal investigator is Dr Andy Eaton of Montgomery Laboratories, with participation from the Metropolitan Water District of Southern California (MWDSC), and West Coast Analytical Services. The second project is a study of arsenic removal by enhanced coagulation and membrane processes. The principal investigator on this project is Dr Janet Hering of the University of California at Los Angeles, with participation from MWDSC.

The goals of the arsenic analytical chemistry project are to improve detection limits, work on analytical methods for speciation, and to develop methods that can easily be used by drinking water utilities. The project will first develop new methods, and evaluate stability, contamination, and sample preservation. These new methods will be tested in an interlaboratory comparison from three labs. Based on the feedback from the three labs, a Statement of Procedure (SOP) will be developed for use by outside labs. The SOP will then be tested by five outside labs, and then further refined based on the feedback from the outside labs.

The goal of the arsenic removal project is to analyze the impacts of water quality on arsenic removal. The impact of the arsenic oxidation state, pH, humic substances, and other substances will be analyzed at bench and pilot scale using synthetic and two source waters.

Conclusions

While several projects are ongoing or underway on arsenic regulatory research, there are still many more questions left unanswered. Much of the results of this research will not be available by the time the arsenic regulation is proposed in November, 1995. The value of future research to answer many of these questions needs to be incorporated into the regulatory development process by US EPA.

Acknowledgements

The author wishes to acknowledge the assistance of Jack Sullivan, American Water Works Association in the review and preparation of this paper.

23 The use of site-specific studies to assess arsenic health risk at a Superfund site

Paul D. Bergstrom[1]*, H.L. Greene[2], Rosalind A. Schoof[3], Catherine Petito Boyce[3], Lisa J.Yost[3], Barbara D. Beck [4] and Peter A. Valberg[4]

[1]*Atlantic Richfield Co.,555 Seventeenth Street, Denver, Colorado 80202, USA*
[2]*Atlantic Richfield Co.,515 South Flower Street, Los Angeles, California 90071, USA*
[3]*PTI Environmental Services,15373 SE 30th Place, Suite 250, Bellevue, Washington 98007, USA*
[4]*Gradient Corporation, 44 Brattle Street, Cambridge, Massachusetts 02138, USA*

Abstract
ARCO has instituted a research program designed to provide site-specific estimates of current and future exposures to arsenic in Anaconda, Montana, that are more accurate than those resulting from use of the U.S. Environmental Protection Agency's default exposure assessment assumptions. These studies include a comprehensive community exposure study, an investigation of children's soil ingestion rates in Anaconda, and studies of the bioavailability of arsenic from Anaconda soils, including detailed mineralogic analyses. Additional information regarding bioavailability and soil ingestion rates is being obtained by comparing environmental data with urinary arsenic levels observed in exposure study participants. As studies are completed, site-specific exposure parameters derived from the results will be used to develop more accurate estimates of exposure. Additional ARCO research efforts focus on delineating and reducing uncertainties associated with current toxicity factors used to assess the risks of ingested arsenic.

Introduction

Assessment of current and future arsenic health risks following US Environmental Protection Agency (EPA) Superfund guidance relies on many default assumptions in the absence of site-specific data. Such default assumptions tend to err on the side of conservatism whenever there is uncertainty regarding the actual values. Consequently, the resulting risk estimates are likely to be based on overestimates of

---

*To whom correspondence should be addressed.

exposure that actually occur at a site. Such uncertainties are most significant where site conditions differ from those assumed in developing the default parameters and where the parameters significantly affect the risk assessment results and subsequent remedial action decisions. In such cases, site-specific studies can be used to develop alternative parameters that better reflect actual site conditions and to provide a better understanding of actual exposure sources and pathways at a site. Use of such parameters will ensure that site investigation and remediation efforts are focused on materials that most significantly contribute to site-related exposures. In light of these considerations, ARCO has instituted a research program to more accurately estimate current and future exposures and risks associated with arsenic in soils related to historic smelting activities at a Superfund site in Anaconda, Montana. This paper provides an overview of these efforts.

The basic model used to characterize site risks and to develop site-specific cleanup levels incorporates exposure and toxicity data to calculate risk estimates as follows:

$$\text{Exposure} \times \text{Toxicity} = \text{Risk}$$

Exposure is typically calculated using the following general equation:

$$\text{Intake} = \frac{\text{Media concentration} \times \text{Contact rate} \times \text{Exposure duration} \times \text{Exposure frequency} \times \text{Bioavailability adjustment}}{\text{Body weight} \times \text{Averaging time}}$$

Some of the key parameters that significantly influence risk estimates and that may vary significantly from the default assumptions are site-specific soil ingestion rates, exposure frequency, and bioavailability. In addition, because EPA default assumptions typically compound the conservative nature of the resulting risk assessments, use of site-specific data provides more accurate estimates of risk, resulting in the derivation of remedial goals that are both realistic and protective of public health.

In addition to the issues raised by the use of default assumptions, the appropriateness of the toxicity values (*i.e.*, reference dose [RfD] and carcinogenic slope factor [CSF]) used by EPA to evaluate health risks associated with arsenic are the subject of increasing scientific concern. Concerns include the degree of arsenic exposure actually experienced by the Taiwanese study populations investigated in epidemiologic studies (Tseng *et al.*, 1968; Tseng, 1977) (which were used to derive toxicity values EPA uses), the potential inapplicability of the ingested arsenic CSF to populations in the United States, and the likelihood that a threshold dose exists below which there is no observed increase in internal or skin cancers. Investigations that ARCO has undertaken to resolve some of these issues are discussed further in the toxicity assessment section of this paper.

Exposure assessment

ARCO has initiated research to examine several key aspects of potential arsenic exposures at the Anaconda site. Specific projects include: a comprehensive environmental health study using urinary arsenic as a biomarker for arsenic exposure, a soil ingestion study, bioavailability and mineralogy studies, and an investigation of the degree to which arsenic in vegetables may occur in methylated forms that are less toxic. Findings from these studies can be used to more accurately identify and quantify sources of arsenic exposure in the community. As described above, estimates of arsenic intake are central to estimating risks associated with ingestion of arsenic in soil and dust. Urinary arsenic is a function of arsenic intake from various exposure media (*i.e.*, diet, soil and dust, and water) and of factors influencing intake estimates for those media, such as bioavailability and demographic and behavioral variables. Arsenic intake from soil and dust may be more accurately estimated by analyzing urinary arsenic data in combination with paired soil, dust, and water concentrations for each household, and with site-specific estimates of soil ingestion rates and soil arsenic bioavailability.

Anaconda community exposure study

A comprehensive community exposure study was begun in Anaconda by Dr Robert Bornschein and other researchers from the University of Cincinnati. This study was begun in 1992 and is almost complete; preliminary results are presented by Dr. Bornschein in these proceedings. During the exposure study, researchers collected environmental, biological, demographic, behavioral, and other data from 600 children less than 6 years old and 250 older individuals from 480 families residing in or near Anaconda. The objectives of the study were to determine the extent of childhood arsenic exposure in Anaconda; to evaluate individual exposures in children 2–6 years old, as reflected by urinary arsenic concentrations; to evaluate the associations between arsenic sources and a biological marker of exposure; and to estimate daily arsenic uptake from soil.

The exposure study, in combination with other research efforts, will provide several types of information relevant to assessing arsenic exposures in the Anaconda community. First, the exposure study will yield estimates of the sources, frequency, duration, and magnitude of exposures in Anaconda by combining information from the environmental sampling program, urinary arsenic measurements, and detailed lifestyle questionnaires. Second, complementary information on arsenic exposure will be collected through an investigation by researchers at the University of Massachusetts of soil ingestion rates in a randomly selected subset of children in the Anaconda Community Exposure Study and an evaluation of the bioavailability and mineralogic characteristics of soil arsenic at Anaconda. This information will be used to support the results of the community exposure study.

As described above, the findings of the community exposure study will be used to identify exposure patterns in Anaconda and to develop site-specific exposure parameters to use instead of standard EPA default assumptions for assessing risks in Anaconda. For example, EPA frequently assumes that arsenic concentrations in outdoor soil and indoor dust are equal. However, several comprehensive investigations conducted in mining and milling communities (*e.g.*, Midvale, Utah [Bornschein *et al.*, 1990]; and Butte, Montana [Butte-Silver Bow DOH and UCDEH, 1991]) have shown that mean arsenic concentrations in outdoor soil are generally

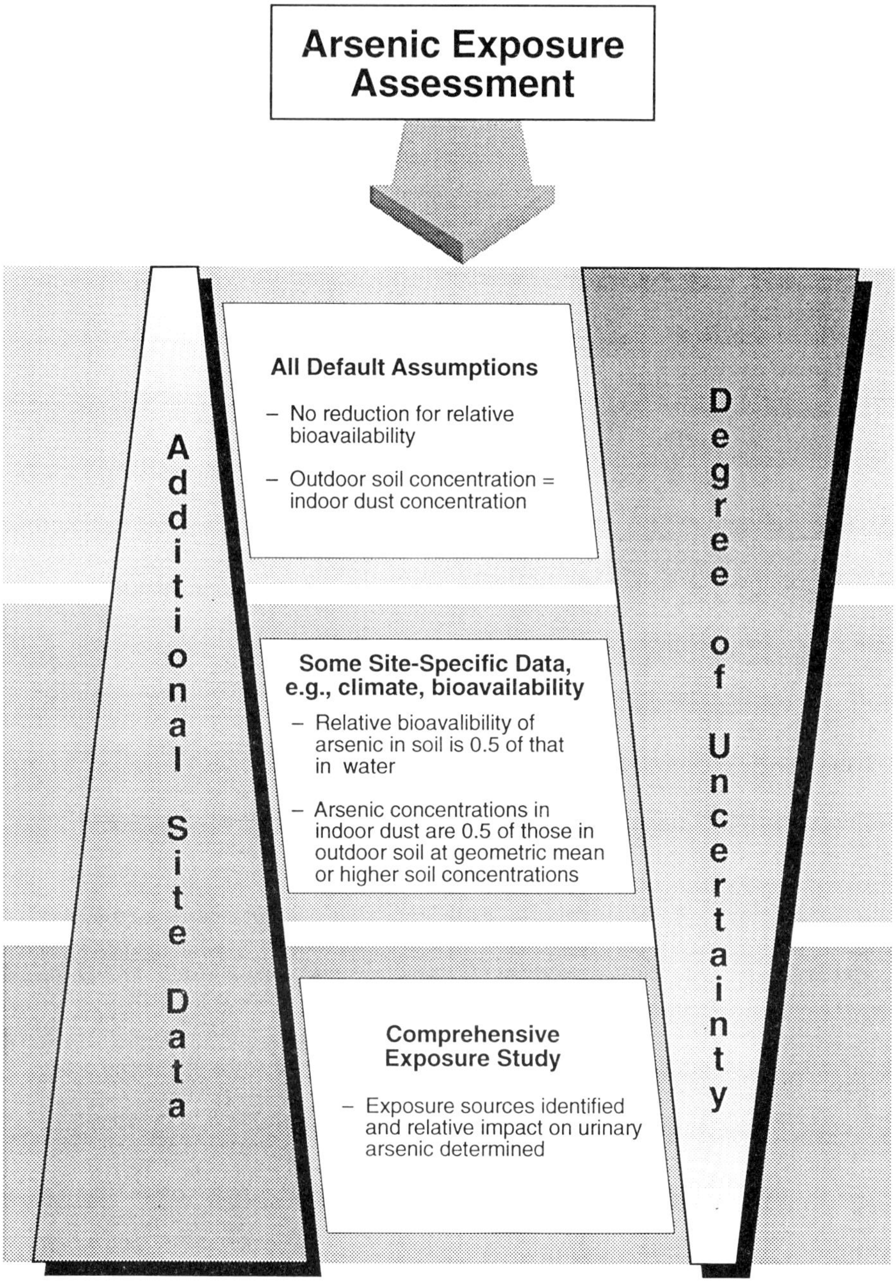

Figure 1 *Reduction of uncertainties through collection of site-specific data.*

higher than those in indoor dust. Results of the community exposure study are consistent with these findings: indoor dust concentrations are generally less than one-half of the concentrations in outdoor soils when outdoor soils are at the geometric mean concentration of 183 mg $kg^{-1.}$

Soil ingestion rate study

The extent to which young children ingest soil is a critical element in estimating arsenic intake from soil. The most reliable data for estimating soil ingestion rates are studies in which the mass of individual elemental tracers found in soil are compared with measurements of tracer mass in feces of children or adults. The selected tracers must be relatively unique to soil and poorly absorbed in the gut. Two tracer studies (Binder *et al.*, 1986; Clausing *et al.*, 1987) were the primary basis for EPA's current default soil ingestion ratc of 200 mg/day for children 1–6 years old (Porter, 1989; US EPA, 1989). However, these studies, and subsequent studies conducted by Calabrese *et al.* (1989), Calabrese *et al.* (1990), Calabrese (1990), Davis *et al.* (1990), and van Wijnen *et al.* (1990), are recognized by the researchers as having limitations. The Anaconda soil ingestion study design was intended to correct the limitations of the previous studies.

To develop site-specific data for the soil ingestion rate, a subset of 64 children, 1–4 years old, was randomly selected from the community exposure study for inclusion in the soil ingestion study conducted by Dr Edward Calabrese and other researchers at the University of Massachusetts. The study was conducted over a 1-week period in August 1992, a period of high outdoor activity. In addition, a validation study, using 10 healthy adult volunteers in Massachusetts, was conducted to estimate the soil ingestion detection capacity of each tracer element measured for the children and to provide quantitative estimates of soil ingestion in the adult subjects. During the ingestion studies, investigators collected duplicate samples of all food consumed by study participants during the study period, all fecal materials, and soil and dust samples from each subject's home. All samples were analysed for eight tracers: aluminum, cesium, chromium, silicon, titanium, vanadium, yttrium, and zirconium. Arsenic was also analysed to obtain an estimate of arsenic bioavailability.

The approach used in the Anaconda soil ingestion study reflects refinements of that used in a previous investigation conducted in Amherst, Massachusetts (Calabrese *et al.*, 1989), and will provide more precise soil ingestion estimates than previous studies. These refinements include an increase in the time period of the study to allow for better matching of ingested and excreted tracer amounts. Analyses of the eight tracers will be useful in differentiating intake from soil from that of indoor dust and in determining the proportion of arsenic intake from dust versus soil. The results of this study can be used to replace the EPA default soil ingestion value in the risk model.

Bioavailability and mineralogic studies

Risk assessments conducted using EPA default assumptions typically assume that gastrointestinal absorption of arsenic in soil is equal to that of arsenic in water (the medium in the studies [Tseng *et al.*, 1968; Tseng, 1977] that formed the basis for

EPA's toxicity factors). Available site-specific data suggest, however, that arsenic in Anaconda soil is at least 50 percent less bioavailable than arsenic in water, and therefore, the associated risk is significantly reduced.

To provide site-specific information regarding bioavailability, Freeman *et al.* (1993) investigated the bioavailability of soil arsenic from a residential area in Anaconda, Montana, by testing five groups of New Zealand white rabbits. Three groups were fed varying arsenic dosages via capsules of arsenic-containing soil, one group was administered intravenous sodium arsenate solution, and one group was administered sodium arsenate solution via oral gavage. Fecal and urinary arsenic levels were monitored for 5 days after arsenic administration in all groups.

To determine soil arsenic bioavailability in the experiment, the authors compared the urinary excretion from three soil dose groups and the oral gavage group against those from the intravenous group. By comparing absorption of arsenic from soil with absorption of dissolved arsenic from water, an overall relative bioavailability of arsenic in soil of less than 50 percent was derived. Recently released findings from a study comparing urinary excretion in monkeys after administration of soil or dust from Anaconda found much lower relative bioavailability than previously estimated in studies using rabbits. Three monkeys were administered four different treatments using a crossover design in which each monkey served as its own control. Each monkey was administered a single dose of arsenic-containing soil or dust, soluble sodium arsenic by gavage,or sodium arsenate solution by the intravenous route in a random order with at least a 2 week washout period between treatments. Relative bioavailability of 20% and 29% were reported for animals fed arsenic in soil and dust, respectively based on comparison with the gavage group. This reduced absorption factor should be used in deriving risk estimates for arsenic from Anaconda soils.

The residential soils used in the bioavailability study were also evaluated as part of a mineralogic investigation designed to characterize the geochemistry of area source materials and residential soils. This investigation, conducted by PTI Environmental Services, focused on evaluating the homogeneity of the mineral distribution within the area soils and on the implications of the observed geochemical characteristics for source identification and bioavailability of ingested arsenic. The forms of arsenic present in each of 21 residential soil samples and 5 samples of process waste materials from the Anaconda area were determined by electron microprobe, as were the particle size and mineralogic associations of each particle analyzed.

The relative occurrence of arsenic-bearing particles in the Anaconda residential soils followed the order: slag > iron-arsenic oxide > arsenic phosphate >manganese-arsenic oxide > iron-arsenic sulfate > metal-arsenic oxide. Reductions in the bioavailability of arsenic from smelting wastes may result from encapsulation of arsenic in insoluble matrices (*e.g.*, the vitreous slag matrix or quartz), formation of insoluble alteration or precipitation rinds on arsenic phases (*e.g.*, iron-hydroxide or silicate rinds on arsenic phosphate grains), and the presence of relatively insoluble arsenic mineral phases (such as metal arsenic sulfides or sulfates). These geochemical limitations to arsenic solubility, and hence bioavailability, help explain the observed reduced bioavailability of arsenic from these soils.

Toxicity assessment

ARCO is also conducting research on the toxicity values used by EPA to estimate human health risks in populations exposed to arsenic. Within ARCO's research program, investigators are developing a preliminary physiologically-based pharmacokinetic (PB-PK) model of arsenic metabolism, performing a meta-analysis combining several epidemiologic studies of skin cancer prevalence and arsenic exposure in the United States, recalculating the oral CSF for arsenic based on revised arsenic intake assumptions for the Taiwanese study population evaluated in deriving the arsenic toxicity values, and reanalysing the contributions of dietary arsenic to the total arsenic intake in the Taiwanese population. These investigations are designed to address concerns regarding the EPA-derived CSF for arsenic ingestion, which have led EPA to acknowledge that risk estimates derived using the CSF may need to be reduced by as much as an order of magnitude for relevant comparison with cancer risk estimates for other chemicals.

Physiologically-based pharmacokinetic model for ingested arsenic

Mammals metabolise arsenate (AsV) to arsenite (AsIII), and methylate AsIII to monomethylarsenic (MMA) and dimethylarsenic (DMA) acids. The methylated forms of arsenic are less toxic and more rapidly excreted than inorganic arsenic. The PB-PK model being developed by Dr. Dan Menzel and others at the University of California at Irvine is designed to take into account these metabolic steps following both acute and chronic oral exposures. Based primarily on arsenic metabolism data from published studies in mice and humans, the PB-PK model is based on flow-limited models of chemical distribution in the body. Dr. Dan Menzel's description of his preliminary model and the research program underway to close remaining data gaps appears in an accompanying paper.

The completed PB-PK model will be used to estimate individual and population variations in internal organ arsenic burden and to plan human studies of arsenic metabolism in populations potentially exposed to arsenic. The model findings will also be used to determine if arsenic methylation was likely to have been saturated in the Taiwanese population used to derive the CSF for ingested arsenic. If detoxification via methylation was saturated to some degree in the Taiwanese population but is not expected to be saturated in Anaconda residents, an alternate CSF should be calculated based on estimates of inorganic arsenic dose in the Taiwanese populations and should be applied to estimates of inorganic arsenic dose in US populations. This calculation is likely to result in significantly lower risk estimates in US populations.

Meta-analysis of skin cancer prevalence in the USA

Although a number of areas in the United States have high levels of arsenic in drinking water, no epidemiologic study of these communities has reported an association between skin cancer prevalence and arsenic in drinking water. Gradient

Corporation is combining data from several individual epidemiologic studies using meta-analysis techniques, to increase the power to detect associations and to determine if the negative results in the US epidemiologic studies are consistent with predictions based on EPA's CSF for ingested arsenic. Three epidemiologic studies in US populations exposed to concentrations of arsenic in drinking water were identified where health outcomes were clearly defined (Southwick *et al.*, 1983; Vig *et al.*, 1984; Harrington *et al.*, 1978). Despite exposure to arsenic concentrations in drinking water of up to 401 μg L^{-1}, no skin cancers were observed in these populations. Data from these three study populations were combined, and the observed skin cancer prevalence was compared with that predicted by the Armitage-Doll model (US EPA, 1988) used by EPA to calculate the arsenic CSF, with adjustments for the study population arsenic doses, age distribution, and baseline skin cancer incidence. The possibility that arsenic exposure was not causing excess skin cancers (*i.e.*, the null hypothesis) was determined to be about 2.2 times more likely than the current CSF. However, the precision and outcome of the meta-analysis are affected by several sources of uncertainty, including the age distribution and duration and classification of exposure of the study populations. A more refined assessment may be conducted as data on additional arsenic-exposed populations or additional information on populations used in this analysis become available.

Recalculation of the skin cancer CSF with exposure assumptions used to calculate the new RfD

Although the oral RfD and the CSF for arsenic are based on the same epidemiological studies conducted in Taiwan (Tseng *et al.*, 1968; Tseng, 1977), the exposure assumptions used by EPA to estimate daily arsenic doses in the experimental population differ for the two values. Specifically, in calculating the RfD, EPA assumed greater daily intake rates for water than were assumed in calculating the CSF. The RfD derivation also considered contributions to total arsenic intake from water used in cooking and some dietary sources of arsenic that were not considered by EPA when deriving the CSF (Abernathy *et al.*, 1989; US EPA, 1993).

When these more accurate arsenic intake assumptions are used in deriving the toxicity factors, a higher daily arsenic intake is associated with adverse effects than would have resulted from use of EPA's previous assumptions. As a result, if the exposure parameters assumed in the RfD derivation were applied in deriving the CSF, the CSF would be lower. This lower value would result from a greater daily arsenic intake being associated with a given increased cancer risk (*i.e.*, a lower carcinogenic potency) than assumed in the current CSF. When Gradient Corporation recalculated the CSF for arsenic with the exposure assumptions used by EPA in deriving the RfD, the current CSF of 1.75 $(mg/kg\text{-}day)^{-1}$ was reduced by 35 percent, resulting in an adjusted CSF of 1.13 $(mg/kg\text{-}day)^{-1}$. Risk estimates derived using the revised CSF would correspondingly be reduced by 35 percent.

Dietary sources of arsenic intake in the studies used to develop EPA's oral toxicity values

As described above, in deriving the RfD for arsenic, EPA did consider arsenic intake from food. Specifically, EPA assumed that the main foods consumed by the Taiwanese study population were rice and yams. It was assumed that rice would contain 0.03 mg kg^{-1} arsenic (dry weight) and that 65 percent of the arsenic occurring in rice would occur in organic forms and, as such, would be essentially nontoxic. EPA also assumed that the arsenic contribution from yams would be insignificant because 95 percent was expected to occur in organic forms. However, supporting data for EPA's assumptions appear to be very limited. Data on arsenic concentrations in Taiwanese rice and yams were apparently not available to EPA. Furthermore, it appears that the assumption that 95 percent of arsenic in vegetables is in organic forms conflicts with available data indicating that only low levels of methylation of arsenic occur in plants. If food sources consumed by the Taiwanese study populations had a higher proportion of inorganic arsenic relative to organic arsenic than assumed by EPA, this would represent an additional source of inorganic arsenic exposure. Accounting for additional exposures to inorganic arsenic would reduce the CSF and increase the RfD for arsenic ingestion, thus lowering risk estimates for both carcinogenic and noncarcinogenic effects of arsenic ingestion.

PTI is currently assisting ARCO in designing and conducting an investigation of arsenic concentrations in yams and rice consumed by populations residing in the areas of Taiwan where the epidemiologic studies used to derive the oral toxicity values for arsenic were conducted. These studies will include analyses of total arsenic in yams, rice, and soil and will seek to establish total concentrations of organic arsenic in vegetable samples. Results from these studies may be used to recommend revisions to the arsenic intake estimates used by EPA in calculating the RfD and the CSF.

Conclusions

Through collection of site-specific data in Anaconda, ARCO has begun to address concerns associated with risk estimates for the site given the overly conservative nature of EPA's default risk assessment assumptions. For example, whereas EPA typically assumes in risk assessments that concentrations of arsenic in indoor dust are equivalent with those in outdoor soil, a comprehensive site-specific environmental study conducted in Anaconda has indicated that indoor dust concentrations are less than one-half those in outdoor soil, where outdoor soils are at geometric mean or higher arsenic concentrations. Similarly, the EPA default assumption that bioavailability of arsenic in soil is the same as that of arsenic in the studies used to derive the toxicity values (*i.e.*, arsenic in water) has been demonstrated to be at least a 2-fold overestimate based on site-specific studies of arsenic in Anaconda soils. Use of site-specific data such as these can greatly improve the reliability of risk estimates and should allow any remedial measures to be more appropriately designed to protect public health with less disruption to the community.

References

Abernathy, C.O., Marcus, W., Chen, C., Gibb, H. and White P. 1989. *Report on arsenic work group meetings.* US Environmental Protection Agency, Office of Drinking Water, Washington, DC.

Binder, S., Kaye, W., Forney, D. and Paschal, D. 1986. *Report on the March 1985 Anaconda, Montana urinary arsenic survey.* Centers for Disease Control, Atlanta, GA.

Bornschein, R., Clark, S., Pan, W. *et al.* 1990. *Midvale community lead study. Final Report.* University of Cincinnati, Department of Environmental Health, Cincinnati, OH.

Butte-Silver Bow DOH and UCDEH. 1991. *The Butte-Silver Bow County environmental health lead study.* Butte-Silver Bow Department of Health, Butte, Montana, and University of Cincinnati, Department of Environmental Health, Cincinnati, OH.

Calabrese, E.J. 1990. Personal communication (conversation with J. Dinan, Toxics Integration Branch, EPA/OSWER/OERR, on October 24, 1990). University of Massachusetts, Amherst, MA.

Calabrese, E.J., Barnes, R., Stanek, E.F., Pastides, H., Gilbert, C.E., Veneman, P., Wang, X. Lasztity, A. and Kostecki, P.T. 1989. How much soil do young children ingest: an epidemiologic study. *Regul. Toxicol. Pharmacol.,* **10**, 123–137.

Calabrese, E.J., Stanek, E.J., Gilbert, C.E. and Barnes, R.M. 1990. Preliminary adult soil ingestion estimates: results of a pilot study. *Regul. Toxicol. Pharmacol.,* **12**, 88–95.

Clausing, P., Brunekreef, B. and van Wijnen, J.H. 1987. A method for estimating soil ingestion by children. *Arch. Occupat. Environ. Health*, **59**, 73–82.

Davis, A., Waller, P., Buschom, R., Ballou, J. and White, P. 1990. Quantitative estimates of soil ingestion in normal children between the ages of 2 and 7 years: population-based estimates using aluminum, silicon, and titanium as soil tracer elements. *Arch. Occupat Environ. Health,* **45**, 112–122.

Freeman, G.B., Johnson, J.D., Killinger, J.M., Liao, S.C., Davis, A.O., Ruby, M.V., Chaney, R.L., Lovre, S.C. and Bergstrom, P.D. 1993. Bioavailability of arsenic in soil impacted by smelter activities following oral administration in rabbits. *Fundam. Appl. Toxicol.,* **21**, 83–88.

Harrington, J.M., Middaugh, J.P., Morse, D.L. and Houseworth, J. 1978. A survey of a population exposed to high arsenic in well water in Fairbanks, Alaska. *Am. J. Epidemiology*, **108**, 377–385.

Porter, J.W. 1989. Memorandum to Regional Administrators regarding interim final guidance on soil ingestion rates. OSWER Directive. US Environmental Protection Agency, Office of Solid Waste and Emergency Response, Washington, DC.

Southwick, J.W., Western, A.E., Beck, M.M., Whitley, T., Isaacs, R., Petajan, J. and Hansen, C.D. 1983. An epidemiological study of arsenic in drinking water in Millard County, Utah. Lederer, W.H. and Fensterheim, R.J. (eds.). Van Nostrand Reinhold, New York. In: *Industrial Biomedical Environmental Perspectives. Proc. of the Arsenic Symposium.* pp. 210–225.

Tseng, W.P. 1977. Effects and dose-response relationships of skin cancer and Blackfoot disease with arsenic. *Environ. Health Perspec.*, **19**, 109–119.

Tseng, W.P., Chu, H.M., How, S.W., Fong, J.M., Lin, C.S. and Yeh, S. 1968. Prevalence of skin cancer in an endemic area of chronic arsenism in Taiwan. *J. Natl. Cancer Inst.,* **40**, 453–463.

US EPA. 1988. *Special report on ingested inorganic arsenic. Skin cancer: nutritional essentiality.* EPA/625/3-87/013. U.S. Environmental Protection Agency, Risk Assessment Forum, Washington, DC.

US EPA. 1989. *Risk assessment guidance for Superfund. Vol. I: Human health evaluation manual (Part A).* US Environmental Protection Agency, Office of Emergency and Remedial Response, Washington, DC.

US EPA. 1993. *Integrated risk information system: arsenic file dated February 1993.* US Environmental Protection Agency, Office of Health and Environmental Assessment, Cincinnati, OH.

van Wijnen, J.H., Clausing, P. and Brunekreef, B. 1990. Estimated soil ingestion by children. *Environ. Res.,* **51**, 147–162.

Vig, B.K., Figueroa, M.L., Cornforth, M.N. and Jenkins, S.H. 1984. Chromosome studies in human subjects chronically exposed to arsenic in drinking water. *Amer. J. Ind. Med.,* **6**, 325–338.

24 The process for setting drinking water regulations under the Safe Drinking Water Act*

Janet Auerbach and Heather L. Shank-Givens

Office of Ground Water and Drinking Water, Drinking Water Standards Division (4603), US Environmental Protection Agency. 401 M Street, SW, Washington, DC 20460, USA

Abstract
The Safe Drinking Water Act was enacted in 1974. In 1986, the Act was amended, at which time the US Environmental Protection Agency (US EPA) was given responsibility for promulgating regulations for 83 listed contaminants by June 1989 and then for 25 contaminants every three years after. The Agency did not complete rules for all 83 contaminants on schedule, and was consequently sued by a citizen coalition. As a result, the Agency is under court-ordered deadlines to complete the rulemakings on particular contaminants by dates certain. Arsenic was among the original 83 contaminants, and is now under court-ordered deadline for proposal by November 1995. Arsenic is currently regulated under the 1986 Amendments. EPA is reviewing the existing standard and development of the revised regulation is underway. This paper provides an overview of the Agency's standard process for developing drinking water regulations.

Safe Drinking Water Act (SDWA)

Regulations are developed under section 1412 of the Safe Drinking Water Act ("SDWA"), 42 U.S.C. 300g L^{-1}, which directs EPA to establish Maximum Contaminant Level Goals (MCLGs) and National Primary Drinking Water Regulations (NPDWRs) for drinking water contaminants. MCLGs are non-enforceable, health-based goals which are to be set at the level at which no known or anticipated adverse effects on the health of persons occur and which allows an adequate margin of safety. In setting MCLGs, the Agency considers all available scientific information regarding the adverse effects that a contaminant may have

*By the time of publication of these proceedings, the Agency has made some revisions to the standard regulatory development process.

upon human health. EPA reviews studies on health effects submitted to the Agency and conducts an independent review of published scientific literature. In certain instances, and within available budgetary resources, EPA itself initiates independent studies to generate relevant information on adverse health effects. EPA also studies contaminant occurrence levels in ground waters and surface waters and evaluates the extent of exposure to a contaminant that exists through drinking water ingestion relative to other routes of exposure (*e.g.*, inhalation or dermal contact).

NPDWRs are enforceable requirements consisting of the highest allowable concentration of a contaminant in drinking water (called a Maximum Contaminant Level, or "MCL"), which is set as close to the MCLG as feasible. Where it is not feasible to ascertain the level of a contaminant in drinking water, EPA is authorized to specify a treatment technique, in lieu of an MCL, that will prevent adverse health effects to the extent feasible. The Agency is directed by SDWA section 1412 to establish NPDWRs based upon the best technology, treatment technique or other means that are available, taking cost into consideration. In establishing NPDWRs, the Agency considers issues such as the availability of laboratory methods to measure a contaminant in drinking water, engineering studies of treatment technologies to reduce contaminant levels, and the costs of those technologies. In accordance with the Act, EPA also establishes monitoring, reporting, and recordkeeping requirements that public water systems must follow, as well as requirements applicable to States that are authorized to administer the public water supply program.

Key supporting information

As noted above, drinking water regulations are developed in the context of several key inputs of information: health effects, occurrence, treatment technology, analytical methodology, and regulatory impacts. This information feeds into the workgroup deliberative process and then forms the basis of the support documents for the regulation: Health Criteria Document, Occurrence Document, Methods Document, and Cost and Technology Document. Under government-wide directives, EPA also develops several additional documents which analyze the regulatory impacts of the rule. The Agency prepares a Regulatory Impact Analysis which examines the costs and benefits of particularly costly rules (including analysis by size of public water systems and by alternative regulatory options). EPA also conducts a Regulatory Flexibility Analysis which examines the impacts on small entities and how these impacts were minimized. The Agency also prepares an Information Collection Request which assesses the burden on the regulated community of the monitoring and reporting requirements of the rule.

Two primary pieces of information, health effects and occurrence, generally establish the need for a rule by demonstrating both the risk as well as actual or potential exposure to the population. Health effects information includes an assessment of toxicological effects, carcinogenicity, and exposure from drinking water. Occurrence data demonstrate how widely and at what levels the contaminant is found. With this basis for regulatory action, and based on statutory directives for setting MCLs, EPA conducts evaluations of the availability and performance of removal technologies, including the cost of applying those technologies, the ability of laboratories to measure accurately and consistently with available methods, and the health risk at different MCL levels.

An initial step in developing a regulation involves determining the feasibility of treatment technologies at various MCL levels. To determine the best available technology (BAT) for meeting an MCL, the Agency considers the existence of a given technology, its affordability, verified performance (not necessarily in the field), and geographic flexibility. Consistent with the legislative history of the SDWA, this assessment focuses on the availability and affordability of the technologies for large public water systems. Systems can use any means they want to comply with an MCL; they do not have to employ a technology designated by EPA as BAT.

The feasibility of setting a precise MCL is further determined by laboratory ability to reliably measure the contaminant. EPA derives quantitation levels for analytical methods which reflect levels that can be measured in good laboratories under normal operating conditions, within specified parameters of accuracy and precision. MCL compliance determinations are dependent on the ability to measure to a given level and this ability is critical for taking enforcement actions.

Regulation development process

Drinking water regulations established by EPA under the Safe Drinking Water Act are developed through deliberations on the key information described above, selection of the best regulatory alternative, and thorough internal agency review and consideration before issuance and publication in the *Federal Register.*

Start Action Notice

The submission of a Start Action Notice to the Agency Steering Committee formally begins the rulemaking process. This notice identifies the need for agency regulatory action and requests offices to nominate representatives for the workgroup which will develop the rulemaking package.

Workgroup

The workgroup has primary responsibility for analyzing data and developing regulatory options. The workgroup provides for full consultation and coordination on a rulemaking package among the staff of all EPA organizations that participate in the rulemaking. The workgroup supports the lead office (the Office of Ground Water and Drinking Water, or OGWDW) in its technical and analytical work, identifies and assesses principal policy issues and options, especially those that are cross-media, resolves issues or elevates them to higher EPA management for resolution, and ensures the quality and completeness of the regulatory package being prepared for publication. The headquarters offices typically included in the workgroup for developing drinking water regulations (besides OGWDW) are the Office of Policy, Planning, and Evaluation (OPPE), the Office of Research and Development (ORD), the Office of General Counsel (OGC) and other concerned media offices such as the Waste program and the Prevention, Pesticide and Toxic Substances program. EPA Regional Offices also participate in the workgroups. Represented on the workgroup are the individuals responsible for providing the key supporting information: engineers, analytical chemists, toxicologists, and economists.

Options selection

Options selection is the process by which senior Agency managers determine the principal regulatory requirements in the rulemaking such as the MCL, Best Available Technology designations, and monitoring frequencies. Options are first developed by the workgroup, and then presented to successive levels of management up to the Deputy Administrator and the Administrator, if appropriate. For option selection, the workgroup prepares detailed briefing documents that explain, among other things, the results of the health effects assessments, treatment technology evaluations, cost estimates and economic impacts analysis. The briefing documents also summarize any major issues that have arisen.

Workgroup closure

After options selection, the workgroup "reaches closure" on (*i.e.*, agrees upon) a rulemaking package, including draft preamble and regulatory language. OGWDW prepares and circulates several drafts of the preamble and regulatory language for comment by the participating offices. Issues relating to implementing the selected options are identified and resolved during this process. During this review process, Assistant Administrators and Regional Administrators have the opportunity to raise any outstanding issues for resolution prior to transmittal of the regulation to the Administrator. Issues which remain unresolved by the Assistant Administrators are forwarded to the Deputy Administrator or Administrator for decision. Workgroup closure occurs when the workgroup members, representing senior management of their respective offices,conclude that the major issues presented in the project are fully addressed and that the rulemaking package (preamble on regulatory language), after changes identified at the time of workgroup closure, will be ready for formal consideration by the Agency's senior management. Issues on which the workgroup could not reach agreement are reserved for resolution by senior management. A closure meeting usually follows review and revision of several drafts of a rulemaking package.

Final reviews

Following workgroup closure, several steps must be taken before publication of proposed regulations. These steps generally begin with revision of the preamble, proposed rule and associated documents in response to the comments raised by concerned offices at workgroup closure.

EPA also consults with the Agency's Science Advisory Board (SAB) during development of the proposal (the SAB is an independent group of scientists that advises Agency decision-makers). SAB is tasked with providing comment on the Agency's risk assessment for the rule, and other technical aspects of the proposal, including occurrence and technology data.

During this time, the Agency compiles the Administrative Record (or docket) which includes all the key documents supporting the rule. Upon approval by the

Administrator, the regulation is sent to the Office of Management and Budget (OMB) for review in accordance with Executive Order 12866 and other Presidential directives. The Administrator may also consult with other Federal Agencies as appropriate. OMB review includes not only the preamble and regulatory language but also the Regulatory Impact Analysis, Regulatory Flexibility Analysis, and Information Collection Request.

Publication and public comment

Once the Administrator approves and signs the rulemaking package, the proposal is published in the *Federal Register*, opening the public comment period. During this period, the Agency also holds public hearings on the proposed drinking water regulations. The Agency provides a sufficient period of time, generally 60 to 90 days, for the public to conduct its review of the proposal and to submit its comments. Drinking water regulations typically present complex technical, economic and policy issues that take time to evaluate. Interested persons have the opportunity to review the proposed regulations as well as the extensive record which supports the regulations. Moreover, interested persons sometimes conduct independent data gathering and analyses, which they submit to EPA as part of their evaluation of the Agency's proposal.

At the close of the comment period on the proposed rule, the comments are compiled, organized, reviewed and summarized to identify significant issues and to initiate the preparation of responses to comments. The time and effort required for this task has proven to be significant for EPA's recent drinking water regulatory packages, especially when the Agency receives hundreds, and sometimes thousands, of comments. Pursuant to the Administrative Procedure Act, the Agency responds to all significant comments received during the comment period.

When EPA proposes a regulation, the Agency seeks to ensure that it has a sound data base and that it has considered all the appropriate factors. Nevertheless, serious challenges to at least portions of the rulemaking record are often raised during the comment period. Commenters frequently challenge the adequacy and validity of the data on which the Agency has based the proposed regulation. To evaluate this type of comment, the Agency reviews the data and analyses supplied by the commenter and may need to gather additional data and perform additional scientific, engineering, or economic analyses. To generate additional data EPA usually either conducts further studies or requests that the industry supply the needed information. This process can require a significant amount of time. Responding to comments submitted in a drinking water rulemaking is often an enormous task because of the variety of pollutants typically covered by the proposal, the range of data that may be available on their health effects, the range of treatment technologies that may be required, the array of laboratory methods that the rulemaking proposes to approve for measuring the contaminants, and the number and diversity of parties and citizens affected by the rule. Commenters may also question the Agency's policy decisions and advocate alternative positions.

Ultimately the Agency must decide whether to modify the proposed rule in response to public comments (including data submitted by commenters), new data developed by EPA since the proposal, or revised analyses performed by EPA. If the

Agency decides to change the package following proposal in material ways, or if significant new data or analyses become available, the Agency may re-propose all or parts of a rule in the *Federal Register* or publish a supplemental notice of data availability. Any notices issued between the publication of the rulemaking proposal and the promulgation of the final rule undergo internal review with many of the steps outlined above, and are published in the *Federal Register* for review and comment by the public.

Final rule

The Agency's final rulemaking package reflects appropriate resolution of comments received and issues raised since the proposal. The workgroup is reconvened to develop final regulatory options, and options selection up through the Assistant Administrator again takes place. Sometimes the Deputy Administrator or the Administrator becomes involved in final options selection. After options selection, Agency staff complete the final version of the rulemaking documents. These documents build on the draft documents issued at proposal, but important portions addressing major issues often cannot be completed until senior managers have addressed these issues.

The rule-making documents consist of the regulation, preamble, technical and economic support documents and other documents referred to above, and a detailed document responding to the public comments. These are the documents which carry out EPA's duties to establish sound, understandable requirements, to respond to comments, and to set out a statement of basis and purpose in the *Federal Register* notice. In any legal challenge, the regulations will ultimately either stand or fall on the strength (or weakness) of these documents, the responses to comments, and the remainder of the rulemaking record. Accordingly, it is vital that all necessary care is taken at the final rulemaking stage. After revisions, the final rulemaking package goes through final review as described above and is again submitted to the Office of Management and Budget for Executive Branch consultation. The Administrator may be briefed and asked to resolve outstanding issues at any point in the process. The final regulation is then signed by the Administrator and published in the *Federal Register.* The final National Primary Drinking Water Regulations become effective 18 months after the date of their promulgation by EPA.

25 Arsenic in the Canadian general environment - evaluation of risks to health

K. Hughes* **and M.E. Meek**

Environmental Health Directorate, Health Canada, Tunney's Pasture, Ottawa, Canada K1A 0L2

Abstract

"Arsenic and its compounds" are included on the List of Priority Substances to be assessed under the Canadian Environmental Protection Act (CEPA). Available data were critically evaluated to determine if levels of arsenic present in the Canadian environment pose a risk to the health of the population. Average daily exposure to inorganic arsenic from drinking water, food, air and soil have been estimated for the general population, as well as for populations residing in the vicinity of point sources. There is sufficient weight of evidence in adequate epidemiological studies in several countries that exposure to airborne arsenic compounds in the occupational environment is associated with respiratory cancer and that ingestion of arsenic compounds in drinking water is associated with skin cancer, and possibly cancers of internal organs. Estimates of exposure in Canada were compared to the estimated potency of inhaled and ingested arsenic to induce cancer. Based on comparison of these values with those of other substances assessed under CEPA, the priority to investigate options to reduce exposure to arsenic in the general environment is considered moderate to high.

Introduction

"Arsenic and its compounds" were included on the first Priority Substances List, published in February, 1989, of 44 substances to be assessed under the Canadian Environmental Protection Act. For the assessment, available data on human exposure and health effects were evaluated to determine if concentrations of arsenic compounds present in the general environment constitute a "danger to human life or health". This risk assessment phase is preliminary to and separate from the risk management phase, in which the cost and feasibility of implementing options to reduce exposure are investigated, if warranted.

Where possible, this assessment has focused principally on inorganic arsenic, the

* To whom correspondence should be addressed.

chemical species of primary toxicological concern. Due to the nature of available data (i.*e.,* the lack of information on speciation in environmental media and in the critical epidemiological studies), this assessment applies to the group of inorganic arsenic compounds as a whole, rather than to individual compounds within the group.

Methodology

Estimates of intake from environmental media for five different age groups of the general population were based primarily on reported mean concentrations of arsenic in air, water, food, soil and other media in Canada and reference values for body weight and intake of food, drinking water, soil and air, derived principally on the basis of national surveys. (Unless otherwise specified, "arsenic" refers to total arsenic measured in environmental media.) In addition, intake of arsenic was also estimated for populations living in the vicinity of point sources.

A draft of the assessment, based on relevant toxicological and epidemiological data identified prior to April, 1992, was externally peer reviewed for technical accuracy and completeness by identified experts in consulting, academia and government. Following external peer review, the assessment was approved by the Standards and Guidelines Rulings Committee of the Bureau of Chemical Hazards of Health Canada and the Environment Canada/Health Canada CEPA Management Committee.

Results

Population exposure

Estimated average daily intake of inorganic arsenic by five discrete age groups in the general population in Canada, and the assumptions upon which these values are based, are presented in Table 1. Most of the arsenic in drinking water, air and soil was assumed to be in the inorganic form (Johnson and Braman, 1975; Mukai *et al.*, 1986), while the proportion or arsenic in food which is inorganic ranged from 0% in saltwater fish to 75% in dairy products, beef and pork, based on limited data (Weiler, 1987).

The estimated total amount of inorganic arsenic ingested daily from environmental sources by the general population ranges from 0.2 to 0.7 μg kg^{-1}-bw day^{-1}, with the greatest exposure being in young children. The principal sources of environmental exposure to inorganic arsenic in all age groups are drinking water and food, although soil may contribute significantly to the total daily intake in infants. Only relatively minor amounts of inorganic arsenic are inhaled daily. Smoking is estimated to contribute an additional 0.01 to 0.04 μg kg^{-1}-bw day^{-1} in teenagers and adults.

Intake of inorganic arsenic may be higher, however, in communities near point sources, such as smelters, landfill sites, geological sources and areas with a history of high arsenical pesticide application. Estimated intakes of inorganic arsenic for populations living near such point sources, based on monitoring data for drinking water, soil and ambient air from various geographical locations in Canada, are presented in Table 2.

Based on the estimates presented in Table 2, contamination of drinking water from ground water supplies due principally to geological sources may greatly increase the total amount of inorganic arsenic ingested daily for all age groups. The intake of inorganic arsenic via inhalation near industrial point sources could be increased by as much as 200- to 250-fold that of the general population, based on available data.

Health effects

For assessment under CEPA, carcinogenicity was considered to be the critical effect for the evaluation of risks to the general population associated with exposure to arsenic. Although non-neoplastic effects have been reported in human populations exposed to arsenic, these effects are generally only observed at levels of exposure associated with estimated large excess risks of cancer. Epidemiological studies in human populations were considered most relevant to the assessment since they obviate the need for interspecies extrapolation and due to the lack of a good animal model for arsenic-induced cancer or skin lesions.

An association between inhaled inorganic arsenic and respiratory cancer has been observed in numerous case reports and analytical epidemiological studies of occupationally exposed populations. Sufficient information on exposure was provided in three cohort studies in occupationally exposed populations to serve as a basis for quantitative estimation of carcinogenic potency. These studies involved workers at the Tacoma smelter in Washington, the Anaconda smelter in Montana and the Ronnskar smelter in Sweden.

Enterline *et al.*, (1987) investigated the mortality as of 1976 in a cohort of 2,802 workers employed for at least one year between 1940 and 1964 at the Tacoma smelter. The standardized mortality ratios (SMRs) for respiratory cancer increased with increased cumulative exposure to arsenic (based on actual monitoring data and conversion of levels in urine to airborne concentrations). In an earlier study of these workers, exposure to sulphur dioxide was not associated with excess in respiratory cancer mortality (Enterline and Marsh, 1982). Although no data on the smoking habits of these workers were available, the large excesses observed (*i.e.*, up to more than 4-fold) are unlikely to be attributable to smoking alone.

The mortality as of 1981 of a cohort of 8,044 Anaconda smelter workers was investigated by Higgins *et al.*, (1986). There was an exposure-response relationship for excess mortality due to respiratory cancer and three measures of exposure (time-weighted average exposure, thirty day ceiling exposure and cumulative exposure), based on actual measurements and extrapolation from existing data to fill gaps. No data on the smoking histories of these workers or their exposure to sulphur dioxide were available. However, in an earlier study of 1,800 employees at this smelter, Welch *et al.* (1982) found little difference between the SMRs for respiratory cancer for smokers and non-smokers, although only a small proportion of the cohort was comprised of non-smokers (*i.e.*, 18.4%).

Jarup *et al.*, (1989) examined the mortality due to lung cancer in a cohort of 3,916 employees at the Ronnskar smelter who had worked for at least three months from

Table 1 *Estimated average daily intake of inorganic arsenic by the general population in Canada.*

Medium	Estimated daily intake ($\mu g\ kg^{-1}$-bw day^{-1})				
	0 - 0.5 yr[a]	0.5 - 4 yr[b]	5 - 11 yr[c]	12 - 19 yr[d]	20 - 70 yr[e]
Water[f]	0.08	0.3	0.2	0.1	0.1
Food[g]	0.2	0.3	0.2	0.1	0.08
Air[h]	0.0003	0.0004	0.0004	0.0004	0.0003
Soil/Dirt[i]	0.03-0.08	0.02-0.05	0.006-0.02	0.002-0.005	0.001-0.004
Total ingested	0.3-0.4	0.6-0.7	0.4	0.2	0.2
Total inhaled	0.0003	0.0004	0.0004	0.0004	0.0003
Tobacco smoking[j]	---	---	---	0.01-0.04	0.01-0.03

[a] Assumed to weigh 6 kg, breathe 2 m^3 of air per day and drink 0.1 L of water per day (Environmental Health Directorate, 1988). Amount of soil ingested per day is estimated to be 35 mg, based on data from van Wijnen *et al.* (1990), in which 0 to 1 year olds ingested approximately 70% as much soil as 1 to 4 year olds (50 mg day^{-1}).

[b] Assumed to weigh 13 kg, breathe 5 m^3 of air per day, drink 0.8 L of water per day (Environmental Health Directorate, 1988) and ingest 50 mg of soil per day (based on average of values reported by Binder *et al.*, 1986; Calabrese *et al.*, 1989; Clausing *et al.*, 1987; van Wijnen *et al.*, 1990).

[c] Assumed to weigh 27 kg, breathe 12 m^3 of air per day and drink 1.1 L of water per day (Environmental Health Directorate, 1988). Due to insufficient data, soil intake estimated to be midpoint between value for 1-4 year olds (50 mg day^{-1}) and that for adults (20 mg day^{-1}), *i.e.* 35 mg day^{-1}.

[d] Assumed to weigh 55 kg, breathe 21 m^3 of air per day, drink 1.1 L of water per day (Environmental Health Directorate, 1988) and ingest 20 mg of soil per day (assumed to be similar to adults).

[e] Assumed to weigh 70 kg, breathe 20 m^3 of air per day, drink 1.5 L of water per day, and ingest 20 mg of soil per day (Environmental Health Directorate, 1988).

[f] Based on a mean concentration of 5 $\mu g\ L^{-1}$; levels in most Canadian surface drinking water supplies are considerably less than this value, although concentrations in groundwater often exceed 5 $\mu g\ L^{-1}$ (Environment Canada, 1989a,b,c,d; Ontario Ministry of the Environment, 1989; Manitoba Environment, 1989).

[g] Based on mean concentrations detected in various foodstuffs in a total diet study in six Canadian cities (mean concentrations ranged from 0.4 ng g^{-1} in applesauce to 3048 ng g^{-1} in saltwater fish) (Dabeka *et al.*, 1993), food consumption patterns data for various age groups in Canada (Nutrition Canada, 1977) and limited information on the percentage of arsenic in various food groups which is inorganic (ranging from 0% in saltwater fish to 75% in dairy products, beef and pork). Insufficient data were identified to estimate intake of arsenic by infants in breast milk.

[h] Based on the mean airborne concentration of arsenic of 0.001 $\mu g\ m^{-3}$ (range <0.0005 to 0.017 $\mu g\ m^{-3}$) in eleven Canadian cities and one rural site surveyed from 1985 to 1990 (Dann, 1990).

[i] Based on the range of mean levels of arsenic in various "uncontaminated" Canadian soil types of 4.8 to 13.6 ppm or mg kg^{-1} reported in a survey of available data (Kabata-Pendias and Pendias, 1984).

[j] Based on estimated arsenic content of mainstream cigarette smoke ranging from 40 to 120 ng per cigarette (US Department of Health and Human Services, 1989) and 20 cigarettes smoked per day.

Table 2 *Estimated daily intake of inorganic arsenic by populations living near point sources in Canada.*

Medium	Estimated daily intake (μg kg^{-1}-bw day^{-1})				
	0 - 0.5 yr[a]	0.5 - 4 yr[b]	5 - 11 yr[c]	12-19 yr[d]	20-70 yr[e]
Water[f] -	<0.08-8.3	<0.3-31	<0.2-20	<0.1-10	< 0.1-1
Food[g]	0.2	0.3	0.2	0.1	0.08
Air[h]	0.003-0.07	0.003-0.085	0.004-0.1	0.003-0.08	0.0025-0.06
Soil/Dirt[i]	0.02-3.0	0.01-1.9	0.004-0.6	0.001-0.2	0.0009-0.1
Total ingested[j]	<0.3-12	<0.6-33	<0.4-21	<0.2-10	<0.2-11
Total inhaled	0.003-0.07	0.003-0.085	0.004-0.10	0.003-0.08	0.0025-0.06

[a] Assumed to weigh 6 kg, breathe 2 m^3 of air per day and drink 0.1 L of water per day (Environmental Health Directorate, 1988). Amount of soil ingested per day is estimated to be 35 mg, based on data from van Wijnen *et al.* (1990), in which 0 to 1 year olds ingested approximately 70% as much soil as 1 to 4 year olds (50 mg day^{-1}).

[b] Assumed to weigh 13 kg, breathe 5 m^3 of air per day, drink 0.8 L of water per day (Environmental Health Directorate, 1988) and ingest 50 mg of soil per day (based on average of values reported by Binder *et al.*, 1986; Calabrese *et al.* 1989; Clausing *et al.*, 1987; van Wijnen *et al.*, 1990).

[c] Assumed to weigh 27 kg, breathe 12 m^3 of air per day and drink 1.1 L of water per day (Environmental Health Directorate, 1988). Due to insufficient data, soil intake estimated to be midpoint between value for 1-4 year olds (50 mg day^{-1}) and that for adults (20 mg day^{-1}), i.e. 35 mg day^{-1}.

[d] Assumed to weigh 55 kg, breathe 21 m^3 of air per day, drink 1.1 L of water per day (Environmental Health Directorate, 1988) and ingest 20 mg of soil per day (assumed to be similar to adults).

[e] Assumed to weigh 70 kg, breathe 20 m^3 of air per day, drink 1.5 L of water per day, and ingest 20 mg of soil per day (Environmental Health Directorate, 1988).

[f] Based on range of concentrations of arsenic of <5 to 500 μg L^{-1} found in drinking water supplies near point sources; 5% of samples from wells near Halifax had concentrations above the upper end of this range (Bergeron, personal communication; Brunswick Mining and Smelting Corporation, personal communication; Meranger *et al.*, 1984).

[g] Based on mean concentrations detected in various foodstuffs in a total diet study in six Canadian cities (Dabeka *et al.*, 1993), food consumption patterns data for Canadians (Nutrition Canada, 1977) and limited information on the percentage of arsenic in various food groups which is inorganic. Insufficient data were identified to estimate intake of arsenic by infants in breast milk. Although it is possible that intake of arsenic in food may be elevated for such populations due to increased uptake by plants and animals, available data were insufficient to estimate intake in food; therefore, it assumed to be similar to that of the general population.

[h] Based on the range of mean airborne arsenic concentrations near industrial point sources in recent years (1986-1991) of 0.0086 to 0.22 μg m^{-3} (Murphy, personal communication; Noranda Mines, 1992; Paquin, personal communication; Teindl, personal communication; White *et al.*, 1986).

[i] Based on range of recent concentrations of 3 to 500 mg kg^{-1} measured in soil, humus, peat or house dust in the vicinity of industrial sources (Azzaria and Frechette, 1987; Binder *et al.*, 1987; Brunswick Mining and Smelting Corporation, personal communication; Hertzman *et al.*, 1991; Murphy, personal communication; Teindl, personal communication; Zoltai, 1988) weighted to account for possible higher concentrations in the vicinity of gold roasting operations (though concentrations of more than 10,000 mg kg^{-1} have recently been reported in soil near two arsenic storage areas in Yellowknife (EBA, 1991), this soil is unlikely to be a source of exposure for the community). It was assumed that all of the arsenic present in soils is inorganic.

[j] It should be noted that these estimates are based on monitoring data from several different geographical locations, and represent a "worst case scenario", in which the individual is assumed to be exposed to contaminated water and soil.

1928 through 1967. Estimates of cumulative exposure were based on monitoring data available from 1950 onwards; earlier exposure was estimated based on limited measurements and extrapolation from later data. Although there appeared to be no exposure-response relationship at the lower exposures, the overall trend was highly significant. When the cohort was subdivided on the basis of average level of exposure and duration of exposure, the excess of lung cancer was more closely related to intensity than to duration. There was no exposure-response relationship between sulphur dioxide and lung cancer. In a nested case-control study within this cohort of smelter workers, Jarup and Pershagen (1991) determined that the interaction between exposure to arsenic and smoking was intermediate between that predicted by an additive and multiplicative model.

The carcinogenicity of ingested arsenic has been investigated in several ecological and analytical epidemiological studies in environmentally exposed

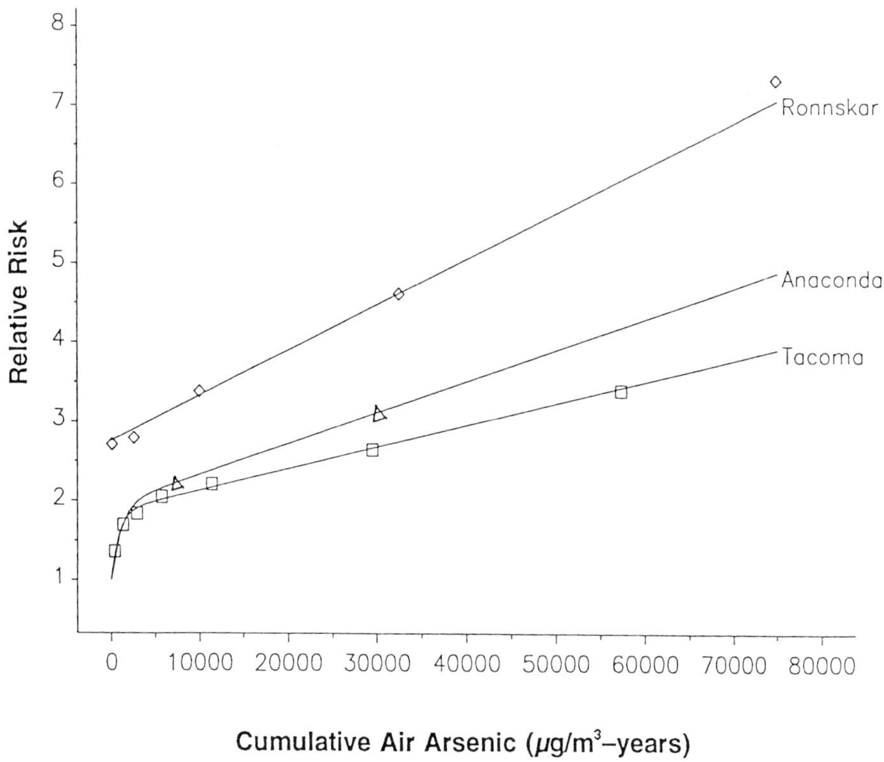

Figure 1 *Exposure-response relationship for inhaled arsenic and respiratory cancer in smelter workers. [Relative risk (Observed/Expected) is plotted against cumulative lifetime arsenic exposure in air (µg m^{-3}-yr^{-1}) for the Ronnskar, Anaconda and Tacoma smelters. Relative risks are denoted by symbols and the predicted curves are denoted by solid lines.]*

populations. In the majority of these studies, conducted in Taiwan (Tseng, 1977; Tseng *et al.*, 1968; Chen *et al.*, 1985), Central and South America (Bergoglio, 1964; Biagini *et al.*, 1974; Cebrian *et al.*, 1983; Zaldivar, 1974, 1977) and England (Philipp *et al.*, 1983), ingestion of arsenic has been associated with an increased prevalence of skin cancer or mortality due to skin cancer. Although such increases have not been reported in smaller populations exposed to arsenic in the US (Southwick *et al.*, 1983; Morton *et al.*, 1976), this may be attributable to shorter duration of exposure, lower concentrations of arsenic in the drinking water or the insensitivity of the study designs.

The largest study was that conducted in Taiwan, where a population of 40,421 had obtained drinking water prior to 1966 from artesian wells which contained up to 1.82 mg L^{-1} arsenic. When the population was divided into three groups based on the arsenic content of their well water, there was a clear exposure-response relationship between concentration and the prevalence of skin cancer (a total of 428 cases) (Tseng, 1977). There were no cases of skin cancer in a comparison population of 7,500 persons whose drinking water contained very low levels of arsenic, although 3 cases would have been expected (Tseng, 1977; Tseng *et al.*, 1968; Yeh, 1973; Yeh *et al.*, 1968).

Ingestion of arsenic-contaminated water in Taiwan has also been associated with increased mortality due to cancer of various internal organs. Increased mortality due to cancer of the bladder, kidney, liver and lung have been reported in several ecological correlational studies, with some indication of an exposure-response relationship (Chen *et al.*, 1985; Chen and Wang, 1990; Wu *et al.*, 1989). In an inherently more sensitive case-control study, elevated odds ratios were reported for deaths due to bladder, lung and liver cancer in those who had used artesian wells for 40 or more years. There was an association between mortality due to these types of cancer and duration of exposure; in addition, the odds ratios were not changed significantly when several other risk factors were taken into account in logistic regression analysis (Chen *et al.*, 1986).

Discussion and conclusions

An association between inhaled arsenic and increased mortality due to respiratory cancer has been consistently demonstrated in available epidemiological studies. Where it was possible to investigate the influence of possible confounders, such as concomitant exposure to sulfur dioxide or cigarette smoke, these factors have not explained the observed excess mortality. In addition, ingestion of inorganic arsenic in drinking water has been consistently associated with an increased prevalence of skin cancer in exposed human populations. Therefore, based on the weight of evidence of carcinogenicity in humans by more than one route of exposure, the group of arsenic compounds as a whole has been classified as "carcinogenic to humans" based on the classification scheme developed for this endpoint for assessment of risk to human health under Paragraph 11(c) of CEPA (Environmental Health Directorate, 1992).

For such substances, where possible, estimated exposure is compared to a quantitative estimate of carcinogenic potency to characterize risk and provide guidance for analysis of options to reduce exposure of the general population. Potency is expressed as the dose or concentration which induces a 5% increase in the incidence or mortality due to relevant tumours ($TD_{0.05}$), based on data obtained in toxicological studies in experimental animals or epidemiological investigations in exposed human populations. In the case of arsenic, $TD_{0.05}$s were developed for exposure by both inhalation and ingestion, based on epidemiological data.

The $TD_{0.05}$ for inhaled arsenic was based on data presented in the three studies of occupationally exposed cohorts described above; the relationships between exposure to arsenic and respiratory cancer in these studies are illustrated in Figure 1. Based on these data, the $TD_{0.05}$s for inhaled arsenic were 7.8, 10 and 51 μg m^{-3} for the Anaconda, Tacoma and Ronnskar smelter workers, respectively. The calculated exposure/potency indices (EPIs), based on comparison on the mean concentration of arsenic in ambient air in several Canadian cities of 0.001 μg m^{-3} to the estimated $TD_{0.05}$s, range from 2×10^{-5} to 1×10^{-4}. However, the calculated EPIs are higher for populations living in the vicinity of industrial sources, such as smelters, where airborne concentrations are reported to range from 0.0086 to 0.22 μg m^{-3}. For these populations (the size of which was not determined for the assessment under CEPA), the EPIs range from 2×10^{-4} to 3×10^{-2}. Therefore, for the general population, exposure in air is approximately 8,000 to 50,000 times less than the $TD_{0.05}$s and for populations residing near industrial sources 35 to 6,000 times less than the $TD_{0.05}$s

For estimation of the $TD_{0.05}$ for ingested arsenic, the quantitative risk assessment prepared by the US Environmental Protection Agency (1988) was modified to predict the dose associated with a 5% increase in the prevalence of skin cancer, incorporating background rates for Canadians. The values were 906 and 844 μg L^{-1} for men and women, respectively. Calculated EPIs for a concentration of 5 μg L^{-1} (concentrations in most drinking water supplies in Canada are below this value) are 5.5×10^{-3} and 5.9×10^{-3} for men and women, respectively. In areas with drinking water supplies containing up to 500 μg L^{-1}, as have been measured in ground water supplies in some areas of Canada, the EPIs are up to 5.5×10^{-1} and 5.9×10^{-1} for men and women, respectively. It should be noted, however, that only a very small proportion of the population in Canada is exposed to such concentrations of arsenic in the drinking water. In summary, therefore, for the general population exposure to arsenic in drinking water is more than 170 times less than the $TD_{0.05}$s and for some communities in Canada about 1.7 times less than the $TD_{0.05}$s .

Based on the calculated EPIs for inhaled and ingested arsenic, the priority for further action (*i.e.*, an analysis of options to reduce exposure to arsenic in the environment), based solely on consideration of potential health effects, is considered moderate to high, in comparison to that of other substances assessed under CEPA.

There are several limitations of the critical studies of exposed human populations which should be borne in mind when interpreting these EPIs. For example, concomitant exposure to sulphur dioxide was only addressed in two of the three occupationally exposed cohorts (Enterline *et al.*, 1987; Jarup *et al.*, 1989), and information on the smoking habits of the workers was not presented in any of these studies. Similarly, the potency estimates for ingested arsenic based on the Taiwanese data must be interpreted with caution, due to several methodological weaknesses, concomitant exposure to other compounds in the water in Taiwan which may have played a role in the etiology of skin cancer, possible dietary deficiencies of the exposed population and intake of arsenic from other sources. It should also be kept in mind that the fatality rate for skin cancer is very low, compared to those for more fatal forms of cancer, although EPIs calculated on the basis of the excesses of internal cancers observed in Taiwan would be similar.

It has been hypothesized that the dose-response relationship for arsenic-induced cancer may be non-linear at low doses. However, available data on mechanisms of action are insufficient to conclude that this is the case. Moreover, available data indicate that the potency of arsenic to induce cancer relative to those of other substances assessed under CEPA is moderate to high. It seems desirable, therefore, to investigate pragmatic options to reduce exposure while additional data on mechanisms of action are generated. Indeed, this is the approach that has been adopted recently in the revision of the Canadian drinking water guideline for arsenic. It has been reduced from 50 to 25 $\mu g\ L^{-1}$ and is considered an "interim value" which is achievable by current technology at reasonable cost and will be reviewed periodically in light of developments in treatment technology and additional data on the risk to human health. Background documentation for the assessment was prepared by BIBRA Toxicology International.

Acknowledgements

The authors wish to acknowledge the assistance of Dr. R. Burnett and Dr. S. Bartlett of the Biostatistics and Computer Applications Division of the Environmental Health Directorate, Health Canada, in the development of the quantitative estimates of carcinogenic potency. Background documentation for the assessment was prepared by BIBRA Toxicology International.

References

Agency for Toxic Substances and Disease Registry (ATSDR). 1989. *Toxicological Profile for Arsenic*. US Public Health Service. ATSDR/TP-88-02. Atlanta, Georgia.

Azzaria, L.M. and Frechette, G. 1987. Natural and industrial sources of trace elements, Rouyn-Noranda-Val d'Or mining area, Québec, Canada. In: Rust,R.W., Davis, T.E. and Augustithis, S.S.(eds.), *The Practical Applications of Trace Elements and Isotopes to Environmental Biogeochemistry and Mineral Resources Evaluation*, pp. 3-26. Theophrastus Publications, Athens, Greece.

Bergeron, A. Personal communication. 1992. Unpublished data supplied by Mineraux Noranda, Inc., Division Horne.

Bergoglio, R.M. 1964. Mortality from cancer in regions of arsenical waters of the province of Cordoba, Argentine Republic. *Pren. Med. Argent.* **51**, 994-998 (cited in US EPA, 1988).

Biagini, R.E., Quiroga, G,C. and Elias, V. 1974. Chronic hydroarsenism in ururau. *Archivos Argentinos de Dermatologia,* **24**(1), 8-11 (cited in US EPA, 1988).

BIBRA Toxicology International. 1990. *The toxicity of inorganic arsenic.* Unpublished report prepared under contract for the Environmental Health Directorate, Department of National Health and Welfare Canada, Ottawa.
Binder, S., Fornery, D., Kaye, W. and Paschal, D. 1987. Arsenic exposure in children living near a former copper smelter. *Bull. Environ. Contam. Toxicol.,* **39**, 114-121.
Binder, S., Sokal, D. and Maughan, D. 1986. Estimating soil ingestion: the use of tracer elements in estimating the amount of soil ingested by young children. *Arch. Environ. Health,* **41**(6), 341-345.
Brunswick Mining and Smelting Corporation. Personal communication. 1992. Unpublished data provided by Brunswick Mining and Smelting Corporation, Belledune, New Brunswick.
Calabrese, E.J., Barnes, R., Stanek, E.J., Pastides, H., Gilbert, C.E., Veneman, P., Wang, X., Lasztity, and Kostecki, P.T. 1989. How much soil do young children ingest: an epidemiologic study. *Reg. Toxicol. Pharmacol.,* **10**, 123-137.
Cebrian, M.E., Albores, A., Aguilar, M. and Blakely, E. 1983. Chronic arsenic poisoning in the North of Mexico. *Human Toxicol.,* **2**, 121-133.
Chen, C.-J. and Wang, C.-J. 1990. Ecological correlation between arsenic level in well water and age-adjusted mortality from malignant neoplasms. *Cancer Research,* **50,** 5470-5474.
Chen, C.J., Chuang, Y.C., Lin, T.M. and Wu, H.-Y. 1985. Malignant neoplasms among residents of a Blackfoot disease-endemic area in Taiwan: high-arsenic artesian well water and cancers. *Cancer Research,* **45**, 5895-5899.
Chen, C.J., Chuang, Y.C., You, S.L., Lin, T.M. and Wu, H.Y. 1986. A retrospective study on malignant neoplasms of bladder, lung, and liver in Blackfoot disease endemic area in Taiwan. *Br. J. Cancer,* **53**, 399-405.
Clausing, P., Brunekreef, B. and van Wijnen, J.H. 1987. A method for estimating soil ingestion by children. Int. Arch. Occup. *Environ. Health,* **59**, 73-82.
Dabeka, R.W., McKenzie, A.D., Lacroix, G.M.A., Cleroux, C., Bowe, S., Graham, R.A. and Conacher, H.B.S. 1993. Survey of arsenic in total diet food composites and estimation of the dietary intake of arsenic by Canadian adults and children. *JAOAC Int.,* **76**(1), 14-25.
Dann, T. 1990. Unpublished data. Pollution Measurement Division, Conservation and Protection, Environment Canada.
EBA, 1991. *Surface Contamination Study Nerco Con Minesite Yellowknife, N.W.T.* Report prepared by EBA Engineering Consultants Ltd, Edmonton, Alberta, for Indian and Northern Affairs Canada, Water Resources Division, Yellowknife, N.W.T., 49 pp.
Enterline, P.E. and Marsh, G.M. 1982. Cancer among workers exposed to arsenic and other substances in a copper smelter. *Am. J. Epidemiol.,* **116**(6), 895-911.
Enterline, P.E., Henderson, V.L. and Marsh, G.M. 1987. Exposure to arsenic and respiratory cancer - a reanalysis. *Am. J. Epidemiol.,* **125**(6), 929-938.
Environment Canada. 1989a. Atlantic Region Federal-Provincial Toxic Chemical Survey of Municipal Drinking Water Sources. *Data summary report - Province of Nova Scotia 1985-1988.* Inland Waters Directorate, Atlantic Region, Moncton, New Brunswick, report no. IWD-AR-WQB-89-154.
Environment Canada. 1989b. Atlantic Region Federal-Provincial Toxic Chemical Survey of Municipal Drinking Water Sources. *Data summary report - Province of Newfoundland 1985-1988.* Inland Waters Directorate, Atlantic Region, Moncton, New Brunswick, report no. IWD-AR-WQB-89-157.
Environment Canada. 1989c. Atlantic Region Federal-Provincial Toxic Chemical Survey of Municipal Drinking Water Sources. *Data summary report - Province of New Brunswick*

1985-1988. Inland Waters Directorate, Atlantic Region, Moncton, New Brunswick, report no. IWD-AR-WQB-89-155.

Environment Canada. 1989d. Atlantic Region Federal-Provincial Toxic Chemical Survey of Municipal Drinking Water Sources. *Data summary report - Province of Prince Edward Island 1985-1988.* Inland Waters Directorate, Atlantic Region, Moncton, New Brunswick, report no. IWD-AR-WQB-89-156.

Environmental Health Directorate. 1988. *Reference values for Canadian populations.* Draft report prepared by the Environmental Health Directorate Working Group on Reference Values. Health and Welfare Canada.

Environmental Health Directorate. 1992. *Approach to Determination of "Toxic" under Paragraph 11c) of the Canadian Environmental Protection Act.* First Edition, 20 November, 1992. Bureau of Chemical Hazards, Health and Welfare Canada.

Hertzman, C. *et al.* 1991. Dept. Health, Care and Epidemiology, University of B.C. (cited in Teindl, personal communication).

Higgins, I.T.T., Oh, M.S., Kryston, K.L. Burchfiel and Wilkinson, N.M. 1986. *Arsenic exposure and respiratory cancer in a cohort of 8044 Anaconda smelter workers. A 43-year follow-up study.* Unpublished report prepared for the Chemical Manufacturers' Association and the Smelters Environmental Research Association. University of Michigan, Michigan.

Jarup, L. and Pershagen, G. 1991. Arsenic exposure, smoking and lung cancer in smelter workers - a case-control study. *Am. J. Epidemiol.,* **134**(6), 545-551.

Jarup, L., Pershagen, G. and Wall, S. 1989. Cumulative arsenic exposure and lung cancer in smelter workers: a dose-response study. *Am. J. Ind. Med.,* **15**, 31-41.

Johnson, D.L. and Braman, R.S. 1975. Alkyl- and inorganic arsenic in air samples. *Chemosphere*, **6,** 333-338.

Kabata-Pendias, A. and Pendias, H. 1984. *Trace Elements in Soils and Plants.* CRC Press, Inc. Boca Raton, Florida, pp. 171-177.

Manitoba Environment. 1989. *Manitoba Drinking Water Quality Data Summary.* Winnipeg, Manitoba.

Meranger, J.C., Subramanian, K.S. and McCurdy, R.F. 1984. Arsenic in Nova Scotian groundwater. *Sci. Total Environ.,* **39**, 49-55.

Morton, W. Starr, G., Pohl, D., Stoner, J., Wagner, S. and Weswig, P. 1976. Skin cancer and water arsenic in Lane County, Oregon. *Cancer,* **37**, 2523-2532 (cited in US EPA, 1988).

Mukai, H., Ambe, Y., Muku, T., Takeshita, K. and Fukuma, T. 1986. Seasonal variations in methylarsenic compounds in airborne particulate matter. *Nature,* **324**, 239-241.

Murphy, M. Personal communication. 1991. Unpublished data supplied by New Brunswick Department of the Environment.

Noranda Mines. 1992. Unpublished data provided by Noranda Mines Limited, Noranda, Quebec.

Nutrition Canada. 1977. *Food Consumption Patterns Report.* Unpublished data. Health and Welfare Canada, Ottawa.

Ontario Ministry of the Environment (OME). 1989. Ontario Drinking Water Surveillance Program. Unpublished data. Toronto, Ontario.

Paquin, E. Personal communication. Unpublished data. From the National Air Pollution Survey provided in letter dated January 10, 1992.

Philipp, R., Hughes, A.O., Robertson, M.C. and Mitchell, T.F. 1983. Malignant melanoma incidence and association with arsenic. *Bristol Med. Chir. J.* **98**(368), 165-169 (cited in US Environmental Protection Agency, 1988).

Smith, A.H., Hopenhayn-Rich, C., Bates, M.N., Goeden, H.M., Hertz-Picciotto, I., Duggan,

H.M., Wood, R., Kosnett, M.J. and Smith, M.T. In press. Cancer risks from arsenic in drinking water. *Environ. Health Perspect.,* **97**.

Southwick, J.W., Western, A.E., Beck, M.M., Whitley, T., Isaacs, R., Petajan, J. and Hansen, C.D. 1983. An epidemiological study of arsenic in drinking water in Millard County, Utah. In: Lederer, W.H., Fensterheim, R.J., eds. *Arsenic: industrial, biomedical and environmental perspectives,* pp. 210-225. Van Nostrand Reinhold, New York (cited in U.S. Environmental Protection Agency, 1988 and BIBRA Toxicology International, 1990).

Teindl, H. Personal communication. Unpublished data from Cominco Metals in letter dated January 10, 1992.

Tseng, W.-P. 1977. Effects and dose-response relationships of skin cancer and blackfoot disease with arsenic. *Environ. Health Perspect.,* **19**, 109-119.

Tseng, W.-P., Chu, H.M., How, S.W., Fong, J.M., Lin, C.S. and Yeh, S. 1968. Prevalence of skin cancer in an endemic area of chronic arsenicism in Taiwan. *J. Nat. Cancer Inst.,* **40**, 453-463.

US Department of Health and Human Services (US DHHS),1989. *Reducing the Health Consequences of Smoking, 25 years of progress.* A report of the Surgeon General of the United States. U.S. DHHS, Washington, DC,

US EPA .1984. *Health assessment document for inorganic arsenic.* Report No. EPA-600/8-83-021F. Washington, DC.

US EPA. 1988. *Special Report on Ingested Inorganic Arsenic. Skin cancer, nutritional essentiality.* Washington, D.C., Risk Assessment Forum US EPA. EPA-625/3-87/013.

van Wijnen, J.H., Clausing, P. and Brunekreef, B. 1990. Estimated soil ingestion by children. *Environ. Res.,* **51**, 147-162.

Weiler, R.R. 1987. Unpublished data. Ontario Ministry of the Environment. Report no. 87-48-45000-057. Toronto, Ontario (cited in US EPA 01988).

Welch, K., Higgins, I., Oh, M. and Burchfiel, C. 1982. Arsenic exposure, smoking, and respiratory cancer in copper smelter workers. *Arch. Environ. Health,* **37**(6), 325-335.

White, E.R., Johnson, C.A. and Crozier, R.J. 1986. *Trail air quality - a compilation and synopsis of ambient air quality and industrial emissions data 1975-1985.* Province of British Columbia, Ministry of Environment. Nelson, British Columbia.

Wu, M.-M, Kuo, T.-L., Hwang, Y.-H. and Chen C.-J. 1989. Dose-response relation between arsenic concentration in well water and mortality from cancers and vascular diseases. *Am. J. Epidemiol.,* **130**(6), 1123-1132.

Yeh, S. 1973. Skin cancer in chronic arsenicism. *Human Pathology,* **4**, 469-485 (cited in BIBRA Toxicology International, 1990).

Yeh, S., S.W. How and C.S. Lin. 1968. Arsenical cancer of skin: histologic study with special reference to Bowen's disease. *Cancer,* **21**, 312-339 (cited in BIBRA Toxicology International, 1990).

Zaldivar, R. 1974. Arsenic contamination of drinking water and foodstuffs causing endemic chronic poisoning. *Beitr. Pathol.,* **151**, 384-400 (cited in US EPA, 1988 and Agency for Toxic Substances and Diseases Registry, 1989).

Zaldivar, R. 1977. Ecological investigations on arsenic dietary intake and endemic chronic poisoning in man: dose-response curve. *Zentralbl. Bakteriol. Parasitenkd. Infektionskr. Hyg. Abt. 1: Orig. Reihe B.,* **164**, 481-484 (cited in US Environmental Protection Agency, 1988).

Zoltai, S.C. 1988. Distribution of base metals in peat near a smelter at Flin Flon, Manitoba. *Water, Air,Soil Pollution,* **37**, 217.

26 Impacts of a stringent arsenic drinking water regulation on the western USA

Bruce A. Macler[1] and Maria W. Tikkanen[2]

[1]*US Environmental Protection Agency, Region 9, 75 Hawthorne Street, W-6-1, San Francisco, California 94105, USA*
[2]*Association of California Water Agencies, 910 K Street, Suite 250, Sacramento, California 95814, USA*

Abstract
Orally ingested inorganic arsenic is known to lead to a variety of adverse health consequences, including cancers. A stringent National Primary Drinking Water Regulation for arsenic, defined as a concentration no greater than 0.002 mg L^{-1}, may be required to reduce these risks to the public to an acceptable level. Such a standard would have substantial impacts on drinking water utilities and regulatory agencies. Since arsenic levels of concern have been shown to be predominantly in ground waters of the western states, the major regulatory burdens are anticipated to be felt there.
For utilities, these impacts are economic as well as technical, and arise largely from the limitations of available technologies to easily achieve the desired arsenic level. Current problems include treatment approaches that conflict with those required for other NPDWRs, limited water resources versus high water loss in waste streams, increased monitoring, waste disposal, power availability, and high operations/ maintenance costs associated with geographically-isolated small systems. Identification of appropriate technologies and funding mechanisms is important.
Regulatory agencies will also be impacted by a strict arsenic standard. Local and state regulatory agencies and EPA must consider regulatory costs from increased oversite activities and the impacts of substantial non-compliance. Implementation resources must be developed and made available in a timely manner.

Introduction

The US Environmental Protection Agency (US EPA) is in the process of revising the National Primary Drinking Water Regulation (NPDWR) for arsenic (As). Based on the requirements of the Safe Drinking Water Act (SDWA) and on the stated intent of the US Congress, it is highly likely that the resulting Maximum Contaminant Level (MCL) will be substantially stricter than the existing MCL of 0.05 mg L^{-1}. Ingestion of inorganic As may lead to a variety of human health problems, many of which were unknown or of limited concern when the current MCL was established in 1975. As is discussed more fully by other authors in this volume (Adams, 1994; Chen, 1994: Guo, 1994; Smith, 1994), As is now known to be a human carcinogen, as well as a cause of vascular disease, neurological effects, liver and kidney damage, and other adverse conditions. It occurs widely in both ground and surface source waters. There is thus reason to be concerned with potential health risks from As in drinking water. Regulation to address these risks at the levels where they become of concern is appropriate to protect public health.

In order that drinking water be protective of public health, Section 1412 of the SDWA directs EPA to set a maximum contaminant level goal (MCLG) " at the level at which no known or anticipated adverse effects on the health of persons occur and which allows an adequate margin of safety." EPA policy has been to set the MCLG for carcinogens at zero, assuming that there is no exposure threshold for cancer and that no level above zero for a carcinogen is without risk. Since a non-zero threshold has yet to be established for As-induced cancers, we expect the MCLG for arsenic to be zero. The SDWA further directs EPA to set the legally enforceable MCL as close to the MCLG as is feasible, based on availability of analytical methods and treatment techniques. Current analytical methods suggest a practical quantitation limit for As in the range of 0.002 to 0.005 mg L^{-1}. A variety of treatment techniques exist (Welch *et al.*, 1988), although no Best Available Technology (BAT) has been established. While as of this writing, no MCL for arsenic has been proposed, possible MCLs considered by EPA and others from this regulatory approach may be in the range of 0.002 to 0.005 mg L^{-1}.

The implications of a stringent MCL are significant for not only the impacted utilities, but also for the federal, state and local agencies that would administer it and for the public that will ultimately bear the costs. While the SDWA and prudent public health policy may dictate a stringent As MCL, latitude exists for the NPDWR to be constructed in an implementable form. A thorough consideration of these implications and approaches to their solution should be made by all parties concerned prior to proposal of any As NPWDR. In general, utilities and regulatory agencies at the field level would like to have a regulation for As, regardless of the absolute MCL set, that can be feasibly met. In this sense, feasible includes practical limits on money, time and personnel. The thesis here is that, given As is a drinking water health problem, then we must find workable solutions, ones that are protective of human health and comply with legal requirements, yet are implementable for regulatory agencies and technically and economically feasible for all utilities.

Arsenic is predominantly a western problem

Data is accumulating on occurrence levels of As in source waters. The work of Davis, *et al.*,1994 (Figure 1), in this volume shows that measured As values in untreated source waters are typically below 0.002 mg L^{-1} for systems east of the Mississippi River. They report 30% of surveyed systems exceed 0.002 mg L^{-1}, with essentially all of these systems located in the more arid regions west of the

Mississippi. This survey reached mostly larger water systems, including many using surface waters of generally lower As levels. Ground water systems were an average of four-fold higher than surface water systems. We believe smaller water systems as a class will have higher average As levels, since proportionately more of these systems use ground water sources. Welch *et al.* (1988), summarized the results of several thousand analyses of ground water and surface water samples in the western USA. They found that high As levels are generally associated with specific geochemical environments typical of the west. A preliminary assessment of As occurrence by EPA indicates as well that the preponderence of higher As levels occur in the western part of the USA (US EPA, 1993). While currently available information is incomplete and additional occurrence information is necessary for a true characterization of As levels in drinking water sources, it is at least qualitatively apparent that a more stringent As NPDWR will affect western drinking water utilities to a greater extent than it will their eastern counterparts.

Impacts on drinking water utilities

Surface water systems

Drinking water systems utilizing surface water sources already face NPDWRs for a variety of contaminants. Surface water systems must filter their water, unless they can prove that they are not vulnerable to microbial contamination. All systems must disinfect their water and meet specific requirements for bacteria, parasites and viruses. Regulations for specific inorganic and organic chemicals, including lead and copper, require monitoring and treatment. These current regulations already require substantial imputs of money, time and personnel.

Future regulations of concern to surface water systems will limit the levels of disinfectants used to treat water and the disinfection byproducts formed by such treatment. These regulations are anticipated to have far-reaching implications as these utilities attempt to meet all of these requirements in some consistent, feasible manner. It appears that in some cases, the treatments best suited to achieve compliance with one set of regulations will conflict with those necessary to meet other requirements.

However, a stringent (0.002 mg L^{-1}) As NPDWR may not necessarily be a major problem for the majority of surface water systems. The reasons are two-fold: typical surface water As concentrations are substantially less than those found in ground water, and conventional water treatment approaches utilizing coagulation, sedimentation and/or lime softening necessary to meet filtration criteria under the current Surface Water Treatment Rule may adequately reduce As to below this value. The range of As concentrations reported by Davis, *et al.(1994)*,for surface waters was less than 0.005 mg L^{-1}. Data on As treatment methodologies indicate that conventional treatment can typically yield 90% or greater removal, suggesting final As concentrations <0.001 mg L^{-1} are possible.

Ground water systems

Drinking water systems utilizing ground water are of greater concern. Current regulatory burdens under the SDWA are in general limited for ground water systems.

While regulations for inorganic and organic contaminants apply to ground water systems as well as to surface water systems, many ground water sources do not exceed the MCLs for these contaminants and have minimal monitoring and treatment requirements. Ground water systems are not currently required by EPA to filter or disinfect for microbial pathogens. Because of this, many of these systems have not had to address the wide range of monitoring, treatment and reporting requirements as have surface water systems. They therefore may have minimal operational infrastructure. Another factor of concern is that ground water systems are generally smaller than surface water systems. Smaller systems are less likely to have the strong rate-payer bases available to the larger systems and can be limited in the financial resources available to them to meet regulatory requirements.

Upcoming regulations will in most cases substantially increase the requirements for these systems. EPA's proposed rule for radionuclides includes an NPDWR for radon, which essentially occurs only in ground water, that would affect the majority of ground water systems in the western USA. EPA is also developing regulations requiring disinfection of all ground water sources, unless they are determined to not be vulnerable to bacterial and viral contamination. All distribution systems may be required to have a disinfection residual present. Ground water systems are likely to be more severely impacted by these rules because, unlike the majority of surface water systems that utilize relatively few sources, these systems may have large numbers of wells. Individual wells are often far apart and centralized treatment or blending is not possible. Wells usually feed into the distribution system independently. Therefore, requirements will apply to individual wells, with corresponding increases in monitoring costs, possible increases in the number of treatment systems required and increased costs per treatment unit due to the loss of economies of scale.

For these systems, a stringent As MCL may pose a significant impact. Because of the non-centralised location problem, treatment for arsenic in ground water would most likely be at individual wells. Additionally, ground water systems are usually low in turbidity and do not typically have to filter their source waters. They therefore do not have conventional coagulation and sedimentation systems in place that could remove As in conjunction with treatment for other contaminants. Treatment approaches for As removal more suitable to ground water systems, such as ion exchange or reverse osmosis, may provide little benefit to these utilities with respect to required treatment for other contaminants. The common co-occurrence of high levels of selenium, sulfate and/or other anions with As in ground water may limit the utility of ion exchange or activated alumina treatments. In addition, serious concerns exist with ion exchange and reverse osmosis for the large wastage of water and the generation of contaminated brines or residuals. Many ground water systems are limited by available water resources. Thus, these treatment approaches may need to be further developed before being feasible as BAT for groundwater systems.

Impacts on regulatory agencies

EPA Region 9 anticipates significant impacts on its Public Water Supply Supervision

Program as a consequence of a stringent As MCL. These fall into three intertwined catagories: problems with direct implementation of the rule, issues of non-compliance, and resource requirements necessary to implement the rule.

While states oversee the majority of drinking water systems in the USA, EPA regions directly oversee the majority of Indian tribal water systems. Tribal lands are for the most part found in the western areas of the USA. Their water systems typically have ground water sources. These systems are generally small and limited in financial resources. Frequently, water treatment for contaminants posing relatively minor health risks is preceived as of low priority relative to treatment for those, such as microbial contamination, that are known problems. Water treatment in general is often considered of low priority relative to larger tribal health issues. Tribal drinking water systems tend to be located in isolated areas, such that consolidation of systems to gain financial leverage is not feasible.

For EPA Region 9, implementation of an As NPDWR means responsibility for some 460 tribal drinking water systems. Since the majority of these systems are ground water-based, we predict most of these will be required to treat for As. The regions will expect compliance by the tribal systems, but realize that many such systems are already overwhelmed by exisiting regulations and that a high degree of non-compliance should be expected. It will be important to identify specific regional implementation needs and resources prior to promulgation of a new As NPDWR.

Small tribal systems are not alone in regional concern for the impacts of a stringent As regulation. Many small groundwater systems under all jurisdictions in Region 9 will be impacted and we expect general problems with non-compliance with this rule. Limited resources means that regional focus will be on enforcement, not on support. We are concerned that this will not be an effective approach to improving public health.

Region 9 is also concerned about the possibility that the states in its region, Arizona, California, Hawaii and Nevada and the Pacific Islands, may need more than the 18 months allowed by the SDWA to adopt or implement the As NPDWR as part of their primacy for oversight of public water supply systems. This would mean that the region would be required to implement this rule until the state does so. If implementation requirements were similar to those of the current NPDWR for lead and copper, which has not been adopted by Region 9 states and is currently implemented by the region directly, significant increases in personnel and funding will be required.

Conclusion

A stringent As MCL will pose significant difficulties for implementation. While that does not preclude such an MCL, which must be based on protection of public health and compliance with the SDWA, close and careful attention given in the development of this rule to the resulting implementation issues of the EPA regions, states and affected water systems will support its success.

We have identified several areas that can be useful in enhancing the implementability of this NPDWR. (1) Concern for consistency between an As

NPDWR and other NPDWRs can help simplify state and utility activities. Additional monitoring requirements can be matched with existing requirements. (2) Care in selection of BAT to consider interactions and impacts of other contaminants and treatment requirements is necessary. (3) Resource requirements should be identified at an early stage and directed towards support for compliance. Adequate funding and funding mechanisms should be in place before regulation. Support for cost recovery from water users must be an element of rule implementation. (4) Time lines for implementation must be realistically worked out to allow state regulatory agencies and utilities adequate time to plan appropriately to meet the specific requirements of an As rule, meld those into the overall regulatory framework and acquire the financial resources necessary to carry out the regulatory mandates. (5) Technical support will need to be provided for, prepared in advance and made widely available. (6) Small drinking water systems must be explicitly considered and provided for in the determinations of feasibility for setting the As MCL. Concern for practical issues of economic and institutional feasibility for these systems being able to meet a strict MCL is paramount.

Acknowledgement

The opinions expressed are those of the authors and do not necessarily represent those of the EPA. The authors wish to thank Bill Thurston and Mike Conlon for timely and helpful comments.

References

Adams, M.A., 1994. *Dietary intake and hazards of arsenic.* Science and Technology Letters, Northwood.

Chen, C-J., 1994. *Ischemicheart disease induced by ingested inorganic arsenic: A seven-year follow-up study.* Science and Technology Letters, Northwood

Davis, M.K., Reich, K.D. and Takkanen, M.W., 1994. *Nationwide and California arsenic occurrence studies.* Science and Technology Letters, Northwood.

Guo, H-R., 1994. *Arsenic in drinking water and urinary cancers.* Science and Technology Letters, Northwood.

Smith, A.H., 1994. *Epidemiological study designs to address potential high bladder cancer risks from arsenic in drinking water.* Science and Technology Letters, Northwood.

Welch, A.D., Lico, M.S. and Hughes, J.L. 1988. Arsenic in ground water of the western United States. *Ground Water,* **26**, 333-347.

US EPA. *Treatment and occurrence (of) arsenic in potable water supplies.* Office of Ground Water and Drinking Water, September 1993.

27 Arsenic in drinking water: health effects and regulatory issues in California

A.M.Fan[1], J.P.Brown[1*], M.A.Milea[2] and P.D.Spath[2]

[1]*Pesticide and Environmental Toxicology Section, Office of Environmental Health Hazard Assessment, California Environmental Protection Agency, Berkeley, California, USA*
[2]*Division of Drinking Water and Environmental Management, Department of Health Services, Berkeley, California, USA*

Abstract
Ingestion of inorganic arsenic in drinking water has been shown to be associated with an increased risk of cancers in humans. On the basis of the US Environmental Protection Agency's unit risk estimate of 5.0 x 10^{-5} μg^{-1} L^{-1}(ppb), the current drinking water standard (Maximum Contaminant Level, or MCL) of 50 ppb would present a lifetime skin cancer risk of greater than one in a thousand (2.5 x 10^{-3}). Arsenic induced skin cancers are mostly non-fatal. Recent findings on cancers of the lung, kidney and liver at similar risk levels have generated an increased concern regarding human exposure since these tumors are often fatal. A reduction of the MCL is being considered. The present work addresses toxicology, risk estimates, water monitoring, cost impact and regulatory issues related to the risk assessment and risk management of arsenic in drinking water. According to a recent risk assessment (OEHHA, 1992) a concentration of 2 ppt is equivalent to a cancer risk of one in a million (i.e., negligible risk) based on a lifetime daily water consumption of 2 liters and a relative source contribution of 20 percent. This level is not feasible to achieve because it is far below the existing analytical detection limit. Also current treatment technology removal capacity and an economic costs would be expected to limit reductions in the current standard. The 2 ppt figure may be considered a goal similar to the federal MCL goals (MCLGs) for other carcinogens. Monitoring data in California showed that large water systems in 15 of 48 counties had water sources with arsenic concentrations greater than 50 ppb. A concentration of 20 ppb is approximately the lowest concentration achievable by conventional treatment technology. Some of the wells having water containing 20 ppb arsenic could be brought into compliance but this concentration offers only a marginal reduction in projected risks compared to the existing MCL. Many water systems would be out of compliance if the MCL were set at 2 ppb.

---

* To whom correspondence should be addressed.

Introduction

Ingestion of inorganic arsenic in drinking water has been shown to be associated with an increased risk of cancers in humans based on extensive epidemiological data obtained from Taiwan. Early investigations showed an association of arsenic ingestion with generally non-fatal skin cancers and Blackfoot disease (Tseng *et al.*, 1968; Tseng, 1977). More recent studies demonstrated an association with fatal internal cancers such as cancers of the bladder, lung, liver and kidney (Chen *et al.* 1986, 1988, 1992; Chen and Wang, 1990).

In 1984, the US Environmental Protection Agency (EPA) evaluated the carcinogenicity data on arsenic in humans and experimental animals and estimated the unit risk for arsenic concentrations to be 4.3×10^{-4} at $\mu g^{-1}\ L^{-1}$ (EPA, 1984). In 1988 a new lower unit risk estimate of $5.0 \times 10^{-5}\ \mu g^{-1}\ L^{-1}$ was calculated with some modifications and additional considerations to the earlier assessment (EPA, 1988). On the basis of the EPA risk estimates, the current drinking water standard (Maximum Contaminant Level, or MCL) of 50 ppb would present a lifetime (skin) cancer risk level of greater than one in a thousand (2.5×10^{-3}). The recent findings on internal cancers at similar risk levels (Smith *et al.* 1992; Chen *et al.* 1992) has generated an increased concern regarding human exposure to arsenic in drinking water and a reduction of the MCL is being considered.

The present work provides a brief summary of the toxicology of arsenic, presents current cancer risk estimates associated with water consumption at and below the MCL, reviews the drinking water standard development process for arsenic in the State of California, and provides an overview of recent State monitoring activities related to arsenic in drinking water. Also examined are the potential economic impacts of a reduction of the current MCL, and some of the toxicological and regulatory issues in the risk assessment and risk management of arsenic in drinking water.

Toxicology

The health effects of arsenic have been reviewed in several major documents (EPA, 1984; 1988). Arsenic compounds may be absorbed via the lungs and the gastrointestinal tract. By ingestion, as little as 10 mg of arsenic may be fatal to humans. The most sensitive non-carcinogenic end-points are probably vascular disorders, neurological disturbances and adverse reproductive effects.

Arsenic compounds are fetotoxic and teratogenic in mice, rats and hamsters. Generally, these effects are seen only at dose levels which also result in maternal toxicity. Common developmental effects seen after administration of arsenic compounds to pregnant mammals include malformations of the brain, urogenital organs, skeleton, and ear and small or missing eyes. Conclusive evidence of human reproductive or developmental toxicity following arsenic exposure is lacking.

Subacute and chronic arsenic exposures generally affect the same organs and systems as those affected by acute exposure. These include gastrointestinal tract, circulatory system, skin, liver, kidney, nervous system and heart. Adverse dermal effects, including hyperkeratosis, hyperpigmentation and de-pigmentation, are attributed to high arsenic levels in drinking water. Palmoplantar keratoses are often surmounted by small, corn-like, elevated nodules up to 10 mm in diameter. Blackfoot disease, an endemic peripheral vascular disorder found in southwest Taiwan, is also associated with consumption of arsenic in well water although a causal connection with arsenic is disputed.

The non-carcinogenic adverse health effects noted above are unlikely to be caused by the concentrations of arsenic compounds currently found in drinking water or in ambient air. The EPA has identified a human chronic No-Observed-Adverse-Effect Level (NOAEL) of 0.8 μg/kg-day based on skin and vascular effects. The reference dose based on this NOAEL is 0.3 μg/kg-day, or 10 ppb in drinking water for a 70-kg human consuming 2 L of water/day. Thus even based on non-cancer endpoints there is some rationale for reducing the current drinking water standard from 50 ppb.

Arsenic compounds are genotoxic in that they inhibit DNA repair and induce chromosome aberrations and sister chromatid exchanges. Although arsenic compounds are usually negative in routine *in vitro* tests for mutagenicity, one assay in mammalian cells indicates that arsenic can inactivate genes by damaging chromosomes. More recent findings suggest that inorganic arsenic may induce gene amplification in mammalian cells and possibly affect the later stages of carcinogenesis, specifically the progression from preneoplastic to malignant lesions. However, the mechanism of arsenic-induced cancer, along with its other toxic effects, is largely unknown.

In general, when given orally to rodents, inorganic or organic arsenic compounds fail to exhibit significant carcinogenic activity. The only exception is the study by Knoth (1966), which reported adenomas of the skin, lung, peritoneum and lymph nodes in mice exposed to a total dose of 7 mg/animal in five months (approximately 1mg/kg-day). Ishinishi *et al.* (1983) and Pershagen *et al.* (1984) have demonstrated increased incidence of lung tumors in hamsters given arsenic trioxide by intra-tracheal instillation. However, the tumor incidences were low (*e.g.*, 3/47). The applicability of these studies to human environmental exposure is questionable. The lack of convincing carcinogenicity via the oral route in experimental animals is unique for a proven human carcinogen.

Risk assessment

As previously noted, arsenic causes skin cancer, the early findings being reported by Tseng and coworkers (1968, 1977). During our evaluation of arsenic for development of a new California drinking water standard, the data reviewed revealed that human exposure to arsenic in an area of chronic arsenicism in Taiwan was also associated with an increased risk of death from several internal organ cancers (Smith *et al.* 1990, 1992; Brown and Fan, (in press); OEHHA, 1992). Using the more recent published data from Taiwan (Chen *et al.,* 1990) and adjusting for body weight and water consumption for the U.S. population, the estimated cancer risks for different types of cancer at the MCL level of 50 ppb are shown in Table 1.

Table 1 *Lifetime risks associated with water consumption at 50 ppb arsenic*

Cancer type	Risk	
	Female	*Male*
Skin	7.6/1000	7.6/1000
Lung	11.0/1000	6.1/1000
Bladder	4.7/1000	2.2/1000
Kidney	3.4/1000	1.4/1000
Liver	0.3/1000	0.2/1000

Note: Water consumption = 2 liters per day

The estimated lifetime risk is highest for the lung but is comparable to that for the skin. In the further analyses of Chen and Wang(1990) and Chen *et al.* (1992), the associations of arsenic with these cancers are especially prominent. Based on the data on skin cancer, the incidences of which were more readily quantifiable, a potency factor of 5.3$(mg/kg\text{-}day)^{-1}$ has been derived for assessing carcinogenic risks of inorganic arsenic via ingestion (Smith *et al.*, 1990). This potency factor was used for the development and proposal of a recommended public health level (RPHL) of 2 ppt (OEHHA, 1992; Brown and Fan, in press) as required by the California Safe Drinking Water Act of 1989.

Regulatory programs and requirements

Under the Safe Drinking Water Act of 1974, amended 1986, the EPA has promulgated National Primary Drinking Water Regulations (NPDWR) and established Maximum Contaminant Levels (MCLs) and MCL goals (MCLGs) for various chemical contaminants in drinking water. The State of California has "primacy", *i.e.*, the right to develop and implement the State's drinking water program according to EPA rules and requirements. The State's standards are similarly referred to as Maximum Contaminant Levels, or MCLs. Under primacy, the State must adopt the same or more stringent MCLs and other requirements of EPA within the time period provided by the regulations.

The State Budget Act of 1986 required the State to develop MCLs for over 40 chemicals . The evaluation of the necessary information for MCL development involves health risk assessment and the consideration of costs and technical feasibility. The promulgation of regulations includes public notifications and hearings before final adoption.

In 1989, AB21, also referred to as the Safe Drinking Water Act of 1989, was passed. AB21 which requires the State to develop and adopt strictly health based drinking water standards called recommended public health levels (RPHLs) and to review existing MCLs once every five years. The RPHL is a new drinking water standard which will apply to public water systems supplying greater than 10,000 service connections when the Act is fully implemented. The water suppliers will be required to report annually on their arsenic and other pollutant concentrations in comparison to RPHL's as well as to State MCL's. Water suppliers will also be required to provide economic and technical justification for any pollutant concentration exceeding an RPHL even though the MCL is met.

The responsibility for the establishment of drinking water standards rests with the Office of Environmental Health Hazard Assessment (OEHHA) and California Department of Health Services (DHS). While the OEHHA performs the risk assessments of chemicals and develops numerical values for standard development, DHS has the responsibility for risk management, assessing economic and technical feasibility, promulgating and adopting regulations and routine monitoring and enforcement.

Water monitoring and human exposure

Arsenic occurs primarily in groundwater. In California, it is found frequently in groundwater in the Central Valley and, therefore, affects water systems in that region which are mostly small, having less than 200 service connections, and serving about 600 persons or less. Some large systems (≥200 service connections) are also affected but these mostly have small distributions serving approximately 1,500 persons or less. About 60% of the State population drinks surface water and 40%, groundwater. Of the small systems, about 80-85% use groundwater while those remaining use both groundwater and surface water.

Monitoring for arsenic has been done once every three years for groundwater and annually for surface water. Traditionally, small systems (< 200 service connections) were the regulatory responsibility of the local counties and were not directly regulated by the State until July 1993.

Since the historical reporting limit for arsenic was 10 ppb, relatively little information exists on levels occurring below 10 ppb. However, as a result of improved analytical methodology and recognition of the need to assess occurrence of arsenic at lower levels, the reporting limit was recently lowered to 5 ppb. The analytical detection limit ranges from 1 to 5 ppb.

The monitoring data on arsenic in large water systems updated through April 1993 are shown in Table 2. The data are reported by county (total 48) as the number of wells showing concentrations for ranges of 5-10, 10-20, 20-50, 50-100 and 100 ppb. Most of the wells (710 among total of 741) have detectable levels of arsenic between 5 and 50 ppb. Twenty four wells had levels in the 50-100 ppb range, and 7 had levels 100 ppb. The counties are also listed according to the levels reported as shown in Table 3. In terms of the number of wells, the majority of them are found in the 5-10 ppb and 10-20 ppb ranges. In most counties, the well water contains arsenic within the 20-50 ppb range. Kings county has the highest number of levels exceeding the MCL.

Data on arsenic levels in small water system sources is limited. To obtain qualitative information on small systems, a telephone survey (February 1991) of

Table 2 *Arsenic in California drinking water based on monitoring of large water system sources (1/84-4/93).*

County	No. of water systems	No. of water sources	Arsenic concentration (ppb)				
			5-10	>10-20	>20-50	>50-100	>100
Alpine	2	2		1	1		
Amador	2	2	1	1			
Butte	2	4		3	1		
Calaveras	4	5	5				
Colusa	3	5	4		1		
Contra Costa	2	2	1	1			1
El Dorado	2	10	4	5	1		
Fresno	7	15	2	6	6	1	
Humboldt	2	2		1	1		
Kern	30	102	31	43	25	1	1
Kings	5	32	2	4	14	12	
Lake	2	2	1	1			
Los Angeles	21	45	17	16	9	1	2
Madera	4	6	2	3			
Marin	2	2		1	1		
Mariposa	3	5	4		1		
Mendocino	1	2			2		
Merced	8	19	12	6	1		
Modoc	2	3	1	1	1		
Mono	4	12	4	5	3		
Monterey	11	22	4	13	4		
Napa	1	2			1	1	
Nevada	3	7		1	5	1	
Orange	11	24	18	3	3		
Placer	2	2	1	1			
Plumas	1	1			1		
Riverside	10	26	3	20	1	1	
Sacramento	13	30	15	10	10	1	
San Benito	1	2			1	1	
San Bernardino	16	39	4	19	12	1	1
San Joaquin	10	80	14	33	33		
San Luis Obispo	11	17	10	5	2		1
San Mateo	1	1	1				
Santa Barbara	5	8	2	3	3		
Santa Clara	5	30	25	5			
Santa Cruz	5	13	3	9	1		
Shasta	5	6	5		1		
Siskiyou	2	2		2			
Solano	3	3		2	1		
Sonoma	10	20	5	11	3	1	
Stanislaus	13	42	22	17	2		1

Sutter	4	14	3	8	3		
Tehama	2	2	1	1			
Tulare	8	11	5	2	2	2	
Toulumne	1	1			1		
Ventura	3	8	1	7			
Yolo	2	7	4	3			
Yuba	2	2	2				
Totals	269	699	239	273	158	24	7

selected counties was conducted based on levels of occurrence in large water systems. The results are shown in Table 4.

Based on the available monitoring data, it can be postulated that human exposure occurs primarily at the concentration range of less than 5 to 50 ppb arsenic, as most of the wells having detectable levels are included in this range. The size of the population affected is not known because no related information is available. The data on arsenic levels ranging from 5 to 10 ppb are incomplete since prior to June 1993, only some water utilities report levels at less than 10 ppb.

Regulatory and cost impacts

The establishment of a new Federal and State MCL for arsenic would affect all water systems. Reduction of the existing standard is expected to be limited by economic and technical feasibility factors. The establishment of a RPHL, as noted above, would impact water systems serving 10,000 connections and a compliance evaluation would need to be performed. In addition, a water quality improvement plan might be required from the affected water suppliers which would address alternatives for mitigation, including information on the present situation, cost and technical feasibility. Documentation would be needed to justify any alternative addressed and there could not be a "no-action" alternative strictly based on the cost impact. If no plans of action were feasible, a report describing studies and measures to be taken would be required to be submitted and reviewed by the DHS' Division of Drinking Water.

A RPHL of 2 ppt for arsenic in drinking water is equivalent to a negligible cancer risk of one in a million based on an average consumption of two liters of tap water per day and a relative source contribution of 20%. This level is not feasible to achieve because it is far below the existing analytical detection limit. It may be considered a goal similar to the federal MCL goals (MCLGs) for other carcinogens. The current detection limit in most laboratories is 5 ppb, although more elaborate laboratory procedures can detect arsenic at 1 ppb or less.

A reduction of the MCL would place a major burden on implementation and compliance and is expected to impose a significant impact on the economy. The precise cost impacts have not been determined. As can be seen in Table 2 many water systems would potentially be out of compliance if the MCL were lowered to

Table 3 *Arsenic levels in California drinking water, listed by Range of detected levels reported for large water systems for each county (1/84-4/93).*

Arsenic concentration, ppb				
5-10	>10-20	>20-50	>50-100	>100
Calaveras	Amador	Alpine	Fresno	Contra Costa
San Mateo	Lake	Butte	Kings	Kern
Yuba	Placer	Colusa	Napa	Los Angeles
	Santa Barbara	El Dorado	Nevada	San Bernardino
	Santa Clara	Humboldt	Riverside	San Luis Obispo
	Santa Cruz	Madera	Sacramento	Stanislaus
	Shasta	Marin	San Benito	
	Siskiyou	Mariposa	Sonoma	
	Tehama	Mendocino	Tulare	
	Ventura	Merced		
	Yolo	Modoc		
	Solano	Mono		
	Sutter	Monterey		
	Tuolumne	Orange		
		Plumas		
		San Joaquin		

2.0 ppb. It should be noted that 20 ppb is about the lowest concentration achievable by conventional treatment technology but gives only a marginal reduction in projected risks compared to the existing MCL.

If the MCL were set at 20 ppb, some of the wells having detectable levels of arsenic at concentrations above this level could be brought into compliance. To reduce arsenic to 5 ppb would require non-conventional (ion exchange) treatment at presumably much higher costs. Only 'state of the art' treatment by reverse osmosis would be able to reduce arsenic to 2 ppb and this is probably infeasible for both small and large systems. If the MCL were reduced to 2 ppb, some water systems with surface water sources such as the Metropolitan Water District of southern California which serves more than 16 million people would have to treat the water at significant additional cost.

Summary and discussion

Current risk assessment by OEHHA and USEPA associate chronic consumption of water containing inorganic arsenic with relatively high lifetime risks (0.1-1.0%) of skin cancer and probably more serious internal cancers of the lung, kidney, bladder and liver (Bates *et al.* 1992). Based on current information there is no detectable concentration of arsenic in drinking water (>1 ppb) that can be considered safe for chronic consumption. This conclusion is based on lack of data supporting a

Table 4 *Arsenic occurrence in small water systems in selected counties in California (February 1991).*

Not detectable	Very few	Few	Some	Most	> MCL
Colusa	San Bernardino	Tulare	Fresno	Kern	San Benito
Contra Costa	San Benito		Merced	Kings	San Joaquin
Imperial	Stanislaus		Monterey		Sacramento
Nevada	Sonoma		San Joaquin		Sonoma
Shasta					Kern
Sonoma					Kings

MCL = 50 ppb

threshold for arsenic carcinogenicity in the 1-50 ppb range (Hopenhayn-Rich *et al.,* 1993) and the assumption of linear dose response which is supported by epidemiology (Smith *et al.* 1990). In California 15 counties contain 31 water system sources with arsenic concentrations above the 50 ppb MCL. Perhaps as many as 35,000 people are exposed at these levels. However as noted above, the number of water systems with arsenic concentrations in the 5-50 ppb range is much larger with a potentially much higher total cancer burden. Perhaps as many as 5 million people are exposed. For the purpose of setting a new MCL for arsenic in California, it is important to obtain a more detailed assessment of population exposures to arsenic via drinking water from affected public water systems. Since the risks of arsenic exposure via drinking water have been compared with those of environmental tobacco smoke and radon, two recognized environmental carcinogens of public health concern, it is important to reduce this exposure as much as is economically feasible.

References

Bates, M.N., Smith, A.H., and Hopenhayn-Rich, C. 1992. Arsenic ingestion and internal cancers: A review. *Amer. J. Epidemiol.* **135,**462-476.

Brown, J.P. and Fan, A.M., Arsenic drinking water standard development. *Journal of Hazardous Materials.*(in press)

Chen, C.J., Kuo, T.L. and Wu, M.M. 1988. Arsenic and cancers. *Lancet,* **1**, 414.

Chen, C.J., Chen, C.W., Wu, M.M. and Kuo, T.L. 1992. Cancer potential in liver, lung, bladder and kidney due to ingested inorganic arsenic in drinking water. *Br. J. Cancer,* **66**, 888-892.

Chen, C.J., Chuang, Y.C., Lin, T.M. and Wu, H.Y. 1986. Malignant neoplasms among residents of a blackfoot disease-endemic area in Taiwan: High arsenic artesian well water and cancers. *Cancer Res.,* **45**, 5895.

Chen, C.J. and Wang, C.J. 1990. Ecological correlation between arsenic level in well water and age-adjusted mortality from malignant neoplasms. *Cancer Res.*, **50**, 5470-5474.

Hopenhayn-Rich, C., Smith, A.H., and Goeden H.M. 1993. Human studies do not support the methylation threshold hypothesis for the toxicity of inorganic arsenic. *Environ. Res.*, **60**,161-177.

Ishinish, N., Yamamoto, A., Hisanage, A. and Inamasu, T. 1989. Tumorigenicity of arsenic trioxide to the lung in Syrian golden hamsters by intermittent instillations. *Cancer Lett.*, **21**, 141-147.

Knoth, W. 1966. Arsenbehandlung. *Arch. Klin. Exp. Derm.*, **227**, 228-234.

Office of Environmental Health Hazard Assessment. 1992. *Arsenic Recommended Public Health Level for Drinking Water.* Draft Report. California Environmental Protection Agency, Berkeley, CA.

Pershagen, G., Nordberg, G., Bjorklund, N.E. 1984. Carcinomas of the respiratory tract in hamsters given arsenic trioxide and/or benzo-a-pyrene by the pulmonary route. *Environ. Res.*, **34**, 227-241

Smith, A.H., Goeden, I.H., Shearn, V., Bates, M., and Allen, H. 1990. *Health Risk Assessment for Arsenic Ingestion.* University of California, Berkeley

Smith, A.H., Hopenhagen-Rich, C., Bates, M., Goeden, H., Hertz, I., Allen, H., Hood, R., Kosnett, M., and Smith, M. 1992. Cancer risks from arsenic in drinking water. *Environ. Hlth. Persp.*, **97**, 259-267.

Tseng, W.P. 1977. Effects and dose-response relationships of skin cancer and blackfoot disease with arsenic. *Environ. Hlth. Persp.*, **19**, 109-119.

Tseng, W.P., Chu, H.M., How, S.W., Fong, J.M., Lin, C.S. and Yeh, S. 1968. Prevalence of skin cancer in an endemic area of chronic arsenism in Taiwan. *J. Natl. Cancer Inst.*, **40**, 453-463.

US Environmental Protection Agency. 1984. *Health Assessment Document for Inorganic Arsenic*, EPA 600/8-83-021F. Environmental Criteria and Assessment Office, Washington, D.C.

US Environmental Protection Agency. 1988. *Special Report on Ingested Arsenic: Skin Cancer; Nutritional Essentiality. EPA/625/3-87/013. Washington D.C.*

28 Regulating arsenic in Nevada drinking water supplies past problems, future challenges

Jeffrey A. Fontaine

Public Health Engineering, Nevada State Health Division, 505 East King Street, Room 103, Carson City, NV 89710, USA

Abstract

One-half of Nevada's community water systems (CWS) contain detectable levels of arsenic. Five CWS serving a total of 19,000 persons contain levels of arsenic between 50 and 100 micrograms per litre (g L^{-1}). At least one-thousand private wells contain levels of arsenic in excess of 50 g L^{-1}. The existing Federal Safe Drinking Water Act standard for arsenic in CWS is 50 g L^{-1}. However, the US Environmental Protection Agency (USEPA) is currently considering a revised standard in the range of 2 to 20 g L^{-1}. Compliance with the 50 g L^{-1} drinking water standard in Nevada has been limited, because of the lack of demonstrated adverse health effects, and the high cost of remediation. The natural and widespread occurrence of arsenic in Nevada has further contributed to the popular belief that low levels of ingested arsenic are not a health concern. For example, the City of Fallon continues to provide water which contains 100 g L^{-1} despite public notification, and attempts by the State Health Division and the USEPA to gain compliance with the standard.

A lower arsenic standard will have a significant cost impact to state drinking water agencies and consumers of public water systems. In Nevada, most CWS serve less than 500 people. Alternative water supplies are not usually available and most communities are served by multiple wells which must be treated separately. USEPA must resolve several issues regarding the health effects, occurrence and treatment of arsenic in drinking water. If risk assessment demonstrates that a lower standard is warranted, state drinking water agencies must be given flexibility in applying the standard to small PWS. Also, appropriate treatment technologies for small CWS and private systems must be developed.

Introduction

More than 200 million Americans are served by public water systems which are subject to the federal Safe Drinking Water Act regulations, including the arsenic standard. In Nevada, arsenic occurs naturally in several groundwater basins in Nevada. One-half of Nevada's 350 community water systems contain levels of arsenic in excess of 3 micrograms per litre (g L^{-1}), including five systems which exceed the current 50 g L^{-1} standard. Additionally, the Nevada State Health Division has estimated that as many as 1,000 private wells in Nevada may contain levels of arsenic in excess of 50 g L^{-1}.

The City of Fallon, Nevada, which has a population of 6,000 people is served by a water supply that contains 100 g L^{-1} of arsenic. City officials and residents vigorously opposed treating their water to meet the standard, arguing that the 50 g L^{-1} standard was arbitrarily based on a foreign study. They also argued that there were no demonstrated adverse health effects associated with their drinking water, that they, and their forefathers, had consumed for years. In 1985, 2,000 customers of the Fallon Public Water System signed a petition indicating that they prefer their water to remain as is, without treatment. After repeated attempts by the Nevada State Health Division and the U.S Environmental Protection Agency (EPA) to enforce the arsenic standard, the City of Fallon continues, to this day, to serve water containing 100 g L^{-1}. Recently however, the City of Fallon has indicated that it will comply with the revised arsenic standard.

This example of non-compliance points out the need to establish a credible standard, and communicate the health risk of not meeting the standard, to those impacted. No responsible public official wants to put their community at risk. But public officials must be convinced that they are expending public money to truly protect public health.

Background

State drinking water regulators have a difficult job. Imagine attending a public meeting and telling a community of 1,000 people, that has just closed a school or fire station because of budget cuts, that they will have to spend a significant sum of money to reduce a contaminant in their water supply which, based on a number of assumptions, should prevent one excess cancer case in-a-million.

In preparation for federal Safe Drinking Water Act (SDWA) reauthorization, the Association of State Drinking Water Administrators, (ASDWA) has been discussing possible legislative changes that could help make the goal of providing safe drinking water more achievable. One of the most important issues expressed by ASDWA is the need for strong research in the development of standards. ASDWA believes that EPA must involve the medical and research community more in the development of standards, to assure we are achieving quantifiable public health protection and that the risks associated with not regulating drinking water contaminants are communicated *to,* and supported *by,* the public health community.

The cost for states to implement monitoring, treatment, and public notification requirements for the 83 drinking water contaminants is also rising significantly. States need $304 million to implement these requirements (EPA, 1993), yet Congress appropriated only $59 million for state drinking water programs last year. As most states are experiencing budget deficits, it is unlikely that states will be providing additional funds to implement drinking water requirements.

Many state drinking water programs, including Nevada's, reside in the health department. When compared to the relative risk of other public health programs, such as AIDS, substance abuse, or childhood lead poisoning, state drinking water programs are usually not a priority for scant public resources. The cost of enforcing a stringent arsenic standard in states where there is a widespread occurrence of this contaminant, will add to this shortfall. States must adopt and implement rules or the state loses primacy and EPA administers the drinking water program.

The cost of safe drinking water is also beginning to affect consumers of public water systems. Repairing aged and neglected water system infrastructure, and the cumulative impact of the SDWA, may make water unaffordable for some. Some Nevada water systems now have monthly household water bills of $75 (RCAC, 1993). Water suppliers are making considerable expenditures for remedial improvements, such as replacing leaking distribution pipes, building new facilities to treat water for acute contaminants, such as giardia and viruses, and remediating priority chronic contaminants, such as lead.

In Nevada for example, the water system which serves 57,000 households and businesses in Reno, is spending $91 million to construct a surface water treatment plant to remove microbiological contaminants. This plant will raise monthly household water bills from $29 to $53 to pay for the surface water treatment plant. The water supplier may also have to reduce disinfection by-products and treat groundwater sources for tetrachloroethylene and radon. This will add $8.43 to monthly household water bills. Arsenic treatment will add another $3.47 to monthly water bills (Sierra Pacific Power Company, 1991). As indicated earlier, one-half of Nevada's community water systems contain detectable levels of arsenic, that is 3 g L^{-1}. Depending on the level of the revised EPA standard, some or all of the systems will need to remediate arsenic. The majority of Nevada's community water systems are small, that is they serve less than 100 people. They include mobile home parks, and rural towns. The cost of remediating arsenic in these Nevada systems will be significant for several reasons.

(1)As the most arid state, alternative water sources in Nevada are not readily available.

(2)Arsenic treatment technology is expensive and relatively costly to operate and maintain. The disposal of treatment by-products adds to this cost.

(3)In Nevada, most communities are served by multiple wells, which usually cannot be combined for central treatment or blending, necessitating treatment at each source.

Conclusions

EPA is under a court ordered deadline to propose a new arsenic standard by

September, 1994. Yet, there are still several unresolved issues regarding not only health effects, but also the occurrence and treatment of arsenic:

(1)A lack of information on occurrence levels between 0.5 g L^{-1}, the lowest detection limit, and 5 g L^{-1}, the reporting limit in most national surveys.

(2)A lack of information on whether or not plumbing materials, such as tin and antimony, represent a source of arsenic which could cause a water system to exceed a low standard.

(3)A lack of information about whether arsenic impurities present in drinking water treatment chemicals, such as coagulants, could cause a water system to exceed a low standard?

(4)A lack of information about full-scale on arsenic removal processes?

(5)A lack of information about whether arsenic treatment could affect water quality to the extent that it conflicts with other drinking water regulations? For example, the Lead\Copper\Corrosion Control regulation requires the optimization of water quality to prevent leaching of lead in distribution systems.

EPA is mandated to set the enforceable arsenic standard as close as feasible to the drinking water health goal, taking cost into consideration. EPA must carefully evaluate these impacts based on reliable data, including the cost for states to enforce, and systems to comply with a revised standard. Also, EPA must consider the impact on small water systems which do not have the rate base to pay for treatment, or qualified personnel to operate complex treatment technologies.

ASDWA believes that drinking water contaminants that pose a risk should be eliminated or reduced. However, this takes money. And the regulation of arsenic could cost a significant amount of money. The public looks to EPA and state drinking water agencies to protect their drinking water, yet, more and more they are asking what benefit they achieve from the resources they expend. It is incumbent upon us, as keepers of this public trust, to develop regulations that protect public health, yet are economical and defensible in today's tight fiscal covenant.

In light of the uncertainty of health effects of arsenic at low levels and unresolved issues regarding occurrence and treatment, is EPA comfortable with promulgating a revised arsenic standard that could have a significant fiscal impact? If not, ASDWA encourages EPA to seek additional time to resolve these issues.

References

US Environmental Protection Agency (EPA). 1993. *Technical and Economic Capacity of States and Public Water Systems to Implement Drinking Water Regulations, Report to Congress.* Washington

Rural Community Assistance Corporation (RCAC). 1993. *Nevada Water Rate Survey.* Carson City, NV

Sierra Pacific Power Company, Westpac Division. 1991. *Surface Water Treatment Rule Compliance Plan.* Reno, NV

29 The water industry perspective on the important arsenic regulatory issues

J. Alan Roberson[1] and Howard M. Neukrug[2]

[1]*American Water Works Association,1401 New York Ave. NW, #640, Washington, D.C. 20005, USA*
[2]*Philadelphia Water Department, One Reading Center, 2nd Floor1101 East Market Street, Philadelphia, PA 19107, USA*

Abstract
The US Environmental Protection Agency (US EPA) is currently developing a new arsenic drinking water regulation. The impact to public water supplies could be substantial. The low levels that are under consideration for the Maximum Contaminant Level (MCL) could make this regulation the most costly of any regulation to date. US EPA is wrestling with many difficult issues in the development of this regulation.
For drinking water regulation, US EPA uses a Category system based on carcinogenicity via ingestion. Based on the Taiwan epidemiological data, arsenic will be classified in Category I, a known human carcinogen via ingestion. The exclusive use of this data, and the lack of any American epidemiological data has generated some controversy. The lack of any detailed occurrence data in source waters at the low levels that are being considered for the MCL is another controversial area. Existing occurrence data was analyzed using analytical methods with detection limits an order of magnitude higher than the low levels being considered for the MCL. Scarcity of data on the ability of available treatment technology to achieve low arsenic levels is another controversial area.

Introduction

The 1986 Amendments to the Safe Drinking Water Act (SDWA) require US EPA to develop national drinking water standards for 83 inorganic and organic chemicals. Since 1986, the vast majority of these regulations have been promulgated by US EPA. However, US EPA is wrestling with difficult policy issues for the balance of the contaminants on the original list. Radionuclides, sulfate, aldicarb and its metabolites and arsenic are taxing US EPA's ability to balance the technical and policy issues of these final drinking water regulations. These regulations have been delayed because they are the most difficult ones to develop.

The current interim drinking water standard for arsenic is 50 $\mu g\ L^{-1}$. US EPA is considering lowering the standard substantially, to the 0.2-20 $\mu g\ L^{-1}$ range. Reducing the MCL for arsenic to these levels raises many questions that are difficult to answer, due to a lack of data. Very little health effects data from the United States exist, although, there are several communities with high levels that could be investigated. There are also limited data on treatment technologies for arsenic removal to low levels at full scale plants, and even less occurrence data at low arsenic levels.

The policy questions on arsenic include many debatable philosophical issues. Will this regulation provide effective relative risk reduction? In other words, will the monies spent to comply with this regulation be balanced with health protection from arsenic? How will the arsenic that is removed from the drinking water ultimately be disposed? Will the average drinking water consumer be able to afford the increased cost due to the additional treatment for the removal of arsenic? Due to the competition for scarce public health dollars and the increasing costs of complying with all of the environmental mandates, not just those of the SDWA, the answers to these questions and others may end up being the driving force behind the final arsenic regulation.

US EPA is in the difficult position of attempting to answer these questions, and many others, in a short period of time. Arsenic regulations were to be proposed by US EPA in September 1994. Based on a recent District Court settlement, the arsenic regulation is now scheduled to be proposed by US EPA in November 1995, and promulgated in November 1997. While a detailed research agenda is still being developed, US EPA is required to continue with the regulatory development.

Discussion

Health effects

Due to the uncertainties associated with the epidemiological studies conducted in communities with high arsenic levels in Taiwan, US EPA must investigate the US communities with high arsenic levels (Kenneth G. Brown presentation to the Society for Risk Analysis, December 6, 1993). The ongoing scientific debate on the health effects may take years to ultimately resolve, if it is at all possible to resolve the debate. There may never be enough health effects data to completely answer all of the questions. Meanwhile, US EPA is mandated by the SDWA to finalize an arsenic regulation for drinking water.

Investigations into the mechanisms of cancer, as well as epidemiological studies are traditionally conducted by such agencies as the National Cancer Institute, the Center for Disease Control, *etc.* Nevertheless, water utilities need to be active participants in these studies, and need to encourage these agencies to start these studies as soon as possible.

In addition, the following issues need to be closely examined by US EPA in its development of its health criteria document on arsenic prior to the development of the final arsenic regulation:

-Arsenic as an essential nutrient

-The potential for a threshold for cancer
-Applicability of the Taiwanese epidemiological data and the corresponding lack of US data
-The relative potency of external versus internal cancers
- Non-cancer health effects

Occurrence data
Generally, the existing occurrence data yields the following trends:
-High arsenic levels are located in groundwater supplies west of the Mississippi river
-High arsenic levels can be found in surface waters in the arid Southwest
-Geological strata such as recent volcanic activity can produce extremely high arsenic levels in groundwater
-Very little *national* occurrence data exists

Since the current drinking water standard for arsenic is 50 μg L^{-1}, the current method detection limit using the US EPA approved Method 206.2, using a nickel modifier has remained acceptable at 5-10 μg L^{-1}. While analytical methods using platform with a palladium/argon modifier have lowered the method detection limits in the last few years, the vast majority of existing arsenic data was not analyzed at these low detection limits. This creates two potentially serious problems. US EPA wants to make data extrapolations to low levels without any real data. The potential for a log-normal distribution at low levels could greatly increase treatment costs. Additionally, many water utilities may not think that they have a problem with arsenic due to the current high detection limit. These same utilities may actually be severely impacted by a low arsenic drinking water regulation.

The determination of background arsenic levels in drinking water sources is a critical need. US EPA needs to develop a national distribution of background arsenic levels, broken down by geographic location, groundwater strata, and surface water sources. The arsenic speciation in water supplies, and the linkage by species to health effects is another key component of this national distribution. The current data from the National Inorganic and Radionuclides Survey (NIRS, US EPA, 1984-1986) does not have data for low (5 μg L^{-1}) arsenic levels.

Treatment technologies
US EPA has sponsored several research projects on the removal of arsenic from drinking water. Malcolm Pirnie, Inc., under contract with US EPA, recently completed a review of the various treatment technologies for the removal of arsenic. This review indicates that at bench scale, treatment technologies such as reverse osmosis and electrodialysis can produce an effluent with an arsenic concentration of 1-5 μg L^{-1}. However, these bench scale studies were only performed on one raw water, and the raw water was pre-treated to lower the natural sulfate concentration. While at the bench scale, many of the advanced treatment technologies appear to be feasible, very little operating data exists for full scale arsenic removal.

Under the existing interim arsenic standard of 50 μg L^{-1}, many drinking water utilities have shut down wells and drilled new wells, or have blended sources in order to meet the current standard. Drilling new wells can be an expensive proposition, and some systems can not easily abandon wells due to community growth. These measures can only be viewed as stopgap measures, particularly if the standard is lowered down to the 1-10 $\mu g\ L^{-1}$ range.

The use of reverse osmosis as a treatment technology probably will not be a viable option in the arid southwest, due to raw water loss. Reverse osmosis systems typically recover 50-80% of the raw water. This loss of water resources, in addition to the intensive energy use, probably makes the use of reverse osmosis not feasible in the arid southwest.

Policy issues

As just discussed, there exists a considerable lack of data for the development of a new standard for arsenic. Occurrence, treatability, costs, analytical methodologies and health effects are the main ingredients necessary for proper regulatory development, all of which have some important knowledge gaps relative to arsenic. Yet the significant reduction in the allowable standard for arsenic in drinking water under consideration by US EPA will have a substantial impact on public water suppliers and the consuming public. In fact, the total capital costs of the arsenic regulation may be more than comparable to the $10 billion price tag for this summer's flooding in the midwest (extrapolated by regression analysis on cost figures cited in Table in AWWA, 1993, assuming MCL = 1.5 $\mu g\ L^{-1}$) And, like the flooding costs, the bulk of the arsenic regulatory costs will be concentrated within specific geographic regions of the country.

Given that the costs for compliance are so significant and that the technical information to form an arsenic rule is so limited, public policy issues (which are important in all environmental regulatory development) take on an even more important role with respect to arsenic. Thus a balanced approach, which considers both technical and policy issues, is critical.

The water industry is attempting to raise these policy issues in this forum and others. It is a fine line that one must follow to include these policy debates within the context of regulatory rule development, but it is something that we must attempt to do, while trying not to take on an adversarial role with US EPA and others. Full and open dialogue is essential at this stage of the regulatory development process.

For example, one policy issue is affordability. A greatly reduced standard for arsenic will be particular hardship for small community water systems where significant costs for treatment must be spread over a relatively small rate base. The elderly on fixed incomes and welfare recipients will, in many cases, need to make conscious "life choices" between drinking water and other necessities. The public health consequence of these actions must be considered and weighed in the evaluation of the benefit of a new limits on arsenic in drinking water.

Perception of drinking water quality and communicating the risk of arsenic in drinking water is another major policy issue. It is the public water suppliers who will be on the front lines answering consumer questions on why water rates in some areas may double or triple. US EPA has the obligation to not only set the drinking water standard for arsenic, but to also be able to communicate to the public the benefits they will receive through paying for its removal. As of today, this type of information and understanding is not available to even convince knowledgeable professionals in the drinking water and health fields of the potential risk reduction. Those on the "front lines" – local utility managers, local health officials and local elected officials must be able to communicate the pressing need for the installation of expensive new treatment technologies.

The formulation of an "overall risk reduction strategy" is a key ingredient necessary in the development of all new environmental regulations. As with the recent controversy with radon regulations in drinking water, the issue here is whether, given a $10 billion budget for arsenic exposure reduction, is US EPA getting a reasonable return on its public dollar investment through regulating drinking water? Alternatively, perhaps food sources are a more critical need, or perhaps the money is better spent for limiting the use of arsenate insecticides, or other multi-media exposures?

The implementability of a new drinking water regulation for arsenic is another critical area. Given its costs and the relative limited availability of field-proven technologies and analytical capabilities, will small systems be able to comply with the regulation?

Clearly, US EPA is under a very tight schedule to develop a new regulation for arsenic in drinking water. Yet we must all keep in focus that we are dealing with a potential chronic health threat and develop a rational, implementable, cost-effective method of reducing exposure to arsenic in drinking water.

The first step is really up to the health professionals at this conference. The SEGH Task Force that will review the findings of this conference needs to develop a firm consensus and develop conclusions as to what is known and not known on arsenic health threats from drinking water. If there is insufficient information at this time, the SEGH Task Force needs to recommend to US EPA additional study needs and a time frame to develop the answers.

One potential avenue for additional health effects information is to conduct epidemiological studies in the United States, in communities with high arsenic levels (*e.g.* Fallon, Nevada and Hanford, California). This is clearly an area which needs some leadership from US EPA, in order to provide a clearer understanding of the health benefits to be achieved through increased treatment dollars.

Although several years may be needed to answer the health effects questions, they are at the crux of the policy issues involved in setting this new regulation. It must be remembered that we are discussing a contaminant that is not an acute health threat to the public, even at current MCL levels. While the clock is running and we all want to do the right thing, we should not move forward blindly.

What can US EPA do? US EPA needs to look beyond the regulatory deadlines by providing the maximum relative risk reduction. Perhaps US EPA needs to look at innovative regulatory strategies, such as that under consideration for the Disinfectant/Disinfection By-Products Rule. With respect to arsenic, the concept would be to develop an action level or interim standard to bring our nation's arsenic limits in-line of those used elsewhere, such as the World Health Organization or the Canadian EPA. By setting a lower action level or interim standard, the most significant exposures to arsenic in drinking water will be addressed in a rational, implementable and technicalogically acceptable manner. The portion of the saved public health dollars will then be available to develop a fuller understanding of arsenic exposure, treatability and health effects.

In reality, while the drinking water industry can do almost anything technically, sound drinking water regulations must be based on a policy that includes full scientific knowledge, communication of risk, a full perspective of regulatory implications, and some good ol' common sense.

Conclusions

Congress has mandated through SDWA for US EPA to balance the technical and policy issues in the development of the arsenic regulation. The two sets of issues are intertwined, and cannot be split apart in the regulatory development. US EPA needs to consider innovative regulatory strategies due to the multitude of scientific questions that are still unanswered. US EPA needs to look beyond the constraints of deadlines in order to develop a regulation that provides the biggest bang for the buck. While the drinking water utilities will feel the immediate impact of the regulation, the secondary impacts will spread to other environmental statutes. US EPA needs to incorporate the principles of integrated environmental management into its regulatory development in order to account for all costs resulting from this drinking water regulation.

Acknowledgements

The authors wish to acknowledge the assistance of Tim Chinn, Montgomery Watson, and Jack Sullivan, American Water Works Association, in the review and preparation of this paper.

References

Brown, K.G. and Chen, C.-J. 1994. On the observed dose-response in the blackfoot endemic region of Taiwan. In: Chappell, W.R., Abernathy, C. and Cothern, C.R. (eds), *Proceedings of the International Conference on Arsenic Exposure and Health Effects*,. Science and Technology Letters, Northwood, England.

AWWA. 1993. *Background Information on Arsenic in Drinking Water* (August, 1993), AWWA, Denver.

30 Derivation of the inorganic arsenic Reference Dose

Charles O. Abernathy and Michael L. Dourson

Human Risk Assessment Branch (4304), Office of Water, US EPA, Washington, DC 20460-0001 and Environmental Criteria and Assessment Office, US EPA, Cincinnati, OH 45268, USA

Abstract
The US Environmental Protection Agency (EPA) regulates certain environmental chemicals that can adversely affect human health. To promote consistency among evaluations of noncancer health effects of chemicals after oral/inhalation exposure, the US EPA has developed a Reference Dose (RfD)/Reference Concentration (RfC) Work Group. This Work Group of Agency scientists considers the entire toxicity data base of a chemical, except for carcinogenicity and mutagenicity, determines the chemical's critical effect and the appropriate uncertainty and/or modifying factors to estimate an RfD or RfC. This report will describe the RfD process, discuss the specific derivation of the arsenic RfD, and briefly consider a novel method to calculate risk above the RfD.

Introduction

For regulating chemicals which exert adverse noncarcinogenic effects after oral exposure, the US EPA uses the Reference Dose (RfD). The RfD and uncertainty factor (UF) concepts have been developed by the US EPA in response to some problems associated with Acceptable Daily Intakes (ADIs) and safety factors (SFs), as outlined in Barnes and Dourson (1988). A related concept, the Reference Concentration (RfC), is used by the US EPA for inhalation exposures (Jarabek *et al.*, 1990).

The RfD is a dose operationally derived from a no-observed-adverse-effect level (NOAEL) of a chemical's critical effect, or occasionally a lowest-observed-adverse-effect level (LOAEL), by consistent application of generally order-of-magnitude UFs that reflect various types of data sets used to estimate RfDs.

The opinions expressed in this manuscript are those of the authors and do not necessarily reflect the opinions or policies of the US Environmental Protection Agency.

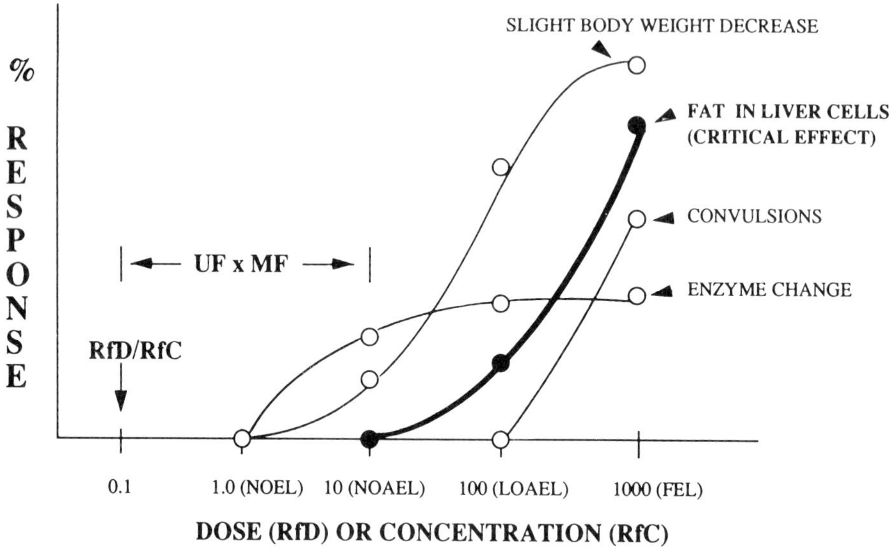

Figure 1 *Examination of the database leads to selection of a principal study. As demonstrated hypothetically from this study, scientific judgement leads to selection of the critical effect* (i.e., *fat in liver cells) and its no-observed-adverse-effect level (NOAEL) and the appropriate uncertainty (UF) and modifying factors (MF). The NOAEL is divided by the UF and MF to derive an RfD or RfC. This figure was adapted from Abernathy* et al., *1993). Other abbreviations - NOEL, no-observed-effect level; FEL, frank-effect level; RfD, Reference Dose; RfC, Reference Concentration.*

For example, a valid chronic animal NOAEL is normally divided by a UF of 100. In addition, a modifying factor (MF), to account for a factor such as differences in absorption, is sometimes used. The total UF and MF are based on the professional judgment of the RfD/RfC Work Group after consideration of the entire data base of the chemical (Figure 1). These factors and rationales for their use are presented in Barnes and Dourson (1988).

Arsenic (As) is a metalloid and exists in valence states of -3, 0, +3 and +5. The -3 form is a gas and the 0 state is poorly absorbed by humans. Accordingly, effects from these forms of arsenic were not considered in deriving the oral RfD for As. Although arsenicals may exert beneficial (medicinal) or harmful (toxic) effects, the RfD process considers only the adverse effects of a chemical, such as As, which is not considered to be an essential trace element for humans (US EPA, 1988).

The Reference Dose (RfD)

The RfD is determined by use of the equation:

RfD = NOAEL (or LOAEL) / (UF x MF)

this equation is the functional equivalent of the one used to calculate the ADI. The RfD is defined as an estimate (with uncertainty spanning perhaps an order of magnitude) of a daily exposure to the human population (including sensitive subgroups) that is likely to be without an appreciable risk of deleterious effects during a lifetime (US EPA, 1993). It is generally expressed in units of milligrams of chemical per kilogram of body weight per day (mg kg^{-1} day^{-1}).

The RfD is also useful as a focal point from which to gauge the potential effects of the chemical at other doses. Usually, doses at or less than the RfD are not likely to be associated with adverse health risks, and are therefore less likely to be of regulatory concern. As the frequency and/or the magnitude of the exposures exceeding the RfD increases, the probability of adverse effects in a human population increases. However, it should not be categorically concluded that all doses below the RfD are "acceptable" (or will be risk-free) and that all doses in excess of the RfD are "unacceptable" (or will result in adverse effects).

EPA has developed a systematic approach to summarising its evaluations, conclusions, and reservations regarding RfDs in a "summary sheet" that may range from a few pages to 20 to 30 pages in length. The summary sheet includes a statement on the confidence (high, medium, or low) that the RfD/RfC Work Group has in the stability of the RfD. High confidence indicates the judgment that the RfD is unlikely to change in the future because there is consistency among the toxic responses observed in different species, sexes, study designs, and/or in dose-response relationships, or that the reasons for existing differences are well understood. High confidence is often given to RfDs that are based on human data for the exposure route of concern, since the problems of interspecies extrapolation have been avoided. Low confidence indicates the judgment that the data supporting the RfD may be of limited quality and/or quantity and that additional information could result in a change in the RfD. These summaries are loaded on EPA's Integrated Risk Information System (IRIS; US EPA, 1993).

Evaluation of the arsenic RfD

The entire systemic toxicity data base of As, except for carcinogenicity and mutagenicity, is considered during RfD deliberations. The acute human LD_{50} for As, estimated from poisoning cases, is in the range of 1 to 4 mg kg^{-1} (Vallee *et al.*, 1960). In experimental animals, the guinea pig (LD_{50} of 9 mg kg^{-1}) is the most sensitive, while the rat is more resistant (LD_{50} of approximately 40 mg kg^{-1}) to the acute lethal effects of sodium arsenite (Lehman, 1951; Fielder *et al.*, 1986). These data suggest that humans may be more sensitive to the acute lethal effects of As than are animals. In humans, hemorrhagic gastroenteritis, vomiting, convulsions and peripheral vascular failure have been observed after acute As poisoning (Buchanan, 1962). Although short-term As exposure affects such organs as the heart, liver, kidney and nervous system (Hamamoto, 1955; Mizuta *et al.*, 1956; Tay and Seah, 1975; Fowler, 1977; Abernathy and Ohanian, 1992; Abernathy, 1994), the RfD is used to estimate a lifetime exposure level that is unlikely to cause adverse health effects. Accordingly, data from acute or relatively short-term exposure scenarios are not usually useful in deriving an RfD for a chemical. They may, however, provide useful toxicological information.

Unlike the data base for most chemicals, there is adequate information on the effects of As on humans under environmental conditions. Most of this data has been collected in populations exposed to drinking water containing As; these reports were examined during the derivation of the As RfD and are summarised below.

The effects of As were evaluated in several California and Nevada communities (Valentine *et al.*, 1984; 1985). In these studies, the concentration of As in the water ranged from approximately 400 μg L^{-1} to less than 1 μg L^{-1}. In one study, they found no differences in neurological, circulatory or skin disorders (Valentine *et al.*, 1984). In the second study, the data suggested that As might decrease nerve conduction velocity, but these effects did not parallel the As body burdens (Valentine *et al.*, 1985). Since some of the people had been exposed to As for only one year and the numbers of participants (14 to 29) in the various groups were low, these studies were not considered adequate for RfD verification purposes.

A clinical and electromyographic (EMG) survey of persons in Canada exposed to As in their drinking water has been conducted (Hindmarsh *et al.*, 1977). The exposed individuals were placed in three groups: (1) high-dose — As above 100 μg L^{-1}; (2) mid-dose —As between 50 and 100 μg L^{-1} ; and (3) low-dose — As less than 50 μg L^{-1}. Clinical findings of As exposure were found in 40% (17 of 42), 16% (8 of 50) and 10% (2 of 21) in the high- to low-dose groups, respectively. The authors reported a positive correlation ($p = 0.0026$) between As exposure and clinical signs. In the EMG studies, the high-dose group, 50% (7 of 14) had abnormal EMGs, while 17% of the mid-dose group (3 of 18) had abnormal results. None (0 of 12) in the less than 50 μg L^{-1} group had abnormal EMGs. This study substantiated the fact that As exposure could affect the nervous system, but the low numbers of subjects in the groups precluded its use in the RfD derivation.

Cebrian *et al.* (1983) studied the effects of As exposure in two Mexican populations. The average concentrations of As in the water were 411 μg L^{-1} and 5 μg L^{-1}, respectively. In the group exposed to 411 μg As L^{-1}, 22% (64 of 296) had hypo- and hyperpigmentation and keratosis, while only 2% (7 of 318) of the low-exposure group had dermal lesions. As there were no data on As in food, the numbers of persons studied were not high (approximately 600), and only one-third of the individuals were over 20 years of age, the RfD/RfC Work Group did not use this study.

Southwick and coworkers (1983) studied the effects of chronic As exposure in Utah. The "controls" consumed water containing an average of 19 μg As L^{-1}, while the two exposed groups drank water containing from 150 to 270 μg As L^{-1}. There was a slight increase in dermal lesions in the exposed group (6.3%; 9 of 144) as compared to those in the control group (2.9%; 3 of 105). Although there were minor differences in nerve conduction velocities between the two groups [exposed — 16% (13 of 83) versus controls — 12% (8 of 67)], the observed differences were not

Table 1 *Summary of the derivation of the oral reference dose for inorganic arsenic*[1].

Critical effect	Experimental doses[2]	UF[4]	MF	RfD
Hyperpigmentation, keratosis and possible vascular complications	NOAEL[3]: 0.009 mg L^{-1} converted to 0.0008 mg kg^{-1} day^{-1}	3	1	3E-4
Human chronic oral exposure	LOAEL: 0.17 mg L-1 converted to 0.014 mg kg^{-1} day^{-1}			
Tseng, 1977; Tseng *et al.*, 1968				

[1]US EPA, 1990.
[2]Conversion factors: NOAEL was based on an arithmetic mean of 0.009 mg L^{-1} in a range of arsenic concentration of 0.001 to 0.017 mg L^{-1}. This NOAEL also included estimation of arsenic from food. Since experimental data were missing, arsenic concentrations in sweet potatoes and rice were estimated as 0.002 mg day^{-1}. Other assumptions included consumption of 4.5 L water day^{-1} and 55 kg bw (Abernathy *et al.*, 1989). NOAEL = [(0.009 mgL^{-1} x 4.5 L day^{-1}) + 0.002 mg day^{-1}] / 55 kg = 0.0008 mg kg^{-1} day^{-1}. The LOAEL dose was estimated using the same assumptions as the NOAEL starting with an arithmetic mean water concentration from Tseng (1977) of 0.17 mg L^{-1}. LOAEL = [(0.17 mg L^{-1} x 4.5 L day^{-1}) + 0.002 mg day^{-1}] / 55 kg = 0.014 mg kg^{-1} day^{-1}.
[3]Abbrev. = NOAEL, no-observed-adverse-effect level; LOAEL, lowest-observed-adverse-effect level; UF, uncertainty factor; MF, modifying factor; RfD, Reference Dose.
[4]The UF of 3 is to account for both the lack of data to preclude reproductive toxicity as a critical effect and to account for some uncertainty in whether the NOAEL of the critical study accounts for all sensitive individuals.

significant. However, the small numbers of individuals did not permit adequate statistical evaluation. Accordingly, this report did not supply sufficient data for RfD derivation.

The epidemiological study selected as the principal study for RfD derivation involved a survey of over 40,000 Taiwan residents (Tseng *et al.*, 1968; Tseng, 1977). This population had various signs of As exposure including hypo- and hyperpigmentation and keratosis and Blackfoot disease (BFD). The BFD, a peripheral vascular disorder causing gangrene of the feet and hands, was not used in the RfD derivation, since Lu (1990) has recently reported that the humic acid, present in the well water of the Taiwanese, could cause gangrene in mice. Accordingly, the critical effects for the RfD derivation were hyperpigmentation, keratosis and possible vascular problems. From these data, a NOAEL of 0.8 μg kg^{-1} day^{-1} and a LOAEL of 14 μg kg^{-1} day^{-1} were calculated for As exposure (Abernathy *et al.*, 1989). The NOAEL of 0.8 μg kg^{-1} day^{-1} was divided by an UF of 3 to obtain an RfD of 0.0003 ($3E^{-4}$) mg kg^{-1} day^{-1} for As (Table I). The RfD Work Group could not reach a consensus on an appropriate UF for As and the matter was sent to the Risk Assessment Council (RAC) of the EPA. The RAC decided on an UF of 3, but stated that any value within the range of 0.0001 to 0.0008 mg kg^{-1} day^{-1} could be used for the RfD, depending on the scientific justification.

Several criticisms of the use of the Taiwanese data for RfD derivation have been listed. The primary one involves whether the data of the Taiwan group are applicable to the US population. There may be nutritional and genetic differences between the two populations (US EPA, 1988) that could affect response to As. Marcus and Rispin (1988) proposed that there is a threshold for As effects depending upon methylation rates, but a recent report (Hopenhayn-Rich *et al.*, 1993) does not support this hypothesis.

Another possible issue in the risk assessment of As is its potential beneficial (essentiality issue) effects (NRC, 1989). The NRC report stated that As may be

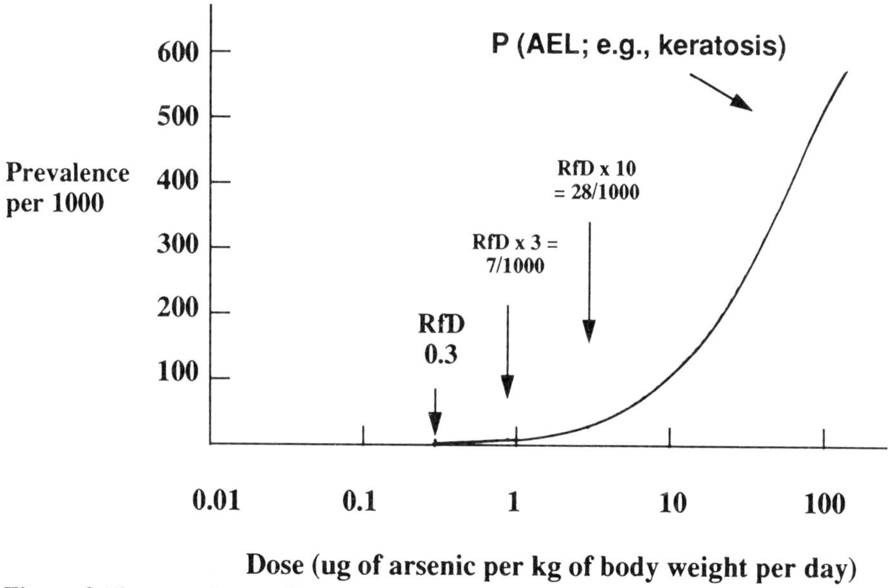

Figure 2 *The prevalence of keratosis/1,000 people, at 70 years of age, versus doses at intervals above the RfD. The 7 and 28 above the X axis represent prevalence of keratosis at 3 and 10 times the RfD, respectively. The values were estimated from Tseng* et al. *(1968) and Tseng (1977) by Farland and Dourson (1992).*

essential in animals, but presently there is inadequate evidence of human essentiality (US EPA, 1988; NRC, 1989) and As is not considered to be an essential trace element in human beings. A recent report (Mayer *et al.*, 1993) has described a correlation between lowered serum As levels in hemodialysis patients and increased risk of CNS injury, cancer and vascular diseases. These data need to be confirmed and extended by other studies. However, the potential beneficial effects were not used in the RfD derivation and will not be considered until As is accepted as an essential trace element in humans. On the basis of current knowledge, the RfD/RfC Work Group decided to utilize the Taiwan data to derive the As RfD, as shown in Table 1.

Risk above the RfD

The EPA has been examining risks above the RfD by using a categorical regression model (Hertzberg, 1989). To illustrate this type of analysis, several toxicological responses for As (*e.g.*, hyperpigmentation, keratosis and Blackfoot disease; Tseng *et al.*, 1968; Tseng, 1977) were translated into four categories of progressive no-effect to effect levels: no effects, non-adverse effects, mild-to-moderate adverse effects, and severe or lethal effects, as defined by Farland and Dourson (1992). These categories were regressed on the logarithm of dose and duration; both parameters were statistically significant indicating that the effects of As are both time- and dose-dependent.

As discussed by Farland and Dourson (1992), some difficulty existed with the interpretation of the Tseng (1977) and Tseng *et al.* (1968) toxicity data. For example, Tseng *et al.* (1968) did not show all necessary data for categorical regression; thus, some data were estimated by Farland and Dourson (1992). Likewise, Tseng and colleagues maintained that Blackfoot disease was attributable to arsenic exposure, but a recent publication disputes this (Lu, 1990). However, Farland and Dourson (1992) used that data set only to illustrate the concept of categorical regression as applied to an RfD derivation. Definitive conclusions on the toxicity of arsenic at levels above the RfD should not be drawn from this example as more analyses need to be done. With this in mind, Figure 2 shows the estimated probability of keratosis after 70 years of exposure. The model shows an increasing probability of adverse effects with dose at doses in excess of the verified RfD. For example, at 3- and 10-times the RfD, the risks of adverse effects are estimated to be 7 and 28 per 1,000, respectively.

References

Abernathy, C.O., Marcus, W., Chen, C., Gibb, H. and White, P. 1989. Office of Drinking Water and Office of Research and Development, US EPA. Memorandum to Cook, P., Office of Drinking Water, and Preuss, P., Office of Regulatory Support and Scientific Management, US EPA. Report on Arsenic (As) Work Group Meetings. February 23, 1989.

Abernathy, C.O., and Ohanian, E.V. 1992. Noncarcinogenic effects of inorganic arsenic. *Environ. Hlth. Geochem.*, **14**, 35-41.

Abernathy, C.O., Cantilli, R., Du, J.T. and Levander, O.A. 1993. Essentiality versus

toxicity: Some considerations in the risk assessment of essential trace elements. *Haz. Assess. Chem.*, **8**, 81-113.
Abernathy, C.O. 1994. Inorganic arsenic: An overview. *Environ. Hlth. Geochem.* (in press)
Barnes, D.G., and Dourson, M. 1988. Reference Dose (RfD): Description and use in health risk assessments. *Reg. Toxicol. Pharmacol.*, **8**, 471-486.
Buchanan, W.D. 1962. *Toxicity of Arsenic Compounds.* Elsevier Publ. Co., Amsterdam.
Cebrian, M.E., Albores, A., Aguilar, M. and Blakely, E. 1983. Chronic arsenic poisoning in the north of Mexico. *Human Toxicol.*, **2**, 121-133.
Farland, W. and Dourson, M.L. 1992. Noncancer health endpoints: Approaches to quantitative risk assessment. In: Cothern, R. (ed.), *Comparative Environmental Risk Assessment*, pp. 87-106. Lewis Publishers, Boca Raton, FL.
Fielder, R.J., Dale, E.A. and Williams, S.D. 1986. *Toxicity Review. 16. Inorganic Arsenic Compounds.* HMSO Publications, London.
Fowler, B.A. 1977. Toxicology of environmental arsenic. *Adv. Modern Toxicol.*, **2**, 79-122.
Hamamoto, E. 1955. Infant arsenic posioning by powdered milk. *Jpn. Med. J.*, 1649, 3-12.
Hindmarsh, J.T., McLetchie, O.R., Heffernan, L.P.M., Hayne, O.A., Eilenberger, H.A., McCurdy, R.F. and Thiobeaux, H.J. 1977. Electromyographic abnormalities in chronic environmental arsenicalism. *J. Anal. Toxicol.*, **1**, 270-276.
Hertzberg, R.C. 1989. Fitting a model to categorical response data with application to special extrapolation to toxicity. *Health Phys.*, **57**(Suppl 1), 404-409.
Jarabek, A.M., Menache, M.G., Overton, J.H., Dourson, M.L. and Miller, F.J. 1990. The U.S. Environmental Protection Agency's inhalation RfD methodology. Risk assessment for air toxics. *Toxicol. Indust. Hlth.*, **6**, 279-301.
Hopenhayn-Rich, C., Smith, A.H. and Goeden, H.M. 1993. Human studies do not support the methylation threshold hypothesis for the toxicity of inorganic arsenic. *Environ. Res.*, **60**, 161-171.
Lehman, A.J. 1951. Chemicals in foods. A report to the Association of Food and Drug Officials on current developments. Part II. Pesticides. *Quar. Bull. Assoc. Food Drug Off.*, **15**, 122-133.
Lu, F.J. 1990. Blackfoot disease: Arsenic or humic acid? *Lancet*, **336**, 115-116.
Marcus, W.L. and Rispin, A.S. 1988. Threshold carcinogenicity using arsenic as an example. In Cothern, C.R., Mehlman, M.A. and Marcus, W.L. (eds.), *Risk Assessment and Risk Management of Industrial and Environmental Chemicals*, pp. 133-158. Princeton Scientific Publ. Co., Princeton.
Mayer, D.R., Kosmus, W., Pogglitsch, H., Mayer, D. and Beyer, W. 1993. Essential trace elements in humans. Serum arsenic concentrations in hemodialysis patients in comparison to healthy controls. *Biol. Trace Elem. Res.*, **37**, 27-38.
Mizuta, N., Mizuta, M., Ito, F., Ito, T., Uchida, H., Watanabe, Y., Akama, H., Murakame, T., Hayashi, F., Nakamura, K., Yamaguchi, T., Miizuia, W., Oishi S. and Matsumura, H. 1956. An outbreak of acute arsenic poisoning caused by arsenic contaminated soy-sauce (shoya): A clinical report of 220 caces (sic). *Bull. Yamaguchi Med. Sch.*, **4**, 131-149.
NRC. National Research Council. 1989. Recommended Dietary Allowances, 10th ed.*Report of the Food and Nutrition Board.* National Academy of Sciences, National Academy Press, Washington.
Southwick, J.W., Western, A.E. and Beck, M.M. 1983. An epidemiological study of arsenic in drinking water in Millard County, Utah. In Lederer, W.H. and Fernsterheim, R.J. (eds.), *Arsenic: Industrial, Biomedical, Environmental Perspectives*, pp. 210-215. Van Nostrand-Reinhold, New York.

Tay, CH., and Seah, C.S. 1975. Arsenic poisoning from antiasthmatic herbal preparations. *Med. J. Austr.*, **2**, 424-428.

Tseng, W.P., Chu, H.M., How, S.W., Fong, J.M., Lin, C.S. and Yeh, S. 1968. Prevalence of skin cancer in an endemic area of chronic arsenicism in Taiwan. *J. Natl. Cancer Inst.*, **40**, 453-463.

Tseng, W.P. 1977. Effects of dose-response relationships on skin cancer and blackfoot disease with arsenic. *Environ. Hlth. Perspect.*, **19**, 109-119.

US EPA. 1988. *Special Report on Ingested Inorganic Arsenic: Skin Cancer; Nutritional Essentiality.* EPA/625/3-87/013. US EPA, Washington.

US EPA. 1990. Integrated Risk Information System (IRIS). Reference Dose for Oral Exposure for Inorganic Arsenic. Online. (Verification date 11/15/90). Office of Health and Environmental Assessment, Environmental Criteria and Assessment Office, Cincinnati, OH.

US EPA. 1993. Integrated Risk Information System (IRIS). Online. Office of Health and Environmental Assessment, Environmental Criteria and Assessment Office, Cincinnati, OH.

Valentine, J.L., Campion, D.S., Schluchter, M.D. and Massey, F.J. 1984. Arsenic effects on human nerve conduction. In: Howell, J. McC., Gawthorne, J.M. and White, C.L. (eds.), *Trace Element Metabolism in Man and Animals,* pp. 409-412. Austr. Acad. Sci., Canberra.

Valentine, J.L., Reisbord, L.S., Kang, H.K. and Schluchter, M.D. 1985. Arsenic effects on population health histories. In: Milk, C.F., Bremmer, I. and Chesters, J.K. (eds.), *Trace Element Metabolism in Man and Animals.* pp. 289-292, Commonwealth Agricultural Bureau, Aberdeen.

Vallee, B.L., Ulmer, D.D. and Wacker, W.E.C. 1960. Arsenic toxicology and biochemistry. *AMA Arch. Ind. Hlth.*, **21**, 56-75.

31 Arsenic and human health:some persisting scientific issues

Paul Mushak

Principal, PB Associates and (adj.) Pathology Department, University of North Carolina-Chapel Hill,Suite G-3, 714 Ninth Street, Durham, NC 27705, USA

Abstract

Over the last several decades, a large volume of scientific information has accumulated on the topic of arsenic (As) and environmental health. Various scientific questions about this topic have arisen and require attention. Concerns and questions include (1) the matter of form-specific As toxicity and newly-identified adverse effects in humans and experimental animals; (2) the nature of the As dose-response relationship for skin and internal cancers at low doses; (3) laboratory measurement of very low levels of total As, measurement of form-specific As, and the question of what should be measured to best reflect bioactive dose; (4) matrix-specific As bioavailability as a factor in human exposure; (5) interactive relationships of As with other bioactive substances; (6) adverse effects of As versus potential nutritional essentiality in humans; (7) relevance of the principal epidemiological data base for human As carcinogenesis to US populations in regions with elevated As exposures from, e.g., *drinking water; (8) importance of the methyl connection –* in vivo *biomethylation and its interrelationships to attenuation of toxicity, to an essentiality role, and to carcinogenesis. The status of these issues is briefly summarised.*

Introduction

A large volume of scientific information exists for the topic of environmental As, particularly with reference to As exposure, toxicity, and health risk assessment. However, various technical aspects of this topic have produced a number of questions which require close attention. This overview identifies and briefly summarises the status of some of these issues. The discussion is organised by headings that reflect the components of risk assessment as set forth by NAS/NRC (1983). These key issues raise some key questions.

Issues in health hazard identification

Form-variable as toxicity

Does As toxicity in humans vary with the element's chemical form? Arsenic toxicity has been conclusively shown to differ with chemical form. The most toxic forms commonly encountered environmentally are the tri- and pentavalent oxyarsenicals (US EPA, 1984a, 1988; WHO, 1981; WHO, 1987).

The trivalent form of inorganic As has generally been considered to be more toxic than As in the pentavalent state. The validity of this comparison, however, is a function of the dosing particulars in the various experimental studies. In acute and subacute poisonings, differential toxicity is detectable. With lower doses administered chronically, the distinction diminishes. Given current knowledge of the *in vivo* behaviour of different As forms, any differences in toxicity of tri- *vs.* pentavalent As is heavily influenced by the relative rates of both systemic/cellular uptake of As(V) and its subsequent reduction to As (III) compared to rates of direct interaction of As(V) with biochemical sites. Most importantly, the phenomenon of *in vivo* reduction in humans establishes that both As valency forms are of concern for As-associated skin and internal cancers (US EPA, 1984a, 1988).

At the other end of the toxicity spectrum, organoarsenicals occurring in seafood appear to be relatively nontoxic and undergo rapid excretion with little or no biotransformation (US EPA, 1984a).

Questions of a causal role for As in certain toxic endpoints

Certain As-associated toxic endpoints remain in dispute with reference to arsenic as exclusive cause. Taiwanese Blackfoot Disease (BFD) is a peripheral artery disease which has progressed to an arteriosclerotic, 'dry' gangrene of the extremities (Chen *et al.*, 1988). BFD has also been claimed to be due to or exacerbated by exposure to other well water contaminants, *e.g.*, humic acid, ergot alkaloids, *etc.* (Lu *et al.*, 1990). Chen (1990) has evaluated those competing claims and argued against a role for humic acid. For example, Chen drew attention to the difference between vasculopathy induced by such substances as humic acid and the histopathological features of BFD. Overall, evidence that As appears to be the main etiological agent in BFD includes the following:

(1) As shows a significant dose-response relationship with BFD (Chen *et al.*, 1988);
(2) Vasculopathy related to BFD has also been reported with As exposures in Chile, Mexico, *etc.*, where the only exposure constant was As (US EPA, 1984a; WHO, 1981);
(3) Humic acid-induced gangrene really does not show the histopathological characteristics of BFD (Chen, 1990);
(4) An *in vitro* cytotoxicity model shows relevant effects of As in venous endothelium (Chen *et al.*, 1990).

Reports of new adverse health impacts of As

A number of reports have now documented that Taiwanese having long-term well water As exposure in an area of endemic arsenicalism show As-associated internal cancers – in bladder, kidney, liver, lung – in both a dose-related and age-related fashion (Mass, 1992; Smith *et al.*, 1992; Chen *et al.*, 1992). The affected organs, in the main, are those associated with As metabolism or excretion. Since these cancers carry mortality risks that are much higher than for skin cancer, the whole risk assessment issue for ingested As carcinogenesis has taken on much more importance for both public health and regulatory policy. At present, the question is not whether there are As-related internal cancer effects but whether the dose-internal cancer response curves are linear or non-linear at low doses.

As toxicity associated with mainly occupational exposures in emerging technology areas is of some concern. For example, As encountered in the manufacture of computer chips has been studied as to toxicity in experimental animals. Such data suggest there is pneumotoxic and immunotoxic potential via inhalation of very small particles of such substances as gallium (III) arsenide (Burns *et al.,* 1991; Webb *et al.,* 1987).

Who is at risk for As toxicity or exposure?

A third issue for As toxicology and epidemiology is the critical question of who is at higher risk for As, either As exposure or toxicity or both.

A number of studies over the years have shown that young children are at higher risk from As exposure as measured by an exposure biomarker, As in urine, or measurement of intake rates on a unit measurement basis (Jensen *et al.,* 1991; Polissar *et al.,* 1990; CDC, 1987; US EPA, 1984a). This occurs because of either ingestion of dusts and soils through early childhood mouthing activity or higher intakes of water on a unit weight basis.

Workers exposed to As in such occupational settings as smelting operations are at highest risk for As-associated lung cancer. Older adults who have had many years or even lifelong elevated As exposure via medical preparations or environmental media are at highest risk for skin and internal cancers (US EPA, 1984a, 1988; Bates *et al.,* 1992; Cuzick *et al.,* 1992; Mass, 1992; Chen *et al.,* 1992).

It is not clear whether high As exposures over a small age band, particularly for children in the preschool period, translate to lifelong increased non-cancer and cancer risks. Are children intrinsically at higher risk for toxic effects relative to adults via more vulnerable developmental biological processes? This question remains open but limited data suggest this as a possibility. Japanese infants accidentally exposed for several weeks to high levels of As in contaminated formula had persisting damage to the central nervous system (US EPA, 1984a).

Dose-response relationships for As

In general, the various carcinogenic and non-carcinogenic adverse effects of As in human populations follow a positive dose-response relationship, at some actually measured range of doses and rates of adverse outcome. This is particularly the case for BFD (US EPA, 1984a, 1988), skin cancer (Tseng *et al.,* 1968; Tseng, 1977; US EPA, 1984a, 1988) and internal cancers (Smith *et al.,* 1992; Chen *et al.,* 1992) as reported for Taiwanese subjects with long-term oral exposures to inorganic As or for patients treated therapeutically with Fowler's Solution containing arsenite (Cuzick *et al.,* 1992; US EPA, 1984a).

The main uncertainty with As dose-response relationships is not their existence, *per se*, but rather whether the relationship remains linear down to very low dose. This question is discussed in the section on risk characterization.

Issues in arsenic exposure assessment

Two areas under the heading of arsenic exposure pose concerns. These include: (1) what should be quantitatively measured and how well can such measurements be done; and (2) dependence of As bioavailability in humans on the chemical and physical nature of the As-containing matrix.

Laboratory measurements and As exposure

Hundreds of papers now exist in the analytical chemical and analytical toxicological literature on the subject of As analysis in environmental and human physiological media (see Irgolic, chapter 5). These methods have been the subject of various critical reviews over the years. The interested reader is referred to the recent comprehensive reviews of Irgolic (1992) and Marita and Edmonds (1992), which collectively cover both total As and As speciation analytical techniques.

Most of the As measurement methods appearing in the literature over the years have dealt with measurements of total As in diverse media. Many of these total element techniques have been limited with respect to such parameters as required sensitivity, applicability to diverse media, and adequate proficiency with respect to accuracy and precision.

Total As methods are still widely used for determining concentrations of As in environmental and workplace media, for identification of sources of potential As exposure in environmental monitoring or occupational hygiene compliance or as part of environmental regulatory actions. Common methods for total As analysis include colorimetric, atomic spectroscopic and electrochemical techniques.

A critical matter at the moment in total As analysis is available methodological capability and laboratory proficiency in determining As at very low levels in drinking water, particularly if a regulatory standard in a range of, say, 0.015-0.15 μmol (1-10 μg) L^{-1} were to be promulgated in future years. Routine determination of compliance at these lower levels by rank-and-file water agencies may be a challenge and quality assurance/quality control (QA/QC) protocols for such measurements will have to be in place (see below).

These is increasing interest in, and wider use being made of, laboratory techniques that quantify specific chemical species of As. Growing interest and use of speciation techniques is due to the recognition that different forms of As have different toxicities. Consequently, ingestion of high amounts of arsenic forms of very low toxicity may be of less concern for public health than intake of lesser amounts of very toxic forms. In addition, As in humans and other mammals undergoes biotransformation to yield metabolites that characterise certain exposure forms. Finally, it may also be necessary to characterize chemical species in various environmental media as part of any attempt to ascertain As bioavailability in humans exposed to certain media.

Arsenic speciation methods as applied to physiological samples from exposed persons or certain As source media, *e.g.*, diet and drinking water, generally consist of two major components: a separation component coupled with an As-specific detection method for the separated species. Separation is commonly achieved by either hydride generation and purge removal or any of a number of chromatographic methods. Detection systems include mass spectrometry (MS) in the inductively coupled plasma-MS (ICP-MS) configuration, flameless atomic absorption spectrometry (GF-AAS), and inductively coupled plasma- atomic emission spectrometry (ICP-AES).

There are a number of concerns with use of speciation methodology. First, there is the general absence of a rigorously standardized method or set of methods to reliably quantitate speciated monomethyl, dimethyl and inorganic As in physiological and environmental media, the three forms of growing interest in terms of human health risk assessment. By and large, speciation techniques as part of epidemiological investigations have generally been those developed by the investigators themselves or those used by others without a full appreciation of how accurate and precise the techniques may be.

Secondly, there is no adequate national or international QA/QC framework for evaluating laboratory proficiency with speciation methods as part of human exposure monitoring. QA/QC protocols for any type of As measurement, in fact, appear to be quite limited. The few available reports of interlaboratory proficiency testing outcomes appear to be confined to clinical toxicology laboratories analysing total As in media at concentrations mainly of interest in forensic analyses (de Zeeuw *et al.*, 1987). In one European Community round robin effort for As in urine involving laboratories in 10 countries, the two concentrations of As used were 2.7 μmol (200 μg) L^{-1} and 8.0 μmol (600 μg) L^{-1} (de Zeeuw *et al.*, 1987).

There appears to be a growing need to develop and promote one or more standardised methods for form-specific quantitative analysis, in support of further epidemiological studies of arsenic exposure and human health impacts. Standardized methodology in turn is best employed within a QA/QC framework for long-term proficiency assessment.

As bioavailability in humans

Arsenic bioavailability in toxicology can be defined as the fraction of As absorbed from receiving body compartments (lung, gut) into the bloodstream and the rate at which such uptake occurs (Mushak, 1991). A variety of factors influence bioavailability and these are biological or physicochemical in nature (Mushak, 1991).

Bioavailability of As is an important element in human health risk assessment and figures as an important factor in regulatory and exposure control actions that may arise from such assessments. We are particularly concerned here with the extent to which the environmental matrix influences bioavailability.

Soluble inorganic As in drinking water, dietary components, medical preparations, home remedies, and particulate As oxides in, *e.g.*, atmospheric samples are generally assumed to be highly bioavailable (US EPA, 1984a, 1988; Ontario Ministry of the Environment, 1986; WHO, 1981). Similarly, the complex organoarsenicals found in seafoods are extensively absorbed (US EPA, 1984a).

Bioavailability of As in other environmental media has not been well studied. In particular, bioavailability of arsenic in dusts and soils is little understood but has drawn increasing attention owing to growing concerns at sites where As-containing wastes pose a risk of environmental contamination. This would include communities affected by sites where mining, milling and/or smelting of As-laced ores have been carried out and sites impacted by such energy production-related waste materials as As-containing fly ash.

It is not very feasible to experimentally study As bioavailability in humans, especially in young children. The question of As bioavailability in humans has mainly been addressed via inferential statistical analyses in epidemiological studies or by use of *in vitro* and experimental animal *in vivo* approaches. *In vitro* studies of As bioavailability are necessarily poor surrogates for what occurs in the human gastrointestinal tract, given the complexity of the GI tract and the relatively crude limits to the *in vitro* studies, *e.g.*, simple solubility parameters (Mushak, 1991).

There are three current concerns regarding As bioavailability in energy-based, high-volume combustion wastes, *e.g.*, coal-fired power plant fly ash, and extractive industry-based wastes, *e.g,* mill tailing piles. First, the question of an appropriate animal model for human As bioavailability from the above types of media, particularly extractive industry waste, remains unresolved (Weis and Lavelle, 1991). Secondly, energy-derived and extractive industry-related waste media are rather heterogeneous, and one might expect that bioavailability would not be uniform when comparing dusts and soils potentially contaminated with fly ash, smelter emission fallout, mill tailings, mine overburden, *etc.* Finally, dosing protocols in experimental bioavailability studies should simulate human exposures with respect to both administered amounts of media and dosing schedule.

A rather large amount of material is appearing in the more current literature on the topic of contaminant behavior in fly ash and is beyond the scope of this overview. Fly ash and other combustion-based As sources are discussed elsewhere in this volume.

Several studies have recently appeared on the topic of As bioavailability in soils impacted by mining (Davis *et al.*, 1992) and smelting activities (Freeman *et al.*, 1993). Arsenic in soils contaminated by mining waste was reported to have 11% total solubilisation in the small intestine of rabbits (Davis *et al.*, 1992). Arsenic in soils impacted by smelter emission fallout was reported to have 25% absolute bioavailability in rabbits (Freeman *et al.*, 1993). In both of these studies, the quantities of administered soil matrix were relatively large, overwhelming gastrointestinal activity. Extrapolated to children's body weight, the corresponding intakes by children would be of the order of a gram or more, compared to typical, non-pica children's soil ingestion rates of about 100-200 mg/day. Bioavailability in rabbits versus a more appropriate species, *e.g.*, non-human primates, requires more study.

Issues in As risk characterization

There are five areas of concern with respect to As and human health risk characterisation: (1) the nature of the dose-response relationship for As at low doses and skin or internal cancers; (2) potential limitations in use of the large Taiwanese well water As data set for estimating cancer risk in the United States; (3) the potential alteration of risk via interactions with other bioactive substances; (4) the interplay of potential micronutrient essentiality of As vs. its toxicity depending on dose; and (5) the multifaceted role of *in vivo* As biomethylation in risk assessment.

Dose-response relationships for As at low dose and cancer: models and mechanisms

A major question for cancer risk characterization of ingested As using data mainly gathered at relatively high As concentrations is whether the overall dose-response relationship remains linear down to zero dose or not. This is of particular concern when dealing with internal cancers with their associated higher mortality rates compared to skin cancer.

Smith *et al.*, (1992), using linear regression analyses for age-adjusted mortality rates for bladder, kidney, liver, and lung cancer from the cumulative Taiwan data set, reported that US subjects consuming one liter water daily with an As content of 0.7 μmol(50 μg) L^{-1} would have a combined lifetime risk of dying from cancer in any of the four organs of 13/1000, a rather high rate. Chen *et al.* (1992) have also reported significant rates (see below) for internal cancers using a potency index based on a dose rate of 10 μg kg^{-1} b.w./d. Brown (see this volume) has questioned whether in fact there is low-dose linearity with the dose-response curve starting upward at the origin, or whether the curve does so at some measurable concentration.

Brown *et al.* (1989) modeled the Taiwanese skin cancer data (Tseng *et al.*, 1968, 1977). Four exposure durations and three dose intervals were employed. Risk increases as a power of three in age and is linear (3×10^{-3}) or quadratic (1.3×10^{-3})in As dose, with a linear coefficient. The authors note that the Taiwanese data-based risk estimates are consistent with data reported for water As effects in Northern Mexico (Cebrian *et al.*, 1983).

Guo (see this volume) carried out an ecological study in the arsenicalism-endemic area of Taiwan with reference to the dose-response relationship for As-associated renal cancers. They found that the association between well water As and urinary cancer incidence could only be observed at very high concentrations, suggesting a non-linear dose-response relationship.

Overall, there is some question as to whether the relationship of As to at least certain internal cancers is linear or not at low to moderate As ingestion rates. Statistical modelling and fine tuning can provide only some of the answers.

Elucidation of the biological mechanism(s) by which As behaves as a skin and internal organ carcinogen would of course help considerably in subsequently determining whether the relationship at low dose is linear or not. Much effort has been expended in studies of As carcinogenesis mechanisms, but questions still remain (see, *e.g.*, Mass, 1992). Complicating such investigations is the widely-known absence of a good animal model for As-associated human cancers. At present, there is considerable evidence from experimental and some epidemiological data to suggest As behaves as a late-stage carcinogen (Mass, 1992; Mazumdar *et al.*, 1989;

Barrett *et al.,* 1989; Brown and Chu, 1986) although isolated information points to arsenic as a co-carcinogen.

Finally, are pre-skin cancer lesions such as hyperkeratoses a histopathological pre-condition for skin cancer? That is, does a threshold in As intake for detectable hyperkeratoses define a threshold for skin cancer? Hyperkeratoses do show a dose-response relationship with As dose and were encountered at significant incidences among the As-exposed Taiwanese as well as among patients treated with Fowler's solution (US EPA, 1984a, 1988). According to Shannon and Strayer (1989), however, a fraction of invasive As-associated squamous cell carcinomas can arise *de novo, i.e.*, need not be rooted in hyperkeratotic lesions and therefore their incidence would not be totally affected by absence or presence of a threshold for hyperkeratosis.

Is there limited applicability of Taiwanese As dose-response data elsewhere?

US EPA's Risk Assessment Forum (US EPA, 1988), US EPA's Science Advisory Board (SAB, 1989) and North (1992) have identified some concerns about the applicability of the cumulative cancer outcomes in the Taiwanese data set to other areas, particularly for western areas of the United States where levels of As in water supplies are moderately elevated relative to most of the nation.

These concerns specifically include such issues as the relative effectiveness of As biomethylation at the As intake levels experienced by the Taiwanese, the impact of likely nutritional deficiencies pervasive in this population on amplification of cancer incidence, adequate accounting of all potential sources of As intake in addition to well water intakes, the relatively higher amounts of As in drinking water in that area of Taiwan compared to most US water supplies with elevated As, and the matter of the background rates of any types of skin cancers in the Taiwanese versus any US Caucasian or mixed exposure populations.

Short of urinary As speciation analysis and comparative assessment of fractional distribution in Taiwanese during the early data gathering, steps which were not done, it would be difficult to define the role of altered biomethylation in the observed carcinogenic responses to their water As consumption. It has been claimed that the relative amounts of inorganic versus methylated species in urine of adult volunteers increases above a daily intake of about 500 µg (Buchet *et al.,* 1981). Evidence for this is quite limited, however, and is based on a single subject exposed for only five days. Furthermore, the percent difference in level of urinary inorganic As was still small when compared to the lower doses. Another criticism of this view is that it does not consider adaptive mechanisms for biomethylation and perhaps excretion at chronically high doses. Vahter and Marafante (1988) have noted the phenomenon of tolerance to As over time in the arsenic eaters of Styria, Austria, and decreased animal tissue levels of As over time with continued dosing.

By contrast, evidence for preservation of a relatively fixed biomethylation efficiency (limited variance in urinary As fractional speciation) over a very broad range of short-term and chronic As exposures is considerable and the human data for occupational and environmental exposures have been critically reviewed (Hopenhayn-Rich *et al.,* 1993).

Any marked nutritional deficiencies in the Taiwanese, particularly for methionine, could affect biomethylation in these subjects and potentiate carcinogenic responses via increased fractions of circulating inorganic As. Experimental data show that dietary restrictions of sources of methyl groups will decrease methylation (Vahter and Marafante, 1988).

With reference to baseline skin cancer rates, sunlight-induced skin cancer in Taiwan is very rare while there are about 500,000 new cases of non-melanoma skin cancer each year in the US. The potential for this high US background to conceal As-associated cancers makes geographic and other comparisons problematic (Mass, 1992). An additional factor for complications in US-Taiwan comparisons are small study population sizes in exposed US communities, which do not provide the necessary statistical power to see associated cancers (Andelman and Barnett, 1986; US EPA, 1988).

Were the Taiwanese exposed to cumulative amounts of As beyond what would be estimated directly from well water concentrations and estimates of volumes that were consumed? There is little evidence to reliably quantify high dose estimates but other sources such as diet components using water for preparation could contribute.

Interactions of As

Arsenic uptake, systemic distribution and toxic effects are affected by the presence or absence of other bioactive substances, particularly nutrients. Available data are mainly experimental in nature.

Various experimental studies document a mainly reciprocal, antagonistic relationship between arsenic and selenium (US EPA, 1984a; Mushak, 1983; Hoffman *et al.*, 1992), and the As-Se interaction is the most widely known. Interactions of As with other substances are less well known. Isolated reports have appeared of the antagonistic effect of protein (Hoffman *et al.*, 1992) and cadmium (Yanez *et al.*, 1991).

Are there documented interactions of As with nutrients that have demonstrable health consequences in human populations? This question cannot be readily answered at present. In Taiwanese showing As vasculopathy, BFD may have a potentiation of effect, particularly by other substances. The incidence of skin and internal cancers among these Taiwanese may have been affected by generally poor nutritional status. Interactions of ingested soil/dust mineral constituents with As may be a factor in As bioavailability from certain wastes.

As toxicity vs. essentiality

Considerable experimental evidence exists to show that As has a nutritional function in at least five test species: rats, hamsters, goats, miniature pigs, and chicks (US EPA, 1988; Uthus, 1992; Nielsen, 1991). Depressed growth and reproductive dysfunctions in the form of reduced fertility and increased perinatal mortality appear to be consistent signs of deficient As intake. The nutritional role of As in these species appears to involve methionine metabolism (alterations in labile methyl, taurine, and polyamine levels). Arsenic deprivation effects resemble those of pyridoxine deficiency and are most pronounced with methionine deficiency or guanidoacetic acid supplementation.

As essentiality has not been demonstrated in humans to date. What, then, is the significance of the experimental nutritional data? It is quite difficult to produce trace element deficiencies in humans, especially where any specific effects of the element's deficiency state in humans are unknown and where the element's presence in the environment is pervasive. It may be that a role for As in human nutrition is only expressed under conditions of manifest physiological stress, rather than in unstressed states.

Of the various trace elements that appear to be both essential and highly toxic, depending on the level of intake, As appears to be unique in that the same forms of As which show experimental essentiality are also implicated in As carcinogenesis, *i.e.*, the tri- and pentavalent oxyarsenical ions. By contrast, ingested soluble chromium (III) at low intakes is a non- carcinogenic, nutritionally essential form of the element (potentiates insulin action in peripheral tissue) but the inhaled, insoluble hexavalent chromate salts are associated with lung cancer in certain occupational settings (US EPA, 1984b; WHO, 1988). Oxyselenium salts are either essential or toxic, depending on dose, but have no carcinogenic potential. Therefore, selenium does not raise the conflicting issues of non-threshold, low-dose carcinogenic response linearity *vs.* nutrient intake requirements.

This uniquely dual character of inorganic arsenic ions raises obvious questions:

(1) Is the daily intake requirement for a potential nutritional role of As in humans also a de-facto threshold for any of its adverse health effects, including skin and internal cancers, or

(2) Is it biologically plausible that low levels of As intake potentially required for some physiologically necessary functions, *i.e.*, with a 100% probability of essentiality across all members of a species, would simultaneously produce some estimable but relatively low cancer risk rates in certain body compartments, or

(3) Do these potentially contradictory roles for As at low dose suggest As is not essential in humans, that estimates of human intake requirements are too high or that cancer risk estimates at low dose using certain risk models are too high?

(4) Do the likely mechanisms for arsenic as a potential nutrient versus its carcinogenic behavior provide any insight to this puzzle?

(5) If essentiality and cancer risk data are in conflict, does a resolution of conflicting interpretations require invoking relative weight-of-the-evidence? That is, is the evidence for low-dose linearity for carcinogenicity dose-response weaker or stronger than the evidence for As functioning as an essential micronutrient?

Several estimates of the amount of daily required intake for arsenic as a putative nutrient have been made. A more current figure of 12 μg As day^{-1} for humans has been estimated (Nielsen, 1991; Uthus, 1992). For a typical adult male of 70 kg body weight (b.w.), this is equal to 0.17 μg As kg^{-1} b.w.day^{-1}.

By contrast, what do other data suggest are the adverse impacts, particularly cancer, at this level of intake? Chen *et al.* (1992), using mortality rates in an area of Taiwan with chronic arsenicalism, estimated drinking water As cancer potency in Taiwanese males at an intake of 10 μg As kg^{-1} day^{-1} for various organs: liver, 4.3×10^{-3}; lung, 1.2×10^{-2}; bladder, 1.2×10^{-2}; kidney, 4.2×10^{-3}. The corresponding linearised risk estimates calculated from these Chen *et al.*, data at the estimated putative nutrient requirement level (0.17 μg kg^{-1} b.w. day^{-1}) are: 7.2×10^{-5}, 2×10^{-4}, 2×10^{-4}, 7×10^{-5}. Smith *et al.* (1992) estimated that at the current US EPA drinking water As standard of 0.70 μmol(50 g)As L^{-1}, individuals consuming one liter would have a combined lifetime risk of dying from liver, kidney, lung, or bladder cancer of 1.3×10^{-2}. At the putative nutrient requirement of 0.17 $\mu g^{-1} kg^{-1}$ day^{-1}, the Smith *et al.*, 50 μg As day^{-1} intake (equivalent to 0.7 μg kg^{-1} day^{-1} for a 70 kg adult male) yields a linearised corresponding lifetime risk of 3.3×10^{-3}.

Brown *et al.* (1989) reported skin cancer lifetime risk estimates for adult males based on the Taiwanese data of 3×10^{-3} (linear in dose) and 1.3×10^{-3} (quadratic in dose) for a drinking water As intake rate of 1 μg kg^{-1} day^{-1}. The corresponding linearised cancer risk estimates at a required nutrient value of 0.17 μg kg^{-1} day^{-1} are 5.0×10^{-4} and 2.2×10^{-4} respectively. It appears that the amount of As required to fill any putative nutrient role, *i.e.*, 100% probability of requirement if essential, is simultaneously associated with relatively moderate to rather high skin or internal cancer risks, risk levels which in most cases would not be palatable with what have been typical risk management cutoff guidelines.

One can also easily calculate the amounts of daily As intake rates associated with typical lifetime cancer risk estimates employed in risk management guidance, *i.e.*, permissible risk rates of 1×10^{-4} for skin cancer and 1×10^{-5} to 1×10^{-7} for internal cancers, and then compare these with the putative nutritional requirement intake for adult males of 0.17 μg kg^{-1} day^{-1}. The As intake rate for the skin cancer risk guideline is 0.08 μg kg^{-1} day^{-1}. For a risk cutoff for internal cancers of 10^{-5}, data of Chen *et al.* (1992) yield 0.008 μg kg^{-1} day^{-1} for bladder and lung, and 0.023 for kidney and liver. The corresponding value from Smith *et al.* (1992) is the lowest intake of all, 0.0005. The corresponding intakes using a 10-fold more restrictive cutoff of 10^{-6} would be 0.0008, 0.0023 and 0.00005 μg kg^{-1} day^{-1} respectively. It is obvious from all of the above calculations that the nutritional intake requirement rate is considerably higher than the As intakes linked to significant cancer risk estimates and there is conflict between the two sets of estimates. How does one use weight-of-the-evidence to resolve this conflict? While we cannot say with scientific certainty that cancer response at low As dose is linear, we also cannot conclusively document that As is nutritionally required in humans.

Methyl connection: detoxification, essentiality, and carcinogenesis relationships
The most mechanistically vexing aspect of the diverse biological activities of environmental As is centered in relationships of such activities to methyl group metabolism.

A number of questions arise with reference to any As-methyl connections. Is As detoxification, the widely accepted purpose of reduction-biomethylation, a biologically programmed role (Mushak, 1983)? Is biomethylation a side effect consequence of other As bioactivity, *e.g.*, As functioning in methionine metabolism? Or, does methylation-detoxification comprise part of a multiple effect on the dose-effect spectrum, *i.e.*, potentially adverse as well as beneficial? How does As function in labile methyl metabolism given that the methyl group forms a strong bond with arsenic?

Several reports have noted the relationship of methylation processes in the context of carcinogenic responses. Mass (1992) has suggested that one mechanism of As carcinogenesis may involve competition for Me groups that are also being allocated for As biomethylation, potentially reducing native DNA methylation patterns. For example, 5-azacytidine inhibits the methyl transferase required for cytosine methylation and is a carcinogen. Hypomethylation is closely associated with gene transcription changes. Mass (1992) also argues that a role for arsenic as a hypomethylator for gene methylation changes and amplification of hypomethylated genes are linked to how As behaves as a carcinogen. Seen in this light, As biomethylation may have multiple biological consequences.

Nielsen (1991) suggests there are linkages of a nutritional role for As to As carcinogenesis via polyamine synthesis or histone methylation. As affects polyamine function while polyamines are involved in controlling proliferation and growth.

References

Andelman, J.B. and Barnett, M. 1986. Feasibility study to relate arsenic to skin cancer in the United States. In: Kopfler, F.C. and Craun, G.F. (eds). *Environmental Epidemiology*, pp. 89-108.Lewis Publishers, Chelsea, MI.

Barrett, J.C., Lamb, P.W., Wang, T.C. and Lee, T.C. 1989. Mechanisms of arsenic-induced cell transformation. *Bio. Trace Elem. Res.*, **21**, 421-429.

Bates, M.N., Smith, A.H. and Hopenhayn-Rich, C. 1992 Arsenic ingestion and internal cancers: A review. *Am. J. Epidemiol.*, **135**, 462-476.

Brown, C.C. and Chu, K.C. 1983. Implications of the multistage theory of carcinogenesis as applied to occupational arsenic exposures. *JNCI*, **70**, 455-463.

Brown, K.G., Boyle, K.E., Chen, C.W. and Gibb, H.J. 1989. A dose-response analysis of skin cancer from inorganic arsenic in drinking water. *Risk Analysis*, **9**, 519-528.

Buchet, J.P., Lauwerys, R. and Roels, H. 1981. Urinary excretion of inorganic arsenic and its metabolites after repeated ingestion of sodium metaarsenite by volunteers. *Int. Arch. Occup. Environ. Health*, **48**, 111-118.

Burns, L.A., Sikorski, E.E., Saady, J.J. and Munson, A.E. 1991. Evidence for arsenic as the primary immunosuppressive component of gallium arsenide. *Toxicol. Appl. Pharmacol.*, **110**, 157-169.

CDC. US Centers for Disease Control. 1987. Reduction of children's arsenic exposure following relocation-- Mill Creek, Montana. *MMWR*, **36**, 505-507.

Cebrian, M.E., Albores, A., Aguila, N. and Blakely, E. 1983. Chronic arsenic poisoning in the North of Mexico. *Human Toxicology*, **2**, 121-133.

Chen, C.-J., Wu, M.-M., Lee, S.-S., Wang, J.-D., Cheng, S.-H. and Wu, H.-Y. 1988. Atherogenicity and carcinogenicity of high-arsenic artesian well water. *Arteriosclerosis*, **8**, 505-507.

Chen, C.-J. 1990. Blackfoot disease. *Lancet*, **336**, 442.

Chen, G.-S., Asai, T., Suzuki, Y., Nishioki, K. and Nishiyama, S. 1990. A possible pathogenesis for blackfoot disease. *J. Dermatol.*, **17**, 599-608.

Chen, C.-J., Chen, C.W., Wu, M.-M. and Kuo, T.-L. 1992. Cancer potential in liver, lung, bladder and kidney due to ingested inorganic arsenic in drinking water. *Brit. J. Cancer*, **66**, 888-892.

Cuzick, J., Sasieni, P. and Evans, S. 1992. Ingested arsenic, keratoses and bladder cancer. *Am. J. Epidemiol.*, **136**, 417-421.

Davis, A., Ruby, M.V. and Bergstrom, P.D. 1992. Bioavailability of arsenic and lead in soils from the Butte, Montana mining district. *Environ. Sci. Technol.*, **26**, 461-468.

DeZeeuw, R.A., Franke, J.P. and Stamm, D. 1987. Interlaboratory survey of the quality of urinary arsenic and thallium assays in the European Community. *J. Clin. Chem. Clin. Biochem.*, **25**, 351-355.

Freeman, G.B., Johnson, J.D., Killinger, J.M., Liao, S.C., Davis, A.O., Ruby, M.V., Chaney, R.L., Lovre, S.C. and Bergstrom, P.D. 1993. Bioavailability of arsenic in soil impacted by smelter activities following oral administration in rabbits. *Fund. Appl. Toxicol.*, **20**, 83-88.

Hoffman, D.J., Sanderson, C.J., LeCaptain, L.J., Cromartie, E. and Pendleton, G.W. 1992. Interactive effects of arsenate, selenium, and dietary protein on survival, growth, and physiology in mallard ducklings. *Arch. Environ. Contam. Toxicol.*, **22**, 55-62.

Hopenhayn-Rich, C., Smith, A.H. and Gaeden, H.M. 1993. Human studies do not support the methylation threshold hypothesis for the toxicity of inorganic arsenic. *Environ. Res.*, **60**, 161-177.

Irgolic, K. J. 1992. Arsenic. *Tech. Instrum. Anal. Chem.*, **12**, 287-350.

Jensen, G.E., Christensen, J.M. and Paulsen, O.M. 1991. Occupational and environmental exposure to arsenic-increased urinary arsenic level in children. *Sci. Total Environ.*, **107**, 169-177.

Lu, F.J., Shih, S.R., Liu, T.N. and Shawn, S.H. 1990. The effect of fluorescent humic substances existing in the well water of Blackfoot Disease endemic areas in Taiwan on thromboplastin time *in vitro*. *Thromb. Res.*, **57**, 747-753.

Marita, M. and Edmonds, J.S. 1992. Determination of arsenic species in environmental and biological samples. *Pure Appl. Chem.*, **64**, 575-590.

Mass, M.J. 1992. Human carcinogenesis by arsenic. *Environ. Geochem. Health*, **14**, 49-54.

Mazumdar, S., Redmond, C.K., Enterline, P.E., Marsh, G.M., Constantino, J.P., Zhou, S.Y.J. and Patwardhan, R.N. 1989. Multistage modeling of lung cancer mortality among arsenic-exposed copper-smelter workers. *Risk Analysis*, **9**, 551-563.

Mushak, P. 1991. Gastro-intestinal absorption of lead in children and adults: Overview of biological and biophysico-chemical aspects. *Chemical Speciation and Bioavailability*, **3**, 87-111.

Mushak, P. 1983. Mammalian transformation processes involving various toxic metalloids and metals. In: Brown, S.S. and Savory, J. (eds), *Chemical Toxicology and Clinical Chemistry of Metals*, pp. 227-245. Academic Press, London.

NAS/NRC. 1983. National Academy of Sciences/National Research Council. *Risk assessment in the Federal government: Managing the process*. National Academy Press, Washington, D.C.

Nielsen, F.H. 1991. Nutritional requirements for boron, silicon, vanadium, nickel, and arsenic: current knowledge and speculation. *FASEB J.*, **5**, 2662-2667.

North, D.W. 1992. Risk assessment for ingested inorganic arsenic: a review and status report. *Environ. Geochem. Health*, **14**, 59-62.

Ontario Ministry of the Environment. 1986. *Criteria document for arsenic standards*. Hazardous Contaminants Coordination Branch, Toronto, Canada.

Polissar, L., Lowry-Koble, K., Kalman, D.A., Hughes, J.P., VanBelle, G., Covert, D.S.,

Burbacher, T.M., Bolgiano, D. and Mottet, N.K. 1990. Pathways of human exposure to arsenic in a community surrounding a copper smelter. *Environ. Res.*, **53**, 29-47.

Science Advisory Board. 1989. Letter report to the Administrator, US Environmental Protection Agency. Publication No. US EPA-SAB-89-038, Washington, D.C.

Shannon, R.L. and Strayer, D.S. 1989. Arsenic-induced skin toxicity. *Human Toxicology*, **8**, 99-104.

Smith, A.H., Hopenhayn-Rich, C., Bates, M.N., Gaeden, H.M., Hertz-Picciotto, I., Duggan, H.M., Wood, R., Kosnett, N.J. and Smith, M.T. 1992. Cancer risks from arsenic in drinking water. *Environ. Health Perspect.*, **97**, 259-267.

Tseng, W.P. 1977. Effects and dose-response relationships of skin cancer and Blackfoot Disease with arsenic. *Environ. Health Perspect.*, **19**, 109-119.

Tseng, W.P., Chu, H.M., How, S.W., Fong, J.M., Lin, C.S. and Yeh, S. 1968. Prevalence of skin cancer in an endemic area of chronic arsenism in Taiwan. *JNCI*, **40**, 453-463.

US Environmental Protection Agency. 1984a. *Health assessment document for inorganic arsenic.* Publication No. EPA-600/8-83-021F. Research Triangle Park, NC.

US Environmental Protection Agency. 1984b. Health assessment document for chromium. Publication No. EPA-600/8-83-014F. Research Triangle Park, NC.

US Environmental Protection Agency. 1988. Risk Assessment Forum. *Special report on ingested arsenic. Skin cancer, nutritional essentiality.* Publication No. EPA-625/3-87-013.

Uthus, E.O. 1992. Evidence for arsenic essentiality. *Environ. Geochem. Health*, **14**, 55-58.

Vahter, M. and Marafante, E. 1988. *In vivo* methylation and detoxification of arsenic. In: Craig, P.J. and Glocking, F. (eds), *Biological Alkylation of Heavy Elements: Special Publication No. 66*, pp. 105-119. Royal Society of Chemistry, UK.

Webb, D.R., Wilson, S.E. and Carter, D.E. 1987. Pulmonary clearance and toxicity of respirable gallium arsenide particulates intratracheally instilled into rats. *Am. Ind. Hyg. Assoc. J.*, **48**, 660-667.

Weis, C.P. and LaVelle, J.M. 1991. Characteristics to consider when choosing an animal model for the study of lead bioavailability. *Chemical Speciation and Bioavailability*, **3**, 113-119.

World Health Organization, 1988. *Chromium. Environmental health criteria 61*. Geneva.

World Health Organization, 1987. Arsenic, Ch. 17. *Air Quality Guidelines for Europe*, pp. 171-181. Regional Office for Europe, Copenhagen.

World Health Organization, 1981. *Environmental Health Criteria. 18: Arsenic*. International Programme in Chemical Safety, Geneva.

Yanez, L., Carrizales, L., Zanatta, M.T., deJesus Mejia, J., Batres, L. and Diaz-Barriga, F. 1991. Arsenic-cadmium interaction in rats: Toxic effects in the heart and tissue metal shifts. *Toxicology*, **67**, 227-234.

Index